Plant Biochemistry

Techniques and Procedures

NIPA® GENX ELECTRONIC RESOURCES & SOLUTIONS P. LTD.
New Delhi-110 034

Plant Biochemistry

Techniques and Procedures

G. Nagaraj

Ex. Principal Scientist (Biochemistry)
Directorate of Oilseeds Research
Rajendranagar, Hyderabad - 500 030
India

NIPA® GENX ELECTRONIC RESOURCES & SOLUTIONS P. LTD.
New Delhi-110 034

NIPA® GENX ELECTRONIC RESOURCES & SOLUTIONS P. LTD.

101,103, Vikas Surya Plaza, CU Block
L.S.C. Market, Pitam Pura, New Delhi-110 034
Ph : +91 11 27341616, 27341717, 27341718
E-mail: newindiapublishingagency@gmail.com
web: www.nipabooks.com

For customer assistance, please contact
Phone: + 91-11-27 34 17 17
Fax: + 91-11- 27 34 16 16
E-Mail: feedbacks@nipabooks.com

ISBN: 978-81-19254-06-4

Composed and Designed by NIPA®.

Preface

Wide diversity exists within and across the species of the plant kingdom. Plants, in general, have extensive utility in our day to-day lives and for the maintenance of congenial environment. The utility of the plants and their products depend on their composition and quality. Useful components like sugars, proteins, fats and other nutrients like minerals and vitamins are present in varying proportions in the plants. Additionally, antinutrients and toxins which are harmful to the humans and animals are present. However, some of the components present are useful as medicine to the living beings, as also, animals. Some chemicals used in growing and increasing the production of crops, apart from having a positive influence, have negative effects on the health and environment. Hence assay of the plants and their products is essential to understand their quality and utility. This book deals with various methods and procedures of analysis of the beneficial and harmful constituents of the plant.

Attempts have been made to compile highly useful and more practical procedures in analysis of the components. The book deals with theoretical aspects of the analytical techniques like chromatography, U.V., I.R., N.M.R., Mass and other photometric techniques. Even radiochemical and radio- immunoassay methods find a place. Traditional methods like volumetry, gravimetry, conductometry, polarimetry, etc are also included. Methods of analysis have been grouped into sugars, fats, proteins, enzymes, nucleic acids, antinutrients, toxins, pesticide residues, alkaloids, plant nutrients etc. Wherever available more than one procedure is listed. The analyst can follow any procedure depending on the convenience and facility available in the laboratory. Each method gives information relating to the principle, reagents and equipment needed, actual procedure stepwise and required calculations.

The author has an experience of around five decades as a researcher and teacher in various Indian Agricultural Research Institutions. He was a practical analyst and compiled a lot of standard analytical procedures during his career. He also had training in use of Radioisotopes, G.C., H.P.L.C. techniques, Mycotoxins, Oil Technology, Electronic Instrumentation and related areas. The same background has been utilized in compiling and writing the present book. This book should be useful to all the students, researchers and scientists in the Pant Science. Plant biochemists should find it most useful to them. In spite of best efforts and intentions limitations and lacunae may exist. Constructive suggestions are welcome, which will be utilized in improving the content and quality of this book.

G. Nagaraj, Ph.D.

Retired Principal Scientist and Head, Crop Production
E mail: guttarla@ hotmail.com; guttarla@ yahoo.co.in

Contents

Part II. Analytical Methods and Procedures

Part I: Analytical Techniques

1

INTRODUCTION

Analysis of materials or samples is essential to understand their nature and quality. In recent years it has become mandatory to display the composition of food as well as non-food items. This is because of the need for their safety from the point of view of their day to day use. Environmental awareness has further imposed many restrictions on items of wide ranging nature. Analytical information gained from a laboratory is highly useful in knowing the qualitative and quantitative composition of the material.

Since times immemorial human being has been evaluating the material around him based on their colour, texture, taste and flavor. These are highly personal or subjective. Lot of variations may crop up with such an evaluation. Material with lot of variation may appear similar from a physical angle. Hence the need for objective analyses and characterization of the components in a sample has arisen. This is more so because of their utility as food, medicine, feed, fuel etc.

Analytical methods

The choice of a method for a given determination depends upon many factors including the purpose for which the analysis is being carried out. Lot of analytical methods are available. They are physical, chemical and biological. Accuracy, reproducibility and related requirements decide the specific method that needs to be followed for analysis. The method selected should be such that a large number of samples at the least cost could be handled.

Methodologies

We have two options before an understanding of the sample is looked into. Initially we can understand the nature of the sample by its appearance, texture or taste or flavor. For example we can say, the sample is wet (moist, or watery) or dry. Also we can easily say it is spicy or bland or pungent etc. If it is sweet, it is likely to contain sugars or if it is sour it can be an acid etc. Such an information can give a qualitative information. But through some simple fractionations we can know the compounds that make up a material, like proteins or sugars or oil or minerals (Ca, Mg, K, C,or S) etc. Basically such a knowledge can only be called qualitative and it is called a qualitative report. It can indicate that a sample contains a substance or a group of substances without specifying the complete composition

Knowing the basic compounds that are present in a sample does not satisfy us. We must know the actual quantity of the constituents. For example water content can be higher in a fresh vegetable or a fruit. But water or moisture content in a grain that needs to be stored for some time should have only 5-8% moisture. This is very critical from safe utility of a food item. Hence we need a method to quantify the moisture content or for that matter any other constituent of importance in a sample. This aspect is dealt with by the quantitative analysis. A quantitative report should specify the accuracy and reliability or reproducibility of the method used.

Methods of analysis

It is obvious that the analysis has two wings, namely qualitative and quantitative analyses. Different backgrounds are used for analysis of samples. They are physical, chemical or biological based on the property of the sample. There are some physical properties based on which a sample can be analyzed with good precision. Mass and volume can be measured easily with very simple apparatus. Properties like specific gravity, viscosity and surface tension etc. can also be determined with ease. Physical properties like absorption, emission, fluorescence, turbidity etc. are highly useful in quantitation of groups as well as specific compounds. Radioactivity and electrical properties are also useful in quantifying certain compounds in and around us.

Quantitative analysis is based on a few aspects. One of them is related to the quantity of reagent needed to complete a reaction and measuring the reaction product. The second one is based on electrical measurements, while the third one is based on spectroscopic properties (absorption properties). Yet another one is dependent on movement of the substance through a medium under specified conditions. The analyses are based on a single property or a combination of one or two of the above techniques (Ex. G.C.- M.S.).

Conventional methods which involve titration (volumetry) and precipitation (gravimetry) are wet chemistry methods. They are called conventional quantitative methods and involve preparation of many reagents and standards. However, these analyses can be carried out in an ordinary laboratory with easily available chemicals, a balance and some glassware etc. They are mostly macro or semi- micro or meso analytical methods. Accuracies up to 0.1g or 1mg level can easily be achieved. These are the basic methods based on which other instrumental or physical methods are standardized.

The macro or meso methods consume a lot of time and chemicals. It is difficult to handle large number of samples. A qualified analysts' attention is needed at most stages of the analysis. The invention of spectroscope towards the final stages of 19th century started the revolution in analytical chemistry. It was initially used for qualitative purposes only. Slowly colorimetric and turbidimetric quantitative methods were developed. Further developments in electronics mostly after 1930's helped develop many more instrumental methods of analysis. Today it is these instrumental methods that dominate the analytical chemistry. They have wide ranging applications in chemistry, medicine, biology, biotechnology etc.

Instrumental methods

Physical properties of elements and compounds are utilized in the instrumental methods of analysis. Mixtures are initially subjected to separation in chromatography. Later, identification of individual components are carried out based on their physical properties. Simultaneously their quantification is done again based on properties like absorption or emission of light (of specific wavelength), fluorescence, mass, heat, radioactivity, refraction etc. The precision is higher in addition to reproducibility in the case of physical methods of analysis. A large number of samples can be handled.

The instruments initially cost more but once installed the cost per sample becomes cheaper. It is necessary to carry out a calibration operation with instrumental methods using standards of known quality and quantity. Reagents of high purity (A.R. – Analytical Reagent grade) are essential. The main advantages with instrumental methods are 1. Detection up to nanogram (10^{-9} g) levels can be achieved and 2. Sample volumes needed are much lower than that of wet chemical methods.

Spectroscopic methods involve measurement of radiant energy absorbed by the sample or that emitted by the sample. Absorption methods are classified into visible, ultra-violet or infrared spectrophotometry. Visible photometry is also called colorimetry. There is another physical property based measurement called nuclear magnetic resonance spectroscopy. Atomic absorption spectroscopy involves atomizing the specimen and absorption of radiation. Light scattering is the basis in turbidimetry and nephelometry. Mass spectrometry is an analysis where separation is based on masses of the components.

Emission methods are based on measurement of light emitted by the sample when subjected to heat. Flame photometry is one such simple technique. Fluorimetry involves excitation of sample using U.V. or Visible radiation and measurement of the emitted lower energy radiation. Chromatographic techniques involve separation of samples between two phases under a support or background. Electrophoretic separation is similar to chromatography, but under an electric potential. Ion exchange is based on exchange of anions or cations present in the sample using resins.

Choice of methods

Analysis of samples for qualitative identification as well as their quantification involves use of one or two methods. Initially empirical identification is carried out. Comparison with standards may be used for confirmation. Two or more techniques may be needed for correct identification. The next step is quantification again based on the standard graphs or internal standard elution etc. Based on all these information, the composition and quality of a sample can be presented and understood.

The available methods vary much with respect to precision and sensitivity. Costs of analyses also differ. Inspite of using the same method, sample handling differs with respect to ordinary or radioactive backgrounds. The selection of an analytical method depends on the fastness or the time or the facilities available for the analysis in addition to the cost.

About this book

This book is aimed at helping the analyst with respect to analysis of samples, mainly, plant material. The book is divided into two parts. The first part deals with the different techniques involved in the biochemical analysis. The theoretical aspects of the same are presented in some detail. The mechanism and some information on the build up of the equipments are furnished. The need and scope of the instruments and their utility are also presented.

The second part deals with specific methods of analysis. Analytical methods of sugars, amino acids, proteins, fats, enzymes and their components are presented. Nucleic acid, vitamin, antinutrient (toxicants) and mineral analytical methods also have been included. Information on alkaloid and pesticide residue analytical methods *etc*. have also been furnished. When available two or more methods have been given. They have been given in such a way that they include reagents, apparatus and step wise procedures. Final calculations needed for arriving at the concentration of the constituents have also been included.

2

CHROMATOGRAPHY

Chromatography is a general technique available for the separation of closely related compounds in a sample. The term chromatography is derived from Greek where *chroma* means colour and *graphein* means to write. It is a very useful technique for both qualitative (preparative) and quantitative (analytical) analyses. The separation is effected by the differential distribution of the components between two immiscible phases, namely, stationary and mobile phases. The various constituents of the mixture travel at different speeds, causing them to separate. The separation is based on differential partitioning (partition coefficient) between the mobile and stationary phases. The stationary phase is a porous medium through which the sample mixture percolates under the influence of a moving solvent (mobile phase).

Chromatography was developed by the Russian scientist Mikhail Tswett in the 19th century. He worked for the separation of plant pigments such as chlorophyll, carotenes, and xanthophylls. Since these components have different colors (green, yellow, and orange, respectively) they gave the name chromatography to this technique. New types of chromatography have been developed during the 1930s and 1950s. The technique became highly useful for many separation processes of closely related compounds belonging to a homologous series etc.

Some materials appear homogenous, but are actually a combination of substances. For example, green plants contain a mixture of different pigments. The black ink in the pens is a mixture of different colored materials. In many instances, we can separate these materials by dissolving them in an appropriate liquid and allowing them to move through an absorbent matrix, like paper. Chromatography is a method used by scientists for separating organic and inorganic compounds so that they can be analyzed and studied. By analyzing a compound, a scientist can figure out what makes up that compound. Chromatography is a great physical method for observing mixtures and solvents. Chromatography is such an important technique that two Nobel prizes have been awarded to chromatographers. Over 60% of chemical analysis worldwide is currently done with chromatography or a variation thereon.

Chromatography is based on differential migration. The solutes in a mobile phase go through a stationary phase. Solutes with a greater affinity for the mobile phase will spend more time in this phase than the solutes that prefer the stationary phase. As the solutes move through the stationary phase they separate. This is called chromatographic development.

Chromatographic process

In chromatography there is a mobile phase and a stationary phase. The stationary phase is the phase that doesn't move and the mobile phase is the phase that does move. The mobile phase moves through the stationary phase picking up the compounds to be tested. As the mobile phase continues to travel through the stationary phase it takes the compounds with it. At different points in the stationary phase the different components of the compound are going to be absorbed and are going to stop moving with the mobile phase. This is how the results of any chromatography are obtained, from the point at which the different components of the compound stop moving and separate from the other components.

In paper and thin-layer chromatography the mobile phase is the solvent. The stationary phase in paper chromatography is the strip or piece of paper that is placed in the solvent. In thin-layer chromatography the stationary phase is the thin-layer cell. Both these kinds of chromatography use capillary action to move the solvent through the stationary phase.

The retention factor, R_f

The retention factor, R_f, is a quantitative indication of how far a particular compound travels in a particular solvent. The R_f value is a good indicator of whether an unknown compound and a known compound are similar, if not identical. If the R_f value for the unknown compound is close or the same as the R_f value for the known compound then the two compounds are most likely similar or identical.

The retention factor, R_f = distance travelled by the solute (D1) divided by the distance travelled by the solvent front (D2). R_f = D1 / D2 .

Chromatographic procedures differ based on the interaction between the sample and the stationary phase. They are adsorption, partition, ion exchange, molecular exclusion, gel permeation, and affinity chromatographies. Based on the mobile and stationary phases used, chromatographic procedures can also be classified. For example, they are liquid chromatography(LC) and gas chromatography (GC) based on the mobile phase. They are paper (PC), thin layer (TLC), column (CC), gas- liquid (GLC) and high pressure (performance) liquid (HPLC) chromatographies based on the stationary phases.

3

PAPER CHROMATOGRAPHY

Paper chromatography (P.C.) is the easiest to perform and requires simple and ordinary apparatus. It easily provides qualitative information since components of a mixture can be separated based on their mobilities or molecular weights. It also can provide quantitative information if some special attention and calculation procedures are followed. Paper chromatography is a technique that involves placing a small dot or line of sample solution on to a strip of chromatography paper. The paper is placed in a jar containing a shallow layer of solvent and sealed. As the solvent rises through the paper, it meets the sample mixture, which starts to travel up the paper with the solvent. This paper is made of cellulose, a polar substance, and the compounds within the mixture travel farther if they are non-polar. More polar substances bond with the cellulose paper more quickly, and therefore do not travel as far.

In paper chromatography, components of the mixture are carried along with the solvent up the paper to varying degrees, depending on the compound's preference to be adsorbed onto the paper versus being carried along with the solvent. The paper is composed of cellulose to which polar water molecules are adsorbed, while the solvent is less polar, usually consisting of a mixture of water and an organic liquid. The paper is called the stationary phase while the solvent is referred to as the mobile phase. Performing a chromatographic experiment is basically a three-step process: 1) application of the sample, 2) "developing" the chromatogram by allowing the mobile phase to move up the paper, and 3) calculating R_f values and making conclusions.

The individual components in the mixture move forward depending on their mobility. Thus separation takes place inside the mobile phase and stops when the solvent movement stops at a particular length/ distance (namely, the top end of paper). The paper serves as the stationary/ inert phase. After drying the paper, the component spots are visualized by spraying a colour reagent.
An important characteristic used in identification of the spots is the R_f value. Individual components usually have specific R_f values.

$$R_f = \frac{\text{Distance moved by the sample spot}}{\text{Distance travelled by the mobile phase}}$$

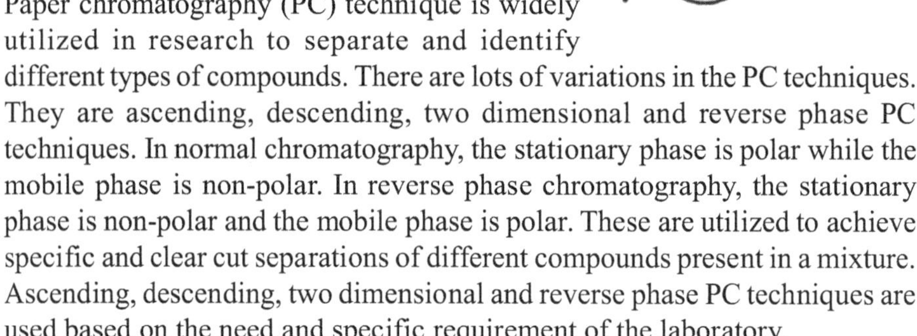

Paper chromatography (PC) technique is widely utilized in research to separate and identify different types of compounds. There are lots of variations in the PC techniques. They are ascending, descending, two dimensional and reverse phase PC techniques. In normal chromatography, the stationary phase is polar while the mobile phase is non-polar. In reverse phase chromatography, the stationary phase is non-polar and the mobile phase is polar. These are utilized to achieve specific and clear cut separations of different compounds present in a mixture. Ascending, descending, two dimensional and reverse phase PC techniques are used based on the need and specific requirement of the laboratory.

Particular mixtures will have chromatographic patterns that are consistent and reproducible as long as the paper, solvent, and time are constant. This makes paper chromatography a qualitative method for identifying some of the components in a mixture. Sometimes quantitaion is carried out based on the intensity or the area of the spots in comparison to that of standards. Densitometry can also be tried. The spots can be cut and extracted with suitable solvents and estimated spectrophotometrically.

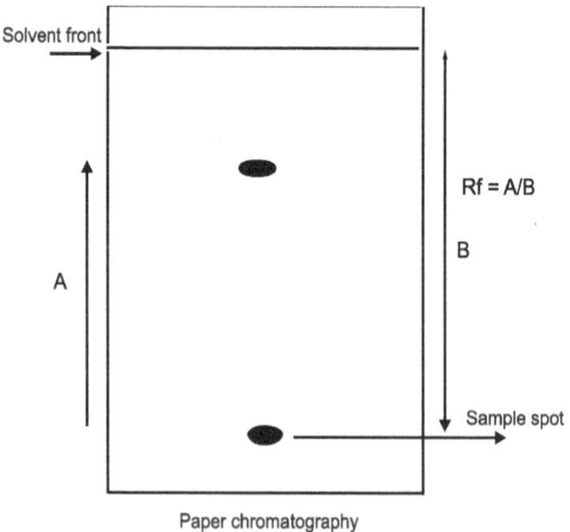

Paper chromatography

4

COLUMN CHROMATOGRAPHY

Column chromatography is a separation technique in which the stationary bed is in a tube. The particles of the solid stationary phase or the support coated with a liquid stationary phase may fill the whole inside volume of the tube (packed column) or be concentrated on or along the inside tube wall leaving an open, unrestricted path for the mobile phase through the tube (open tubular column). Differences in rates of movement through the medium are calculated as different retention times of the sample.

The separation of compounds in the column chromatography takes place based on the phenomenon of adsorption. Here the substance gets attracted by electrostatic forces to the surface of a unit particle. If the substance is in solution and the particle insoluble in the solvent, then part of the substance is adsorbed and part remains in solution. The ratio between the amount adsorbed and the amount in solution is a constant called adsorption coefficient. The separation takes pace based on the differences in the adsorption coefficients.

Adsorbent is taken in a column. The solute dissolved in a solvent is allowed to percolate down. The solute molecules get adsorbed, released and re-adsorbed on the surface of the adsorbent as in counter current distribution. The compound with least adsorptivity moves fast and gets eluted first while the one with maximum adsorptivity travels slowly and gets eluted last along with the solvent. Thus a separation of the different molecules takes place depending upon the adsorptivity and partition between the solvent (mobile phase) and the adsorbent. Different fractions of the eluting solvent can be collected and examined for the components/ solutes. This technique of separation is called column chromatography, since a column is used.

Adsorbents

Solids are generally used as adsorbents. The most widely used ones are alumina, silicic acid, cane sugar, activated charcoal, diatomaceous earth(celite) and magnesium silicate. The choice depends on the specific application. The main condition for selection of an adsorbent is that there should not be any chemical reaction between the solute, the solvent and the adsorbent. The other important factor is the size of the adsorbent particle. The smaller the size, the better it is as an adsorbent. Too small a size will retard the flow of the solvent and care should be taken in having a proper adsorbent particle size.

Solvent

Electrostatic attraction plays a big role in the adsorption phenomena. Polarity of the solvent greatly influences adsorption considerably. Generally adsorption is maximum in non polar solvents. Increased polarity of solvents decreases the adsorption of components. The following is a list of solvents with increased level of polarities. Petroleum ether/ hexane < benzene < CCl_4 < ether < $CHCl_3$ < acetone/ alcohol < water < salt solutions < acid / alkali solutions.

Separation of plant pigments

Take about 5g of flowers and grind them using 20ml of 2:1 benzene and methanol solvent. Add 10ml of water and shake. Remove the methanolic aqueous layer. Separate the benzene layer. Evaporate benzene and re-dissolve in 5ml of benzene. Alumina is used as an adsorbent. A glass column with a stop cock is taken. Glass wool is put at the bottom of the column above the stop cock. 5g adsorbent (dried at 120^{0}C overnight) is taken in a beaker and a small amount of benzene is added. The slurry is carefully poured into the column by avoiding air bubbles. Keep the solvent 1 cm above the adsorbent level. Open the stop cock and when the solvent level touches the adsorbent, add the flower extract, namely, benzene solution. Add more benzene. When 20ml of benzene has run through, add 5ml of 5% acetone in benzene. Acetone concentration can be increased to obtain better separation. Coloured compounds get separated and can easily be seen.

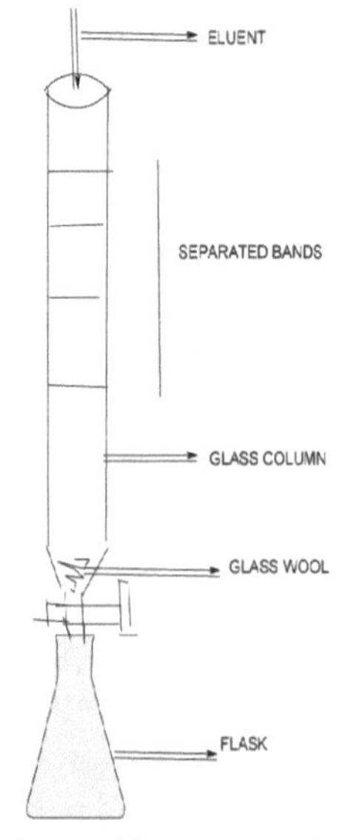

Column Chromtography

Eluents/solvents are used as isocratic (fixed ratio or single solvent) or as gradients (varying solvent composition) in a phased manner. All this is done to obtain efficient separation of solutes.

5

THIN LAYER CHROMATOGRAPHY

Kirchner was the first in 1950 to think of thin layers. But it was Egon Stahl in 1958 who was responsible for bringing out standard equipment and related procedures to popularize thin layer chromatography (TLC). TLC is a modified form of column chromatography. Hence it is also called open column chromatography. The adsorbent is spread over a supporting material (glass or plastic sheets) to form a thin layer of adsorbent. As a binder, plaster of Paris or gypsum is added to the adsorbent material like silica gel (hence silica gel TLC). The adsorbent is spread over the plate as a thin layer of 0.25 or 0.5 mm. The adsorbent to water ratio is 1:2. It is initially allowed to air dry, followed by drying in an oven at 110°C for 1-2hr or overnight (activation of TLC plates) after which the plate is ready for thin layer chromatography.

The sample spots are put at the bottom of the plate. Standard spots are also applied and the plate is allowed to develop in a closed chamber containing a layer of mobile phase. The solvent level should be such that the sample and standard spots should be above its level. The mobile phase is allowed to move up and when it reaches a specified level (1-2 cm lower than the top), the plates are taken out. The solvent is allowed to dry up. The components present in the spot move up as per their mobilities and get separated. The separated spots are visualized by spraying an appropriate reagent. General spraying reagents are sulphuric acid and phospho molybdic acid etc. The organic molecules get charred when dried at higher temperatures and appear as dark spots. The R_f values are calculated and compared with the values of standard spots for identification. The procedure followed is in many ways similar to the ascending paper chromatography.

Quantification is also possible. This is done based on the area of the spots or the intensity of the spots by measuring their densities using a densitometer. The spots prior to their visualization can also be extracted with a solvent and their quantities can be measured spectrophotometrically. For still better resolution and quantification, high-performance TLC (HPTLC) can be used. HPTLC plates are prepared using specially purified silica gel with particle size of 3-5μm. Such plates are available commercially. HPTLC plates have about 5000 theoretical plates and hence provide improved performance. In the absence of densitometer and other facilities, visual quanfication can be carried out. This can be done by comparing with the areas of standard spots.

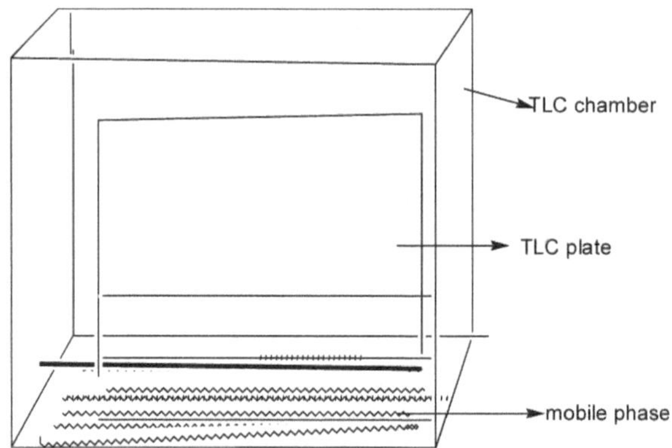

Thin Layer Chromatography Chamber and Development of Plate

Example: TLC of dyes

Prepare a slurry of alumina and spread over the glass plate. Dry the plate at 110°C. Then spot a mixture of methyl red, cresol red, methylene blue etc. on the plate at the bottom and develop the plate by putting in a mobile solvent of butanol saturated with Na_2CO_3. When the solvent reaches the top (2cm below the top) of the plate, it is removed and dried. Since the samples are coloured, the separated spots can easily be viewed.

Advantages of TLC

The main advantages are as follows.

1. This technique needs very simple and cheap apparatus.
2. It is a faster technique than paper chromatography and column chromatography.

3. Better separations are obtained.
4. Smaller quantities can be detected than that in PC.
5. Sample size needed is smaller.
6. Preparation of TLC plates is easy and ordinary workers can prepare them and
7. The TLC combines the advantages of both column and PC and is superior to both of them.

Disadvantages

It cannot handle larger quantities. However, it can be used for small scale preparative work. TLC can be performed as uni- dimensional, two dimensional or for reverse phase chromatography, depending on the requirement and efficiency of separation.

Adsorbents

Lot of adsorbents are available for TLC. They are alumina, aluminum silicate, bauxite (aluminum oxide), bentonite, $CaCO_3$, $Ca(OH)_2$, calcium oxalate, calcium silicate, $CaSO_4$, fullers earth, Kieselgurh, magnesia, magnesium silicate, silica gel, Zn CO_3, cellulose, charcoal, sucrose etc.

Developed TLC Plate

Two dimensional TLC

Two dimensional chromatography is utilized to improve the separation efficiency. The separations are carried out in two different solvent systems. The sample spot is applied on bottom left corner. The first solvent system is run in one direction. After drying, the plate is turned at right angles and run in the second solvent. Visualization of spots are carried out using colour reagent spray etc. Separations are better under two dimensional chromatography.

Preparative chromatography

The TLC technique can be used for isolation of separated components. Such separation is called preparative chromatography. In this case the adsorbent layer is thicker ranging up to 5mm. Larger quantity of sample can be applied on to the plate. After the separation the sample spot is cut out and extracted with a suitable solvent. The sample is applied many times as a streak rather than as a spot. The identification and quantification can be carried out using standard procedures.

Identification

Identification of separated components is carried out using one or the other of following techniques.

1. Spray with 50% sulphuric or phosphomolybdic acid and heat in an oven. The spots get charred and appear as dark spots. This is a destructive procedure. Most organic compounds can be identified with this spray.
2. The plates can be examined under U.V. light. Fluorescent compounds appear as bright spots.
3. Unsaturated compounds reveal their presence when exposed to iodine/bromine vapour
4. Ninhydrin reagent is used (sprayed) to identify amino acids.
5. Autoradiography is used if separated components are radioactive.
6. Specific reagents need to be used to identify various other organic compounds.

6

ELECTROPHORESIS

Electrophoresis is the migration of particles under the influence of a direct electric current. There are two requirements for this process to be carried out. First the particles to be separated should be charged or at least should accept charge. Proteins, amino acids, nucleotides etc. conform to this requirement. Second, the medium used for separating the compounds also should carry charge. The mobilities of the ions depend on the nature of the electrolyte solution, the concentration and temperature. The cations and anions move in opposite directions with the cations moving towards the anode and the anions towards the cathode.

Paper electrophoresis: Samples are applied at the centre of the support, namely, filter paper which is moistened with the buffer of appropriate p^H and connected to buffer reservoir by paper wicks. When current is applied molecules with positive charge move towards the cathode and those with negative charge move towards the anode. When molecules have the same charge, the molecular weight determines their mobilities. Low molecular weight molecules move fast while the heavier ones move slowly. After developing the chromatogram, the dried paper is sprayed with a colour reagent or visualized under U.V. light. Identification and quantifications are carried out in a manner similar to either P.C. or TLC.

High voltage electrophoresis (HVE): This involves application of a potential gradient of 50 volts/cm – 100 volts/ cm. High voltage electrophoresis has some advantages. HVE can separate particles with similar charges including stereo isomers. The time of run is much smaller in HVE. Sharp, discrete spots are obtained . During HVE due to high voltage, heat is generated. Hence cooling of the system is needed.

Capillary electrophoresis (CE): By using capillaries it is possible to minimize zone spreading, improve heat dissipation, shorten separation times and increase efficiencies comparable to that of HPLC. The CE system consists of a fused silica capillary filled with an aqueous buffer electrolyte. The two ends of the capillary are dipped into containers of electrolyte, one holding the anode and the other cathode. The sample is introduced into the anode end of the capillary. The electric field is applied. CE development occurs rapidly.

Electophoretic separation of proteins: The apparatus consists of two parts, chamber vessels and D.C. power supply. Two tanks filled with buffer solutions and platinum electrodes are connected through a wetted filter paper. The electrodes are connected to the +ve and –ve terminals of the D.C. power supply (100 volts).

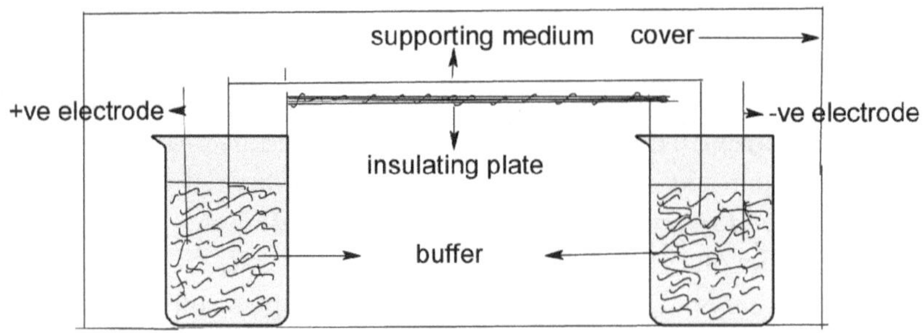

Electrophoresis Apparatus

A protein solution is applied at the centre of the moistened filter paper. A barbitone buffer (p^H 6.8) is kept in the tanks. Switching on the power supply (36volts – 4volts/cm) allows the electrophoresis to go on which can extend for up to 6hr. Then the paper is removed and dried. Soaking the paper in a 1% alcoholic solution of bromophenol blue is carried out next. Excess dye can be removed by soaking in 0.5% acetic acid solution. Protein spots then get visualized.

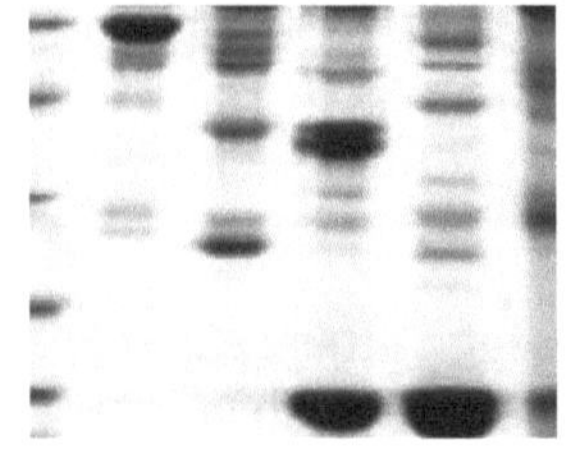

Protein Fractionation

Gel electrophoresis: Gel electrophoresis is an adaptation of electrophoresis. Instead of filter paper, a gel medium is used containing a hydrophobic medium as internal phase and an electrolyte buffer as the external or solvent phase. Gel electrophoresis is used to separate protein or nucleic acid fractions like DNA and RNA etc. DNA fragments are separated according to their size while proteins are separated based on their size and charge. Separation is based on differences in charge density. It is also achieved through molecular sieving (separation based on pore size). Agar gel, starch gel, and acrylamide gel are the commonly used gels. After separation, the gels are stained to locate the spots. Identification, quantification etc. are carried out as in other chromatographic methods.

7

ION EXCHANGE CHROMATOGRAPHY

This technique is used in the separation of ionised and ionisable compounds. It consists of an insoluble matrix to which charged groups are covalently bound. The charged groups are associated with mobile counter ions. These counter ions can be reversibly exchanged with other ions of the same charge without altering the structure of the matrix.

The ion exchanger is complex and polymeric. The polymer carries an electric charge that is exactly neutralized by the charges on the counter ions. These active ions are cations in a cation exchanger and anions in an anion exchanger. A cation exchanger consists of polymeric anions and active cations whereas anion exchanger consists of polymeric cations and active anions. Amberlite IR 120 and Dowex 50 WX are cation exchangers while Amberlite IRA 400 and Dowex 1X are anion exchangers. Ion exchangers should be insoluble in water and hydrophilic to allow the ions to pass through. It should be chemically stable and should be denser than water when swollen.

Deionization of water by ion exchange chromatography: Salt water is deionized by passing through ion exchange columns. When salt water is passed through cation exchanger, it removes ions like Na^+. Later the water is passed through anion exchanger to remove ions like Cl^- etc. The water finally emerges as "deionized water".

Amino acid separation by ion exchange chromatography: The technique of amino acid analysis by ion exchange chromatography plays a central role in the investigation of amino acid composition of proteins and the free amino acids in biological samples. Ion exchanger is complex and is polymeric. The polymer carries an electric charge, which gets neutralized by the moving solute in the mobile phase.

The response of amino acids to p^H is used in separating them by ion exchange chromatography.

1. It is necessary to make sure that the resin is in proper state before it is used for chromatography. If a cationic exchanger is used, it must be in the acid form or must contain H^+ ions. This is achieved by suspending the resin in 4N HCl (4-8 lit/100g) for about 15 min.
2. Resins are available in different mesh sizes. For amino acids, 200-400 mesh resins are used. Suspend 10g of the resin (Dowex-50) in water and allow to settle. The supernatant containing any light particles is decanted. Now add citrate buffer (p^H 3.4) enough to cover the resin. Stir and let stand for 1hr. Use this suspension to set up the column (30x 0.9 cm). Run about 2 column volumes of buffer. Adjust flow rate to 5ml/ min.
3. The amino acid mixture is adjusted to p^H 1.0. Place one ml of sample on top of the column. The resin is allowed to run.
4. The amino acid bound on the column can be eluted by increasing the p^H of the buffer.
5. One ml aliquots of the eluate is collected. They are reacted with ninhydrin reagent. The intensity of colour in the fractions reveal the amino acids present.

The method is highly reproducible. Hence it is regularly used for analysis of amino acids in proteins. A plot of the intensity of colour against time of elution or eluate number gives a better picture of the amino acids eluted. Automated amino acid analyzers adopt the same principle. Protein hydrolysates and buffer gradients are loaded with the help of pumps. The eluates pass through a tube where they get mixed with ninhydrin. The coloured solution passes through a photometer cell. The colour intensity is measured and results are fed to a recorder which gives a graph of intensity against time. The identification is based on the elution pattern of standards.

8

GEL-FILTRATION CHROMATOGRAPHY

Gel-filtration chromatography or molecular sieve chromatography is a method based on molecular dimension. The gel consists of beads which are composed of dextran polymer (sephadex, agarose or polyacrylamide gels). The separation depends upon the different abilities of the various sample molecules to enter pores which contain the stationary phase. Very large molecules, which never enter the stationary phase, move through the chromatographic bed fastest. Smaller molecules, which can enter the gel pores, move very slowly through the column, because they spend a proportion of their mobility, time in the stationary phase. Molecules are, therefore eluted in the order of their decreasing molecular size.

Poly acrylamide gel electrophoresis (PAGE)

It is a powerful technique used to fractionate proteins. A gel matrix carries forward proteins under the influence of an applied current. The matrix is composed of polymers of a small organic molecule, namely, acrylamide ($CH_2 = CH - CO - NH_2$). They are cross linked to form a molecular sieve. A poly acrylamide gel may be formed as a thin slab between two glass plates or as a cylinder within a glass tube.

Once the gel is polymerized, the slab (or tube) is suspended between two compartments containing buffer in which opposing electrodes are immersed. In a slab gel the concentrated protein containing solution is layered in slots along the top of the gel. The protein sample is prepared in a solution containing sucrose or glycerol whose density prevents the sample from mixing with the buffer in the upper compartment. A voltage is then applied between the buffer compartments. Current flows across the slab causing the proteins to move toward the oppositely charged electrodes.

Separations are typically carried out using alkaline buffers, which make the proteins negatively charged and cause them to migrate towards the positively charged anode at the opposite end of the gel. After the PAGE process, the slab is removed from the glass plates and stained.

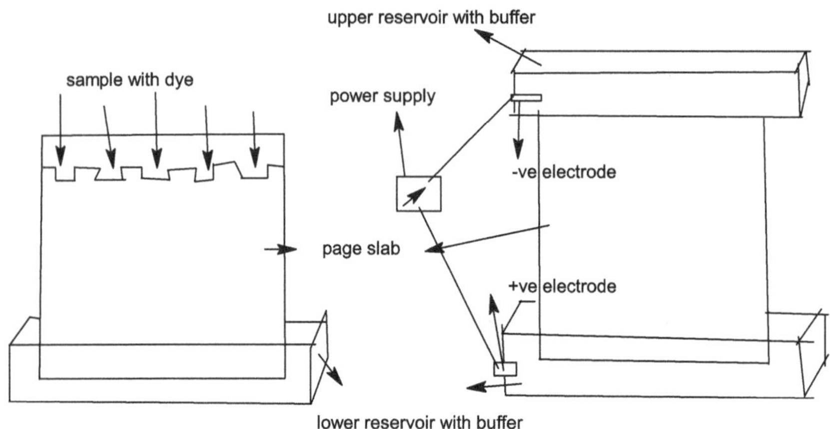

The relative movement of proteins through the gel depends upon the charge density (charge per unit of mass) of the molecules. The greater the charge density the more forcefully the protein is driven through the gel and thus the more rapid its rate of migration. In addition to charge density, size and shape of the molecules also play a role. Poly acrylamide forms a cross linked molecular sieve that entangles proteins passing through the gel. The larger the protein, the more it becomes entangled and the slower migration of protein takes place. Shape is also a factor because compact globular proteins move more rapidly than elongated fibrous proteins of comparable molecular mass.

The concentration of acrylamide used in making the gel is also an important factor. The lower the concentration of acrylamide, the less the gel becomes cross linked and the more rapidly the protein molecule moves or migrates. A gel containing 5% acrylamide will be useful in separating proteins of 60-250 KDa whereas a gel of 15% acrylamide will be useful in separating proteins of 10-50KDa.

Progress of electrophoresis is followed by watching the migration of a charged tracking dye that moves just ahead of the fastest protein. After the tracking dye has moved to a desired location, the current is turned off and the gel is removed from its container. Gels are typically stained with Coomassie Blue or silver stain to reveal the location of proteins. Radioactively labeled proteins can be located using autoradiography.

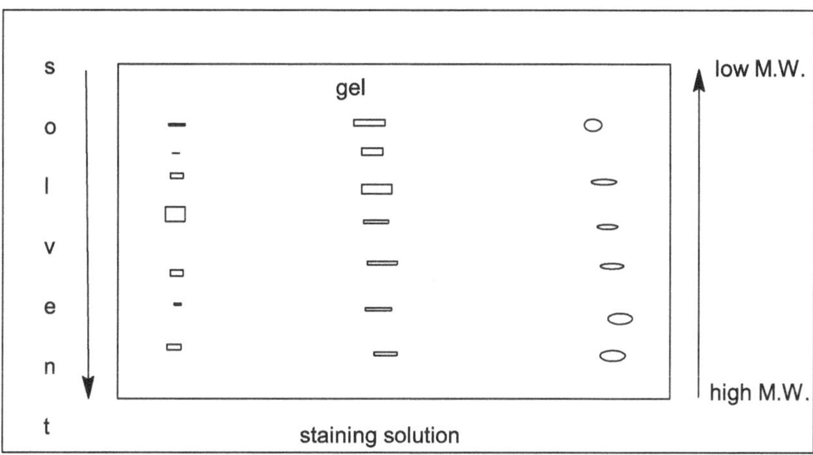

SDS- PAGE: PAGE is normally carried out in the presence of negatively charged detergent sodium dodecyl sulphate (SDS) which binds in large numbers to all types of protein molecules. The electrostatic repulsion between the bound SDS molecules causes the proteins to unfold into a similar rod-like shape, thus eliminating differences in shape as a factor of separation.

The number of SDS molecules that bind to a protein is roughly proportional to the proteins' molecular mass (about 1.4g SDS/g of protein). Consequently each protein species regardless of its size has an equivalent charge density and is driven through gel with the same force. However, because the poly acrylamide is highly cross linked , larger proteins are held up to a greater degree than the smaller proteins. As a result, protein becomes separated by SDS-PAGE on the basis of a single property- their molecular mass. In addition to separating the proteins in a mixture, SDS-PAGE can be used to determine the molecular mass of various proteins by comparing the positions of the bands to those produced by proteins of known mass.

Proteins are separated in a tubular gel according to their *isoelectric focusing*. After separation, the gel is removed and placed on top of a slab of SDS- saturated poly acrylamide subjected to SDS PAGE. The proteins get separated according to their molecular mass. Once separated individual proteins can be removed from the gel and digested into peptide fragments, which can be analyzed by mass spectrophotometry. The resolution is high enough to distinguish most of the proteins. Two dimensional PAGE is ideally suited to detect most of the proteins. The technique is not suitable for high molecular weight proteins that are hydrophobic.

Affinity Chromatography: Poteins carry out their biological activity through binding or complex formation with specific small bio-molecules or ligands as in the case of an enzyme binding its substrate. Protein molecules of interest are immobilized through covalent attachment to an insoluble matrix (cellulose or polyacrylamide) based on specific and reversible molecular interaction between two biologically active substances. The protein of interest, in displaying affinity from its ligands, becomes bound and immobilizes itself. The protein bound to the matrix is then eluted or removed from the matrix by the adddition of high concentration of the free ligand in solution.

9

SOLVENT EXTRACTION

Solvent extraction is mainly used to remove dissolved substances from solutions or soluble material from solids /powders/ solid mixtures. Solvent extraction is explained using distribution law or partition law. The law states that if to a system of two liquid layers made up of two immiscible liquids/ slightly miscible liquids is added a third substance soluble in both layers , then the substance distributes itself between the two layers so that the ratio of concentration in one solvent to the concentration in the second solvent remains constant at a constant temperature.

C_A / C_B = constant (K)

C_A = concentration in layer A, and C_B = concentration in layer B

K = distribution or partition coefficient

Approximately, K is assumed to be equal to the ratio of solubility in the two solvents. Organic compounds are more soluble in organic solvents than in water. Hence they can be extracted from aqueous solutions easily due to salting out effect.

Single extraction is not efficient in removing the dissolved substances in another solvent or solid mixture. Normally at least three extractions are considered better to remove the maximum quantity.

The general equation relating to the quantity remaining after "n" extractions is as follows.

$W_n = W_0 (KV/ K V + S)^n$

W_n = weight in g remaining after n^{th} extraction,

W_0 = weight in g of substance remaining initially,

K = Partition coefficient, V = Volume of aqueous solution and

S = Volume of solvent used for extraction.

The above equation applies strictly to a solvent which is immiscible with water, such as benzene or chloroform or carbon tetrachloride. If the solvent is slightly miscible, for example, ether, the above equation is only approximate. Let us consider a specific example. Extraction of 4g of n-butyric acid in one litre of water with100 ml of benzene at 15^0 C. The partition coefficient for benzene:water is 3 or 1/3 for water: benzene at 15^0 C. For a single extraction with benzene we have

$W_n = 4(1/3 \times 100/100/3 + 100) = 1.0g$

For three extractions with 33.3 ml portions of fresh benzene

$W_n = 4(1/3 \times 100/100/3 +33.3)^3 = 0.5g$

Hence one extraction with 100 ml of benzene removes 3.0 g or 75% of n-butyric acid whilst three extractions removes 3.5g or 87.5% of the total acid. This clears the greater efficiency of extraction obtainable with several extractions when the total volume of solvent is the same. Moreover, the smaller the distribution coefficient between the organic solvent and water , the larger the number of extractions that will be necessary. Several washings with portions of the solvent give better results than a single washing with the total volume of the solvent.

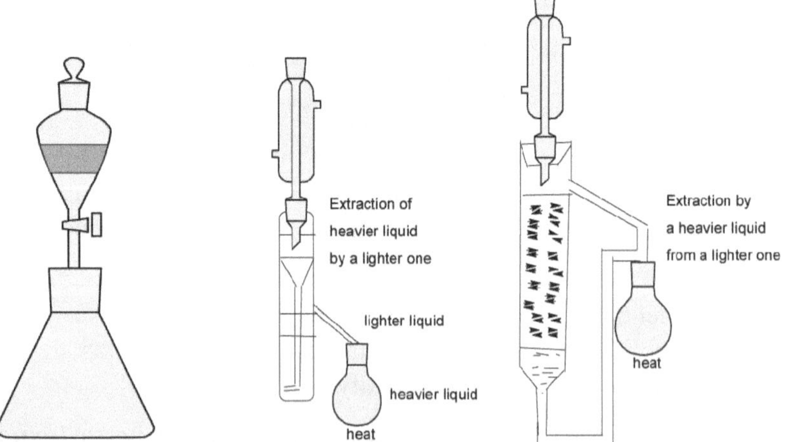

Extraction using a Separating Funnel **Different Extraction Apparatus**

Technique of solvent extraction

One solvent is normally water, while the other solvents are ether, di-isopropyl ether, benzene, chloroform, carbon tetra chloride and hexane. Diethyl ether is the most preferred due its low boiling point, 30^0 C. Separating funnel is mostly used for solvent extraction at the laboratory level.

Extractions are repeatedly carried out. The completeness of extraction can be tested by evaporating a small portion of the solvent and weighing the solute. The organic layers are combined and the clear solution is evaporated to obtain the solute. Emulsions are formed which can be broken up using the following steps.

1. A stream of air can be drawn through the solvents,
2. Saturating the aqueous phase with NaCl,
3. Add a few drops of alcohol or a suitable solvent (generally undesirable).
4. Centrifuge the mixture- most useful to break the emulsions.

10

DIALYSIS, ULTRA FILTRATION AND LYOPHILIZATION

Dialysis

The main purpose of dialysis in biological research is the removal of small molecular weight (M.W.) constituents from biological fluids. This is achieved by dialysing the material in a cellulose tubing with a pore size of 40-80 A^0 (4-8mm) diameter which allows the passage of compounds with molecular weight less than 10,000. The following is the procedure followed: cut the length of the dialysis tubing depending upon the volume to be processed. Open the sides of the tubing with a flow of distilled water and transfer it to a boiling water bath. Do not allow the tubing to dry as it will distort the pore size. Make a bag by tying a knot at one end of the tubing. Transfer the solution to be dialysed into the bag. Tie another knot at the top of the bag. The bag is then put in a beaker containing the solution against which dialysis is to be carried out. In general, one volume of the desired solution is dialysed against 50-100 volumes of buffer. The total period of dialysis could range from 8-48hr. The efficiency of dialysis could be increased by stirring and changing the dialysate solution at frequent intervals.

Dialysis is more familiar to the biologist than to the chemist. The semi-permeable membrane can be cellulose acetate with pore size of 1-5nm. In biological applications one of the aqueous phases contains both ionic species and large biomolecules and the other phase pure water. During dialysis ionic species move out to the pure water phase. The other phase thus gets concentrated with respect to high molecular weight species. Dialysis is used to clean up samples prior to their analysis, like deproteination prior to HPLC fractionation.

Ultrafiltration

Ultrafiltration is a separation process using membranes with pore sizes in the range of 0.1 to 0.001 micron. Typically, ultrafiltration removes high molecular-weight substances, colloidal materials, and organic and inorganic polymeric molecules. Low molecular-weight organics and ions such as sodium, calcium, magnesium chloride, and sulfate are not removed. Because only high-molecular weight species are removed, the osmotic pressure differential across the membrane surface is negligible. Low applied pressures are therefore sufficient to achieve high flux rates from an ultrafiltration membrane. Flux of a membrane is defined as the amount of permeate produced per unit area of membrane surface per unit time. Generally flux is expressed as cubic meters per square meters per day.

Ultrafiltration is used to separate high molecular weight compounds like proteins, nucleic acids etc. from small molecular weight compounds in biological studies. This technique utilizes semi-permeable membrane to retain large molecular weight compounds and allows small molecular weight compounds to pass through. Centrifugal force or high pressures are used to separate small molecular weight compounds. "Cellophane" membranes are mostly used.

Ultra filtration, like reverse osmosis, is a cross-flow separation process. Here liquid stream to be treated (feed) flows tangentially along the membrane surface, thereby producing two streams. The stream of liquid that comes through the membrane is called permeate. The type and amount of species left in the permeate depends on the characteristics of the membrane, the operating conditions, and the quality of feed. The other liquid stream is called concentrate and gets progressively concentrated in those species removed by the membrane. In cross-flow separation, therefore, the membrane itself does not act as a collector of ions, molecules, or colloids but merely as a barrier to these species.

Conventional filters such as media filters or cartridge filters, only remove suspended solids by trapping these in the pores of the filter-media. These filters therefore act as depositories of suspended solids and have to be cleaned or replaced frequently. Conventional filters are used upstream from the membrane system to remove relatively large suspended solids and to let the membrane do the job of removing fine particles and dissolved solids. In ultrafiltration, for many applications, no prefilters are used and ultrafiltration modules concentrate all of the suspended and emulsified materials.

Lyophilization

This is nothing but freeze drying. It is mostly useful for drying of thermo-labile material. Dehydration takes place under low temperatures and reduced pressures. Hence physical and chemical properties of the material are retained.

Freeze drying is the removal of water at low temperature from a specimen after it has been frozen. The solution can be frozen to the walls of the round bottom flask by rotating the flask in an alcohol bath at - 20°C. The flask is connected to one of the parts of a multiport chamber provided on the lyophiliser through a quick seal valve. The system is closed to the atmosphere and evacuated by a mechanical pump. The condenser coils are pre- cooled to -40^0 C to ensure efficient condensation of water vapour on the condensing coils. When the pressure is 0.3 mm small amount of heat can be applied to the sample initially to enhance the rate of lyophilisation. Residual moisture can be removed by applying a secondary drier such as P_2O_5.

If a freeze-dried substance is sealed to prevent the reabsorption of moisture, the substance may be stored at room temperature without refrigeration, and be protected against spoilage for many years. Preservation is possible because the greatly reduced water content inhibits the action of microorganisms and enzymes that would normally spoil or degrade the substance. Freeze-drying also causes less damage to the substance than other dehydration methods using higher temperatures. Freeze-drying does not usually cause shrinkage or toughening of the material being dried. In addition, flavours, smells and nutritional content generally remain unchanged, making the process popular for preserving food. Freeze-dried products can be rehydrated (reconstituted) much more quickly and easily because the process leaves microscopic pores.

There are essentially three categories of freeze-dryers: the manifold freeze-dryer, the rotary freeze-dryer and the tray style freeze-dryer. Two components are common to all types of freeze-dryers: a vacuum pump to reduce the ambient gas pressure and a condenser to remove the moisture by condensation on a surface cooled to “40 to “80°C. The manifold, rotary and tray type freeze-dryers differ in the method by which the dried substance is interfaced with a condenser. In manifold freeze-dryers a short usually circular tube is used to connect multiple containers with the dried product to a condenser. The rotary and tray freeze-dryers have a single large reservoir for the dried substance.

Rotary freeze-dryers are usually used for drying pellets, cubes and other pourable substances. The rotary dryers have a cylindrical reservoir that is rotated during drying to achieve a more uniform drying throughout the substance. Tray style freeze-dryers usually have rectangular reservoir with shelves on which products, such as pharmaceutical solutions and tissue extracts, can be placed in trays, vials and other containers.

Manifold freeze-dryers are usually used in a laboratory setting when drying liquid substances in small containers and when the product is used in a short period of time. A manifold dryer will dry the product to less than 5% moisture content. Without heat, only primary drying (removal of the unbound water) can be achieved. A heater must be added for secondary drying, which will remove the bound water and will produce a lower moisture content.

Tray style freeze-dryers are typically larger than the manifold dryers and are more sophisticated. Tray style freeze-dryers are used to dry a variety of materials. A tray freeze-dryer is used to produce the driest product for long-term storage. A tray freeze-dryer allows the product to be frozen in place and performs both primary (unbound water removal) and secondary (bound water removal) freeze-drying, thus producing the driest possible end-product. Tray freeze-dryers can dry products in bulk or in vials or other containers. When drying in vials, the freeze-dryer is supplied with a stoppering mechanism that allows a stopper to be pressed into place, sealing the vial before it is exposed to the atmosphere. This is used for long-term storage, such as vaccines.

11

CENTRIFUGATION

When a particle revolves around an axis, a force is developed and acts away from the axis of rotation. This force is called centrifugal force. A particle suspended in a medium takes some time to sediment, which depends upon the size, molecular weight, and viscosity or density of the medium under normal "g" (gravity). To speed up the process of sedimentation, the external field of force is allowed to act on the particle which depends also on the distance of the particle from the axis of rotation in addition to the other factors.

Centrifugation is classified into two types: 1. Analytical and 2. Preparative. Analytical centrifugation is useful in studying the physical properties while preparative centrifugation is useful in separation of the sediment and supernatants. Ultracentrifugation is a process of separation of particles whose molecular weights are very high and which do not get sedimented under fields of force applied in ordinary centrifugation. To achieve ultra speeds, extra protection is offered to the motor by application of refrigeration and vacuum.

Centrifugal force: It is nothing but an extra gravitational force. If a stone tied to a string is swirled around holding the string at the other end, the stone is subject to a centrifugal force, which is acting away from the centre of axis. The faster the speed of rotation (measured as angular velocity or rotations / minute), the greater will be the force. The longer the radius of rotation the greater is the force.

Centrifugal force = (angular velocity)^2 x radius (γ)

Angular velocity depends on rotations/minute

$$\text{Angular velocity} = 2\pi \text{ X } \frac{\text{rpm}}{60} \text{ radians/sec}$$

Centrifugal field is expressed as relative centrifugal field (RCF) in 'g' units.

RCF = $4\pi^2$ (rpm)^2 x r / 3600 x 980, "g" units

Or 1.11 x10^{-5} (rpm)^2 x r, "g" units.

Centrifuge

A centrifuge is an instrument which produces centrifugal force. Basically it has containers fixed in such a way, that they can be rotated around the central axis with the help of electric motors. There are centrifuges which produce speeds up to 20,000 rpm. In these cases the friction of the rotor with air produces so much heat that they have to be run under refrigeration.

Machines with speeds up to 60,000 rpm produce gravitational forces of the order of 1,00,000 'g' are called ultracentrifuges. The rotors here are spun under refrigeration and vacuum to reduce friction. The rotors also need to be made of special metal to withstand the great force. There are two types of rotors: 1. Angle head and 2. Swinging bucket. In the former the samples are kept at 30^0 angle and in the latter they are horizontal. The swinging bucket method produces more gravitational force.

Density gradient centrifugation

Density gradient or zonal centrifugation is widely used and is a versatile procedure for separating proteins and other macromolecules. A continuous density gradient of sucrose is first prepared in a centrifuge tube by mixing concentrated sucrose solution and water in decreasing ratio. The density is more at the bottom and less at the top. The mixture of macromolecules is layered on the top of the tube. Centrifugation horizontally at high speed causes the molecules to sediment down the density gradient. The positions of the macromolecule (protein) bands can be detected optically or by staining or collecting different fractions through a pin hole at the bottom. The plastic centrifuge tube can be frozen and cut into thin slices.

Centrifuges and their applications

Centrifuges can be classified into four types. 1. The small bench centrifuge,

2. Refrigerated centrifuge, 3. High speed refrigerated centrifuge, 4. Ultracentrifuges - preparative or analytical type.

Bench centrifuge: These are the simplest and are available in different types and designs. They are generally cheaper. They are useful to collect small quantities of sediments. The speeds range from 4000-6000 revolutions / min, with relative centrifugal forces of 3000- 7000g. They operate at ambient temperatures. Small centrifuges are available with instant acceleration to speeds of 8000- 13,000 rev/min (10,000g). These centrifuges are useful for sedimenting small volumes (0.25ml- 1.5 ml) of material very quickly (1-2 min). Typical applications include sedimentation of blood samples.

Large refrigerated centrifuge: Maximum speeds of 6000 rev/min (6500g) can be reached in these centrifuges. They have refrigerated rotor chambers. They have carrying capacities of 10, 50, and 100 ml. Large capacity centrifuges are also available (1000ml). Centrifuge tubes and contents should be balanced accurately (<0.25g). These instruments are used to compact or collect substances that sediment rapidly, Ex. erythrocites, coarse or bulky precipitates, yeast cells, nuclei, and chloroplasts.

High speed refrigerated centrifuges: These centrifuges have speeds around 25,0000 rev/min (60,000g). They generally have a capacity up to 1500 ml. These centrifuges are used to collect microorganisms, cellular debris, organelles and proteins (NH_4 SO_4 precipitated). They are not useful to sediment viruses or small organelles (ribosomes).

Continuous flow centrifuges: The rotor through which particles are suspended in the medium flow continuously, is long, tubular and non- interchangeable. As medium enters the rotating rotor, particles are sedimented against its walls and excess clarified medium overflows through an outlet port. It is used to harvest bacteria or yeast cells from culture media. High speed continuous flow centrifuges are also available.

Ultracentrifuge: Speeds up to 80,000 rev/min(6,00.000g) can be attained in the ultracentrifuge. The rotor chamber is refrigerated, sealed and evacuated (to absorb temperature of friction). Centrifuge tube balancing should be up to 0.1g. For safety, the rotor chambers are always enclosed in heavy armor plating. In the analytical centrifuge, light absorption system, detecting changes in refrigerative index of the solution is available.

Determination of molecular weight: Centrifuges are useful in determining the molecular weight of the components being centrifuged. Sedimentation coefficient of the molecule is initially determined by the boundary sedimentation or zonal (band) sedimentation. The particles are uniformly distributed through the solution in the analytical cell at the start. When the ultra centrifuge is operated the particles migrate through the solvent radially outwards from the centre of rotation. The sedimentation rate is measured and utilized in the molecular weight (M.W.) determination.

$$Mr = \frac{RTS}{D(1 - r\rho)} \text{ (Svedberg equation)}$$

Mr = relative mass, D = diffusion coefficient,

r = partial specific volume of the molecule,

ρ = density of the solvent at 20°C, R = molar gas constant,

T = absolute temperature in K, S = sedimentation coefficient.

Differential centrifugation: The method is based on differences in the sedimentation rate of particles of different size and density. Centrifugation will initially sediment the largest particles. Those with the highest density (Ex. peroxisomes with density of 1.23g/cc. in sucrose) will sediment at a faster rate than the less dense particles (Ex. plasma membranes, density of 1.16g/cc).

In differential centrifugation the material to be separated (Ex. tissue homogenate) is divided into a number of fractions by increasing (stepwise) the applied centrifugal field. It is chosen so that a particular type of material sediments. Separation can be improved by re-suspension and repeated differential centrifugation.

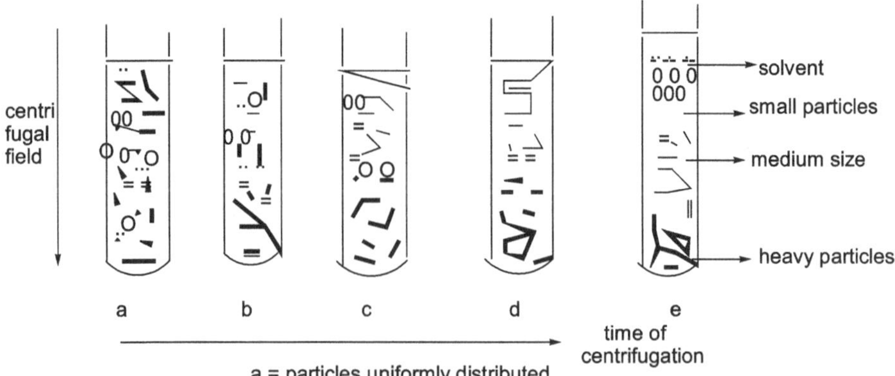

a = particles uniformly distributed,
b to e = sedimentation taking place stepwise

12

GAS LIQUID CHROMATOGRAPHY

Gas chromatography is one of the most valuable techniques available for organic analysis. It has been applied for the separation, identification and analysis of virtually all volatile compounds. In gas chromatography, the mobile phase is gas. When the stationary phase is liquid , it is called gas liquid chromatography (GLC). If the stationary phase is solid, it is called gas solid chromatography (GSC). The sample is transported through a stationary phase with help of an inert carrier gas, namely the mobile phase. The fractionation of the components in the sample mixture takes place based on the partitioning between the stationary and mobile phases.

An inert solid base is coated with a liquid like paraffin oil or silicone oil and packed in long tubes. This column is kept in an oven, which can be heated to the desirable temperatures (60°-250°C) depending upon the components to be separated. An inert gas like helium or nitrogen is passed through as a carrier. The mixture of gases or volatile sample mixture is injected into the carrier gas flowing through the column at one end. The components in the sample get adsorbed, released and re-adsorbed similar to that in the column chromatography. The component with the least adsorptivity gets eluted first followed by the ones with lower adsoptivities. Suitable detectors when placed at the exit end of the column, detect the eluting samples. Recorders and other instruments can capture the signals and quantify the samples.

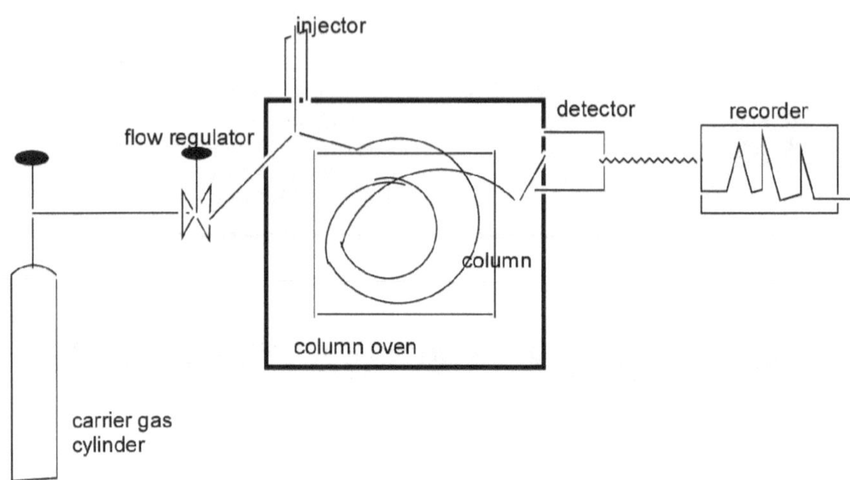

A Schematic view of gas chromatograph

Components of a gas chromatography (GC) set up

The main components of gas chromatography are 1. Carrier gas (mobile phase), 2. Injector port, 3. Column (containing stationary phase), 4. Column oven (to enclose the column), 5. Detector, 6. Recorder and/or Integrator.

Carrier gas: An easily available inert gas which can carry the samples to be analyzed, is generally used in GC. It should not explode and should be cheap. The usual gases are hydrogen, helium, argon and nitrogen. Hydrogen and helium are preferred gases in the TCD (thermal conductivity detector) due to their high thermal conductivity. Nitrogen and helium are used in FID (flame ionization detector). Argon with 5% methane is used in ECD (electron capture detector). The gas is present in the compressed form in a cylinder. The gas is released from the cylinder through a regulator at a specified rate (1-2 ml /min). The flow rate is very important for efficient separation.

Injector: The sample is introduced into the GC column through the injector. A hypodermic syringe is used for introducing the sample at µl levels. A self sealing septum is present at the entrance of the injector. The sample should get vaporized immediately and hence the injector port is heated to a particular temperature, preferably 25^{0} C above that of the column temperature.

Column: It is the heart of the instrument. This is the one which brings about remarkable separations. The columns contain the stationary phase. It is a liquid of zero vapour pressure and low viscosity. Stationary phases are either polar or non-polar. Their use depends on specific purpose of separations and high thermal stability. Different types of columns are available. The main types are 1. Packed columns 2. Capillary colums and 3. Support coated open tubular columns.

1. **Packed columns:** Packed columns (5m x 2-4 mm) are prepared by packing metal or glass tubes with granular stationary phase. For GSC, the columns are packed with size graded adsorbents or porous polymers, while for GLC the packing is prepared by coating the liquid phase (1-15%) over a size graded inert support. Solid supports like activated carbon, silica gel, alumina, porapak etc. are used as solid stationary phases. Liquid phases like paraffin oil, squalene, apiezon L grease, silicone gum rubber, dinonyl phthalate, carbowaxes etc. are utilized as GLC as stationary phases.
2. **Capillary columns:** Fused silica columns, in which the stationary phase is coated on the inner surface are used as capillary columns. The length of capillary columns range from 5m to 100m with 0.1 to 0.75 mm in diameter. The columns can be operated up to 450^0 C. Polyamide plastic is used as a coating on the inert solid support. Capillary columns are more efficient than packed columns.
3. **Support coated open tubular columns:** These have a finely divided layer of solid support material deposited on the inner wall. The stationary phase is then coated over the support material. There is a variation called *wall coated open tubular columns*. In this case the stationary phase is directly coated on to the inner wall of the tubing.

Thumb rules in separation

i) Like separates like. Polar compounds get separated better on polar columns, while non polar compounds get easily separated on non-polar columns.

ii) The longer is the column, the better it is for separation. Generally 25m column is considered as a standard.

iii) The smaller the diameter of the column, the more efficient it is for the separation of components. An internal diameter of 0.1 mm is the most efficient.

iv) Normally 1μm thick films are prepared in GLC. But thicker films by cross linking the inert support and mobile phase can be obtained. Thicker films appear to be more efficient.

Stationary phases used in gas chromatography

Liquid phase	Trade name	Application
Non-polar		
Ploy dimethyl siloxane	OV-1,OV-0,SPB-1SE-30, RSL150, HP-1	Amines, fatty acid methyl esters, waxes
Poly (5% diphenyl-95% diphenyl siloxane)	DB-5, SPB-5, OV-73, SE-54	Fatty acid methyl esters, tocopherols
Intermediate polar		
Poly (6-50%) cyano propyl (14-94%)demethyl siloxane	HP1310,DB624, HP1701, OV-17,SP-2250	Alcohols, pesticide residues, glycols
Polar		
Poly ethylene glycol	DB- WAX, CP-WAX, SP2330	Fatty acid methyl esters, aromatic compounds, positional geometric isomers

Detectors

There are many types of detectors in GC. The most important ones are thermal conductivity detector (TCD), flame ionization detector (FID), and electron capture detector(ECD).

Thermal conductivity detector: The principle of operation of TCD is based on the heat loss phenomenon. The rate of loss of heat from a body depends on thermal conductivity of the surrounding gas. It is a function of the composition of the body. Thus the rate of loss of heat can be used to decide on the composition of the body. In the TCD detector, two pairs of matched filaments in opposite arms of a bridge are surrounded by the carrier gas while the other two are surrounded by the effluent from the GC column.

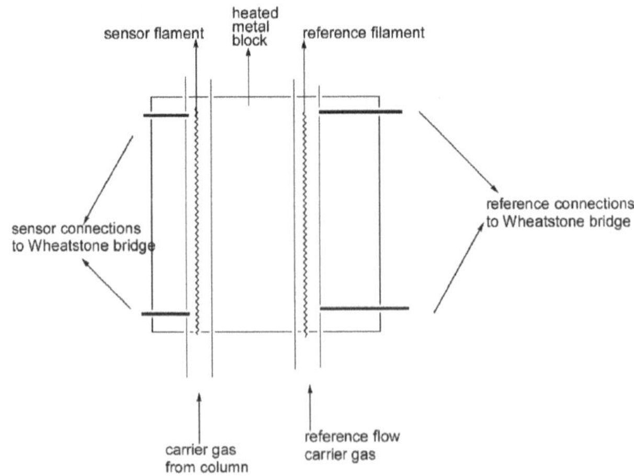

Thermal conductivity detector

When pure carrier gas passes over both the reference and sample filaments the bridge is balanced, but when the vapor from the column emerges, the rate of cooling of sample filaments changes and the bridge becomes unbalanced. The extent of imbalance is a measure of the concentration of vapor in the carrier gas. This is measured using a recorder producing a chromatogram.

Flame ionization detector: A tiny flame of hydrogen is maintained at the exit end of the column. Air or oxygen is introduced through a side inlet for supporting the combustion. Column effluents are led into the flame wherein ionization of components takes place. An electrode system closely located picks up the current, which is then amplified and suitably fed to a recorder to obtain the chromatogram. Inert gases like nitrogen, argon, helium, neon etc. are utilized as carrier gases. In the absence of eluent the current is adjusted to zero. FID is the most popular GC detector due to its high stability and sensitivity. It can be used up to a temperature of 400^0 C.

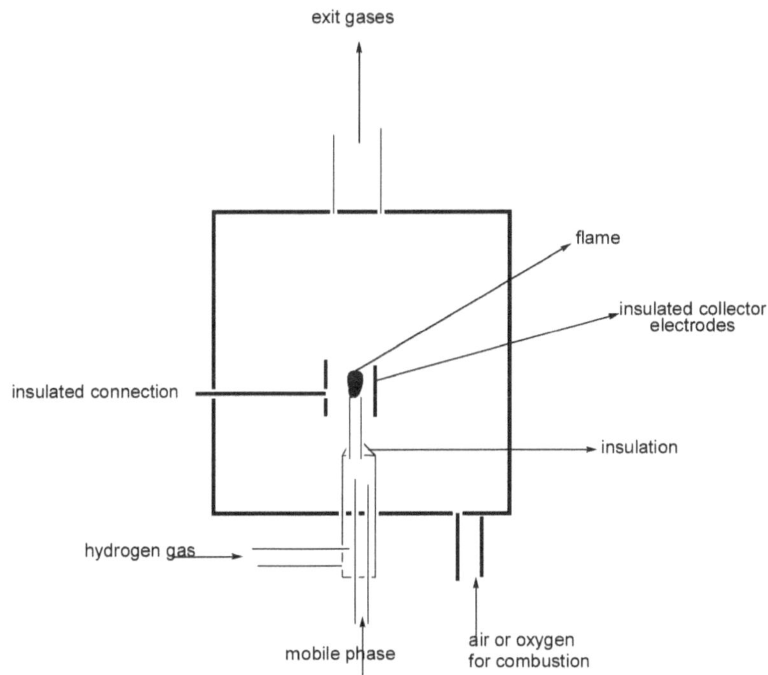

Flame Ionization Detector

Electron capture detector: ECD is a specialized detector. It responds only to those compounds whose molecules have an affinity for electrons, Ex. chlorinated compounds, alkyl lead etc. A γ ray source (Tritium or Ni^{63}) is used to generate slow electrons by ionization of the carrier gas (N_2) flowing through the detector. These electrons migrate to the anode under a fixed potential and give rise to a steady base line current. When an electron capturing gas (namely eluate) emerges from the column and reacts with an electron, it reduces the current flow. This loss of current is traced by the recorder as a peak. The pesticide residue analysis is best carried out by this ECD.

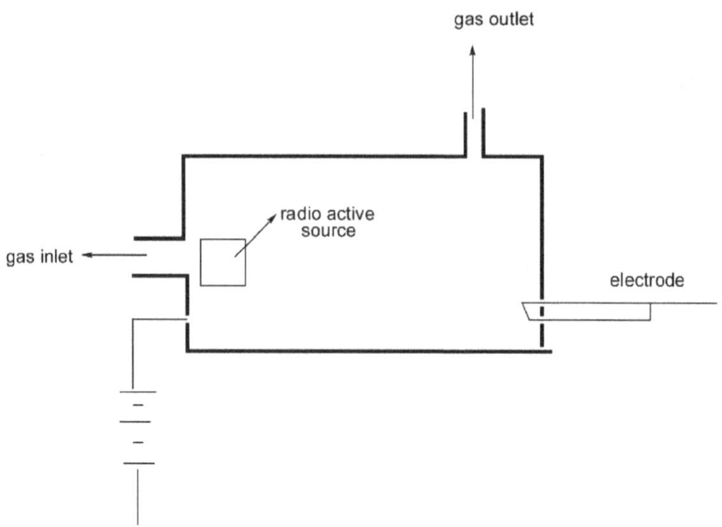

Schematic Diagram of ECD

More detectors: Other important detectors are nitrogen phosphorus detector (NPD), flame photometric detector (FPD), and photo ionization detector (PID). NPD is a sensitive detector and gives strong response to organic compounds containing nitrogen and or phosphorus. FPD is most useful for measuring the content of compounds containing sulfur and phosphorus, which emit radiation at characteristic wave lengths. The emitted light is measured using a photomultiplier detector. PID utilizes the properties of photoluminescence. PID has a U.V. lamp to emit photons. PID is useful to detect volatile organic compounds, heterocyclics, organosulphurs and organo-metallic compounds.

Recorders, integrators and identification: Other items like recorder and integrator are used to quantify the eluting peaks. These days the computer takes care of these objectives. Identification of the eluate peaks is carried out by comparing R_T(retention time) values of standards with unknowns. Internal standards are run for better quantification of peaks. R_T value is the retention time of the sample peak with reference to injection point. Columns are maintained at isothermal or programmed temperatures to achieve better separation of components depending upon the need. The areas of the peak are either calculated manually or with the help of integrators or computers. The area of the peak as percent of total of all the peaks is considered as percent concentration(of the component).

Some details of detectors and their utility

Detector	Type	Support gases	Compounds analysed	Detectability
FID	Mass flow	Hydrogen and air	Most organic compounds	100pg
TCD	Concentration	Reference	General purpose	One ng
ECD	Concentration	Make-up	Halides, nitrates, nitriles. organometallics	50fg
NPD	Mass flow	Hydrogen and air	Nitrogen, phosphorus	10pg
FPD	Mass flow	Hydrogen and air, also O_2	S,P, Sn, B, Ar, Ge, Se, Cr Volatile organics, heterocyclics, organophosphorus	100pg
PID	Concentration	Make-up	Organometallics	2pg

ng = 10^{-9}g, pg = 10^{-12}g, fg = 10^{-15}g

13

HIGH PRESSURE LIQUID CHROMATOGRAPHY

High pressure (or performance) liquid chromatography (HPLC) is more versatile than GC since it is not limited to volatile compounds. The choice of mobile and stationary phases is wider. HPLC has high resolving power, separations are speedier, quantifications are more precise and it is in general, more reproducible. Liquid chromatography basically involves separation of components due to differences in the equilibrium distribution of sample components between the mobile (liquid) and stationary phases.

To effect separations the sample mixture is dissolved in a mobile phase and introduced through an injection port into the column containing the stationary phase. The sample moves ahead under pressure with the help of a pump. The separated components moving out of the column containing stationary phase is fed to a detector and the concentration of the eluates are measured using a detector and a recorder (or computer). The normally used detectors are U.V.-Visible spectrophotometer and refractometer.

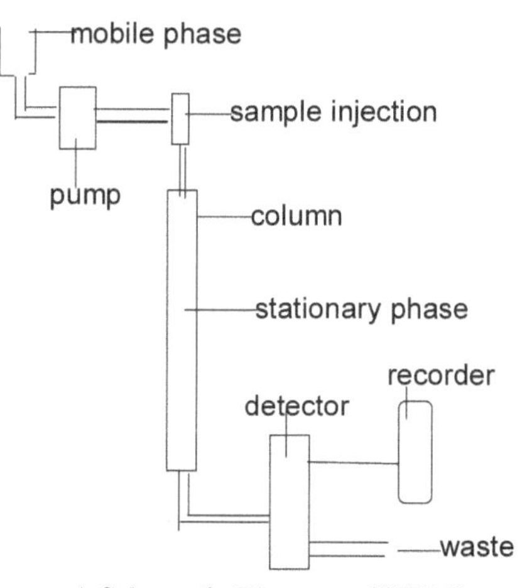

A Schematic Diagram of HPLC

Liquid chromatography involves separation of components due to differences in the equilibrium distribution of sample components between two phases, a liquid mobile phase and a stationary phase. Because of the narrow bore of the column and the tight packing of the stationary phase (adsorbent), very high pressures are necessary to drive the solvent/ mobile phase. The separations are better, namely, the instrument performance is better. Hence the chromatography is called high pressure/ high performance liquid chromatography. The important parts of HPLC are 1. mobile phase, 2. stationary phase, 3.column, 4.injector and 5.detector.

Mobile phase: The mobile phase acts as a carrier for the sample solution. There are two types of mobile phases. The first one isocratic and the other one is gradient solvent. In the isocratic elution the compounds are eluted using a constant mobile phase and constant flow. In the gradient flow, the solvent composition and flow rates are programmed for better elution and separation. The different components in the sample migrate depending on the covalent interactions of the component with the column matrix. The enormous number of mobile phases along with gradient possibilities provides a great opportunity for separations of complex mixtures in regard to resolution and time.

Normally isocratic separation is preferred over gradient evolution since it is less expensive. In GC, the separation factor (S, the ratio of retention times), for a pair of sample components depends only on the nature of the stationary phase. In LC, it depends on both the mobile and stationary phases.

Stationary phase: This is the solid support contained within the column over which the mobile phase continuously flows. As the sample solution flows with the mobile phase through the stationary phase in the column, the components migrate according to the non-covalent interactions of the compounds with the stationary phase. There are different types of stationary phases namely, normal phase, reverse phase, ion exchange, size exclusion and affinity types. Two types of columns, namely, normal phase and reverse phase columns are widely used. In the former one a polar stationary phase and a non polar mobile phase are used. In the latter type, a non polar stationary phase and an aqueous or moderately polar mobile phase are used.

Columns: Different types of columns, namely, analytical, semi-preparative and preparative type columns are the most widely used columns in HPLC. Analytical columns have diameters in the range of 3.0 to 4.6 mm while semi-preparative columns have 4.6 to 19.0 mm diameter and preparative columns have 7.5 to 50 mm diameter. Analytical columns are the most common. Narrow bore and capillary columns are also available.

Injectors: The injection port of the HPLC commonly consists of an injection valve and the sample loop. The sample is dissolved in the mobile phase and injected into the loop with the help of a syringe. Volumes ranging from 10μl to 500μl are injected manually or through an automatic injector.

Detectors: There are many types of detectors that are used with HPLC. The most common ones are 1. Refractive index (RI), 2. Ultra-violet(UV), 3. Fluorescent (FL), 4. Evaporative light scattering (ELSD) and 5. Mass detector (MS).

Refractive index (RI) Detector: The property of refractive index, namely, the ability of sample molecules to bend or refract light is utilized in the RI detector. It monitors the difference in refractive index between the pure mobile phase and the sample eluent continuously. This detector is useful under isocratic elution. It is highly sensitive to changes in temperature. Volatile samples are rarely used. Detection up to 10μg is possible. RI detectors are used in lipid analysis.

U.V. Detector: UV detector measures the ability of samples to absorb UV light. They have a sensitivity of 10^{-8} to 10^{-9} g/ml. The most widely used solvents are hexane, isopropanol, acetonitrile, methanol and water. Solvents which absorb in the range of 200-210 nm should be avoided.

Fluorescent (FL) Detector: The fluorescence property is utilized in the FL detector. Detection up to 1.0μg can be achieved with FL detector. The sensitivity is 100 times better than U.V. detector. Those samples which do not exhibit fluorescence need to be converted to a derivative which exhibits fluorescence.

Evaporative light scattering (ELSD): The detection principle of ELSD involves nebulization of the column effluent to form an aerosol, followed by sample evaporation. The particles enter the light scattering cell and the scattered light by particles serve to quantify the eluent. This detector has wide applicability as it responds to many solvents. The instrument gives excellent results under gradient elution conditions. The sensitivity is comparable to that of RI detector. Mobile phase, temperature and flow rates do not affect the sensitivity.

Mass detector (MS). The MS detector is useful in quantification as well as structural elucidation. It is more sensitive than UV detector. MS detector is highly useful in identification of components in a mixture. An MS detector consists of three main parts, namely, a) ionization source, b) the mass analyzer and c) the electron multiplier. Again three types of ionization techniques are available. They are a) electro-spray ionization(ESI), b) atmospheric pressure chemical ionization(APCI) and c) matrix assisted laser sorption (MALDI).

Applications: Even non volatile compounds can be fractionated in HPLC. Thus it is better than GC. Recovery of eluents is better and hence there is scope for their structural elucidation.

14

GC/MS AND LC/MS

The mass spectrometry (MS) is not very well suited to the analysis of mixtures of similar substances because of the multiplicity of fragment ions. The gas chromatography (GC), of course, has no ability to identify compounds on its own. The GC/MS can easily overcome these defects and can identify the components (fractionated by GC). The combined instrument of GC/MS is an unexcelled instrument that can identify the compounds after their fractionation.

The GC-MS is composed of two major building blocks: the gas chromatograph and the mass spectrometer. The gas chromatograph utilizes a capillary column which depends on the column's dimensions as well as the stationary phase properties. The difference in the chemical properties between different molecules in a mixture and their relative affinity for the stationary phase of the column will promote separation of the molecules as the sample travels the length of the column. The molecules are retained by the column and then elute (come off) from the column at different times and this allows the mass spectrometer downstream to capture, ionize, accelerate, deflect, and detect the ionized molecules separately.

The pressure of the gas stream exiting from a GC is essentially atmospheric whereas, that admitted to the ionizing source of the MS cannot be greater than about 10^{-3} Pa (Pascal- unit of pressure, 10^{5} Pa = 1atmosphere) for the electron impact (EI) or 10 Pa for chemical ionization (CI). The amount of a sample component is generally only a small fraction of the eluent from the GC. The carrier gas and the sample are separated using a separator. The carrier gas moving out is pumped away while the heavier sample molecules enter the MS. The mass spectrometer carries out a process of breaking each molecule into ionized fragments and detecting these fragments using their mass-to-charge ratio.

These two components (GC and MS), used together, allow a much finer degree of substance identification than either unit used separately. It is not possible to make an accurate identification of a particular molecule by gas chromatography or mass spectrometry alone. The mass spectrometry process normally requires a very pure sample while gas chromatography using a traditional detector (e.g. Flame ionization detector) cannot differentiate between multiple molecules that happen to take the same amount of time to travel through the column (*i.e.* have the same retention time), which results in two or more molecules that co-elute. Sometimes two different molecules can also have a similar pattern of ionized fragments in a mass spectrometer (mass spectrum).

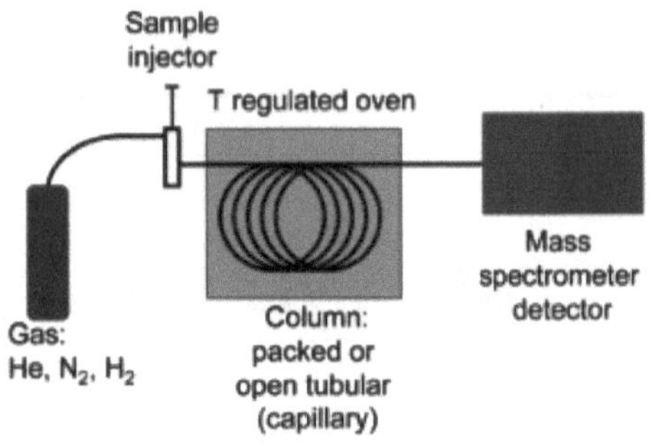

GC-MS Schematic

Combining the two processes reduces the possibility of error, as it is extremely unlikely that two different molecules will behave in the same way in both a gas chromatograph and a mass spectrometer. Therefore, when an identifying mass spectrum appears at a characteristic retention time in a GC-MS analysis, it typically increases certainty that the analyte of interest is in the sample.

Purge and trap GC-MS

For the analysis of volatile compounds, a purge and trap (P&T) concentrator system may be used to introduce samples. The target analytes are extracted and mixed with water and introduced into an airtight chamber. An inert gas such as nitrogen (N_2) is bubbled through the water; this is known as purging. The volatile compounds move into the headspace above the water and are drawn along a pressure gradient (caused by the introduction of the purge gas) out of the chamber. The volatile compounds are drawn along a heated line onto a 'trap'. The trap is a column of adsorbent material at ambient temperature that holds the compounds by returning them to the liquid phase. The trap is then heated and the sample compounds are introduced to the GC-MS column via a volatiles interface, which is a split inlet system. P&T GC-MS is particularly suited to volatile organic compounds (VOCs).

Types of mass spectrometer detectors

The most common type of mass spectrometer (MS) associated with a gas chromatograph (GC) is the quadrupole mass spectrometer, sometimes called "Mass Selective Detector" (MSD). Another relatively common detector is the ion trap mass spectrometer. Other detectors may be encountered such as time of flight (TOF), tandem quadrupoles and magnetic sector mass spectrometer (MS-MS).

Applications

1. **Environmental monitoring and cleanup:** GC-MS is becoming the tool of choice for tracking organic pollutants in the environment. There are some compounds for which GC-MS is not sufficiently sensitive, including certain pesticides and herbicides, but for most organic analysis of environmental samples, including many major classes of pesticides, it is very sensitive and effective.
2. **Forensic:** GC-MS can analyze the particles from a human body in order to help link a criminal to a crime.
3. **Law enforcement:** GC-MS is increasingly used for detection of illegal narcotics, and may eventually supplant drug-sniffing dogs.
4. **Doping and sports:** GC-MS is the main tool used in sports anti-doping laboratories to test athletes' urine samples for prohibited performance-enhancing drugs, for example anabolic steroids.
5. **Security:** Explosive detection systems (GC-MS) have become a part of airports. These systems run on a host of technologies.
6. **Food, beverage and perfume analysis:** Foods and beverages contain numerous aromatic compounds, some naturally present in the raw materials and some forming during processing. GC-MS is extensively used for the analysis of these compounds which include esters, fatty acids, alcohols, aldehydes, terpenes etc. It is also used to detect and measure contaminants from spoilage or adulteration which may be harmful.
7. **Medicine:** Dozens of congenital metabolic diseases also known as Inborn Error of Metabolism are now detectable. GC-MS can determine compounds in urine even in minor concentration. It is now possible to test a newborn for over 100 genetic metabolic disorders by a urine test at birth based on GC-MS.

LC-MS System

The LC/MS is a tougher instrument. There are two ways in which LC/MS is used. The first one involves solvent evaporation. The effluent from the LC is split and a selected fraction (up to 1ml/min) is deposited and IR(infra red) heater evaporates the solvent which is removed. The sample then enters the MS. The second one involves direct injection of the effluent. The solvent is made to play the role of CI reagent in producing ions from solutes.

There are many different mass analyzers that can be used in LC/MS. Single quadrupole, triple quadrupole, ion trap, time of flight (TOF) and quadrupole-time of flight (Q-TOF). The interface between a liquid phase technique which continuously flows liquid, and a gas phase technique carried out in a vacuum was difficult for a long time. The advent of electrospray ionization(EI) improved the situation. The interface is most often an electro-spray ion source or a variant such as a nano-spray source. Sometimes, atmospheric pressure chemical ionization interface is also used. Various deposition and drying techniques have also been used such as using moving belts. However, the most common of these is off-line MALDI deposition. A new approach still under development called Direct-EI LC-MS interface, couples a nano HPLC system and an electron ionization equipped mass spectrometer.

Applications

1. **Pharmacokinetics:** LC-MS is very commonly used in pharmacokinetic studies of pharmaceuticals and is the most frequently used technique in the field of bioanalysis.
2. **Proteomics/metabolomics:** LC-MS is also used in proteomics where again components of a complex mixture must be detected and identified in some manner. Profiling of secondary metabolites in plants or food like phenolics can be achieved with liquid chromatography–mass spectrometry.
3. **Drug development :** LC-MS is frequently used in drug development at many different stages including peptide mapping, glycoprotein mapping, natural products de-replication, bioaffinity screening, *in vivo* drug screening, metabolic stability screening, metabolite identification, impurity identification, quantitative bioanalysis, and quality control.

The GC-MS, LC-MS techniques have both qualitative and quantitative uses. These include identifying unknown compounds, determining the isotopic composition of elements in a molecule, and determining the structure of a compound by observing its fragmentation. Other uses include quantifying the amount of a compound in a sample or studying the fundamentals of gas phase ion chemistry (the chemistry of ions and neutrals in a vacuum). MS is now in very common use in analytical laboratories that study physical, chemical, or biological properties of a great variety of compounds.

15

SPECTROPHOTOMETRY

Spectrophotometry measures how much a chemical substance absorbs light by measuring the intensity of light as a beam of light passes through the sample solution. Chemical compounds normally absorb or transmit light over a certain range of wavelength. Hence spectrophotometry is used to measure the amount of a known chemical substance. It is one of the most useful methods of quantitative analysis in the laboratory and is widely used.

A colorimeter or spectrophotometer is an instrument that measures the amount of photons (the intensity of light) absorbed after it passes through the sample solution. With the spectrophotometer, the amount of a known chemical substance (concentrations) can also be determined by measuring the intensity of light detected. Depending on the range of wavelength of light source, it can be classified into two different types, namely, 1) UV-visible spectrophotometer which uses light over the ultraviolet range (185-400nm) and visible range (400-700nm) of electromagnetic radiation spectrum and 2) IR spectrophotometer which uses light over the infrared range (700-15000nm) of electromagnetic radiation spectrum.

Light consists of radiation to which the human eye is sensitive. Light waves of different wavelengths give rise to light of different colours. A combination of different wavelengths, namely, Violet, Indigo, Blue, Green, Yellow, Orange and Red (VIBGYOR) make up the white light. The visible spectrum ranges from 400-700 nm. The actual electromagnetic spectrum is far wider and ranges from γ (gamma) - rays, 10^{-14} nm to audible frequency, 10^{8} m. The ultra violet light (U.V. – below violet light)) frequency ranges from 200-400 nm while infra red (I.R. – beyond red light) ranges from 800-1500 nm. X- rays and γ- rays have shorter wave lengths than U.V. light while radio and audio waves have longer wave lengths than I.R.

X-rays, gamma rays etc. ←	U.V.	VISIBLE	I.R.	→ Audio, radio fequency etc.
	200	400	750	1500 nm

Wavelength

$$Wave\ no(\gamma) = \frac{1}{\lambda} = \frac{\text{frequency (v)}}{\text{velocity of light(C)}}$$

$$\gamma = \frac{1}{\lambda} = \frac{V}{C} \quad \text{where } C = 10^{10}\ \text{cm/sec}$$

Wave no (γ).= 1/λ, waves/cm

Frequency (η) = C/λ waves/sec

Theory of spectrophotometry

When light either monochromatic or heterogeneous falls on a homogeneous medium, a portion of it is reflected, another portion is absorbed and the rest is transmitted.

$$I_0 = I_r + I_a + I_t$$

I_0 = initial light, I_r = reflected light, I_a = absorbed light, I_t = transmitted light

For air-glass interfaces, about 4% of light only is reflected, and hence it is usually neglected/ eliminated using a control.

$$I_0 = I_a + I_t$$

The absorption of light by a substance can be quantized for practical use with the help of two laws. They are Beer's law and Lambert's law.

Beer's law: When a beam of monochromatic light enters an absorbing medium, the intensity of light coming out decreases exponentially with the increase in concentration of light absorbing constituent in the medium.

Lambert's law: This law states that under similar conditions, the intensity of light coming out decreases exponentially with the increase in length of the medium through which the light passes.

$$A \text{ (Absorbance)} = \log I_0/ I_t = \log 100/T = 2 - \log T$$

T= transmittance

$A = € \times c \times l$

where c = concentration, l = length in cm through light passes

and € = extinction coefficient.

When c = 1 mol/ lit and l = 1 cm, then $€ = €_M$ or molar extinction coefficient.

$€_M$ (molar extinction coefficient) is a constant for a particular compound

The transmittance is simply the fraction of light in the original beam that passes through the sample and reaches the detector. In most applications, the amount of light absorbed is correlated to the concentration of the absorbing molecule. It turns out that the absorbance rather than the transmittance is most useful for this purpose. If no light is absorbed, the absorbance is zero (100% transmittance).

Based on the above laws and the derived equations, the concentration of a particular compound can be determined. Absorbency or transmission of light passing through them is measured and the concentration is calculated. The general term used to measure the absorption of radiation is absorptiometry. Colorimetry is the term used when absorption of radiation in the visible region is used. Colorimeter is used to measure the absorption of visible light while a spectrophotometer is used to measure U.V. light and visible light. When infra red light is involved, I.R. spectrophotometer is used.

Coloured solutions absorb radiation (light) characteristic of the solute or compound. For example Cu^{++} ion absorbs yellow (complementary colour) and is transparent to blue(thus it appears blue). Hence the concentration of copper in a solution is measured based on the intensity of absorption of yellow light. Generally the quantity of any coloured compound is measured like this based on the absorption of it's complementary colour.

Complementary colours

Wave length(nm)	Colour	Complementary colour
400-430	Violet	Yellowish green
400-480	Blue	Yellow
500-560	Green	Purole
580-595	Yellow	Blue
595-610	Orange	Greenish blue
610-750	Red	Bluish green

Colorimeter

Colorimeter is an instrument used to measure the intensity of light after it has passed through an absorbing medium. The colorimeter consists of a light source, a filter and a photo cell to measure the light intensity. A test tube (cuvette) containing a solvent (control) is placed on the light path. The solvent used should not absorb light at the particular wavelength of measurement. The photo cell measures the light I_0. Now another tube containing the coloured substance dissolved in the same solvent is placed in the light path. The length of the light path is 1cm. The photo cell measures the intensity of the light transmitted through the coloured solution placed in the test tube, namely I_t. The measured value of optical density (O.D.) is directly proportional to the concentration of the substance, since' l' the length and E the molar extinction coefficient are constant.

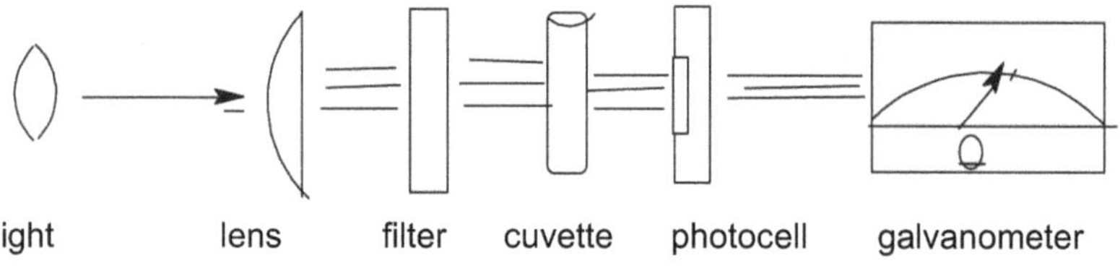

Schematic view of a Colorimeter

Standard graph: A graph is plotted using absorbance against different concentrations of standard solution. A straight line graph is obtained. Later the absorbance of an unknown(sample) is measured. It is extrapolated to read the concentration from the standard graph. Further calculations are made based on the quantity of sample weighed, volume of solvent used for extraction and dilutions made prior to reading the absorbance from the colorimeter.

Applications: Colorimetry has very wide applications. It is easy to measure the colour of coloured substances. A colorless substance can be converted to a coloured derivative and then the colour is measured.

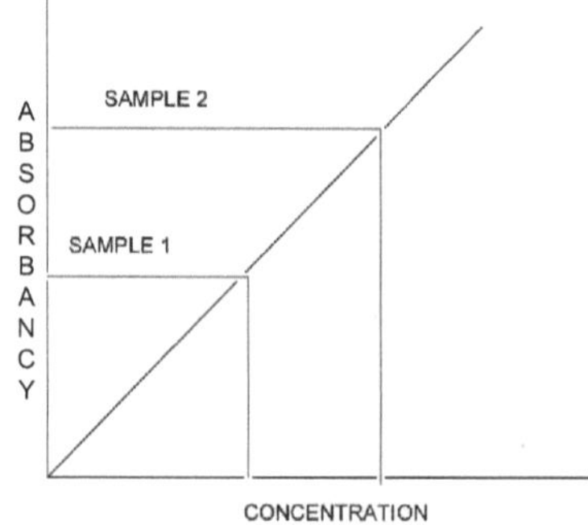

Deviations to Beer's law: Beer's law holds over a wide range of concentrations. Discrepancies arise when the coloured substances ionize, dissociate or associate in solution. The law doesn't hold when the solute forms complexes. Discrepancies also arise when monochromatic light is not used. Whenever discrepancies occur it is better to use a series of calibration curves prepared with two or more standards. Photo cell, cuvettes, etc. should be clean and chemicals used should also be highly pure.

Spectrophotometer

The spectrophotometer is an improved version of the colorimeter. In this equipment, instead of filter a prism or grating is used to split the light to a desired wavelength and made to pass through the sample cell. A spectrophotometer is employed to measure the amount of light that a sample absorbs. The instrument is also useful in recording the absorption spectrum of the test substance.

The spectrophotometer operates by passing a beam of light through a sample and measuring the intensity of light reaching a detector. The beam of light consists of a stream of photons. When a photon encounters an analyte molecule (the analyte is the molecule being studied), it gets absorbed. This absorption reduces the number of photons in the beam of light, thereby reducing the intensity of the light beam.

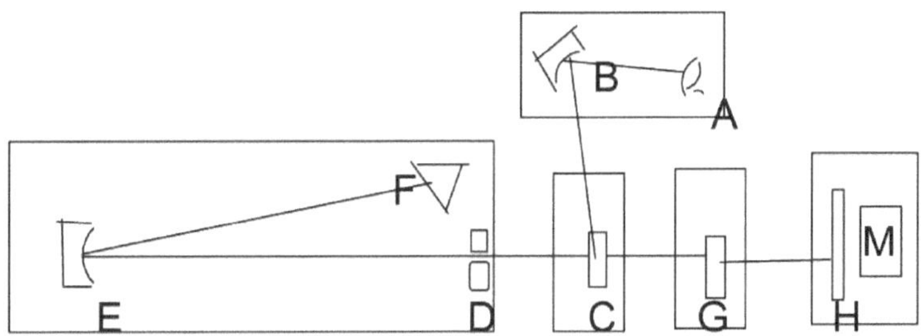

A = light source, B = mirror, C = mirror, D = entrance slit, E= collimating beam, F= prism, G = cuvette/cell, H=Photocell, M = meter

Schematic Diagram of a Spectrophotometer

Single beam spectrophotometer: The light from a source, namely tungsten halogen lamp, is focused on the prism using a mirror. The refracted light of selected wave length passes out and made to pass through the absorption cell or cuvette. Using instrumentation (electronics), the intensity of transmitted light is measured. Lot of automation exists and the absorbency or transmittance or calibrated concentration can be measured using a modern spectrophotometer.

Double beam spectrophotometer: The monochromatic light from tungsten or deuterium Lamp is divided into two identical beams, one of which passes through the reference cell while the other passes through the sample cell. The absorption at the reference cell automatically gets subtracted from that of the sample cell. Adjustments to dark current also is made by the photocell. Stray light and other interferences are also minimized. Double beam measurements are hence more accurate.

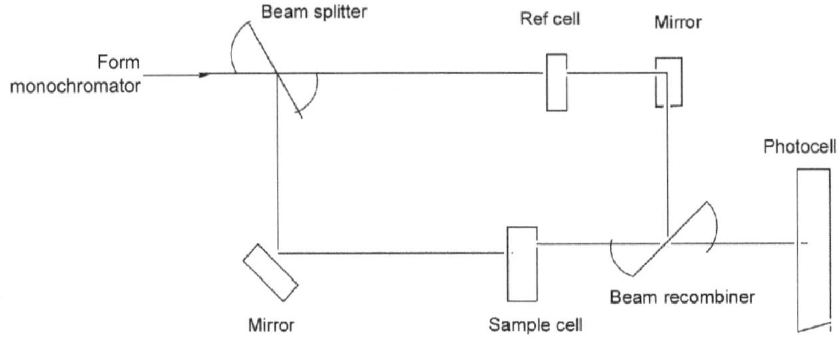

Schematic Diagram of Double Beam spectrophotometer

16

INFRARED SPECTROSCOPY

Infrared light (800-1500nm) interacts with molecules at the level of atomic vibrations. An infrared spectrum gives valuable information about the molecular structure. It is rarely used for quantitative purposes and hence concentration measurements are not possible. Infrared light detects vibrations characteristic of functional groups. For example the C=O (carbonyl) group absorbs at 1700 cm^{-1} (wave number). The main region of interest for analytical purposes is from 2.5- 25μm or 4000- 400 cm^{-1}. At wavelengths below 25μm the radiation has sufficient energy to cause changes in the vibrational energy levels of the molecule, and these are accompanied by changes in the rotational energy levels. The pure rotational spectra of molecules occur in the far infra red region (50-1000 μm) and are used for determining molecular properties.

Infrared spectroscopy is very important in organic chemistry. It is a routine tool for detecting functional groups, identifying compounds and analyzing mixtures. Instruments which record infrared spectra are commercially available and easy to use on a routine basis. Identification of inorganic compounds is not easy because their solvent, namely, water absorbs strongly beyond 1.5 μm. Also inorganic compounds have broad absorption bands.

I.R. spectra are used to identify pure compounds or impurities in samples. The I.R. spectrum of a compound is a sort of “fingerprint” of that compound. For identification of a pure compound, the spectrum of that compound is recorded and compared with that of unknown compound. When there is a match, the identification is complete. With a modern double beam spectrophotometer,

I.R. spectrum from 2.5 - 15 μm is scanned and recorded on a chart. Samples can be vapour, liquid or solid phases. CCl_4 and CS_2 are the chosen solvents as they are transparent over a wide range of wavelengths and can be used safely in cells made with rock salt.

The spectrum of a mixture of compounds is the sum of the spectra of the individual components (provided they do not react between themselves). To detect an impurity, the spectrum of the impure sample should be compared with that of a pure one. The impurity can cause an extra band in the spectrum. It will be advantageous if the impurity has a characteristic grouping not present in the pure sample.

Most groups such as C-H, O-H, C=O and C ≡ N, give rise to infra red absorption bands which vary only slightly from one molecule to another depending on other substituents. Many of these absorptions occur in the "fingerprint" region of the spectrum namely 6.5 to 14μm. It is easy to locate the absorption band associated with the presence of a functional group in the infra red spectrum.

Instrumentation

Lenses are never used in I.R. instruments. But, concave mirrors are used as they do not have chromatic aberrations. The absorption cells (cuvettes) are made of mostly NaCl or KBr. These cells must be carefully protected from moisture. A tungsten lamp with a glass or silica envelope is used as a source of near IR. A simple coil of nichrome wire can also be used as an IR source.

Infrared detectors are of two types. 1. The thermal detectors (thermocouple) and 2. The photon detectors (photoelectric)

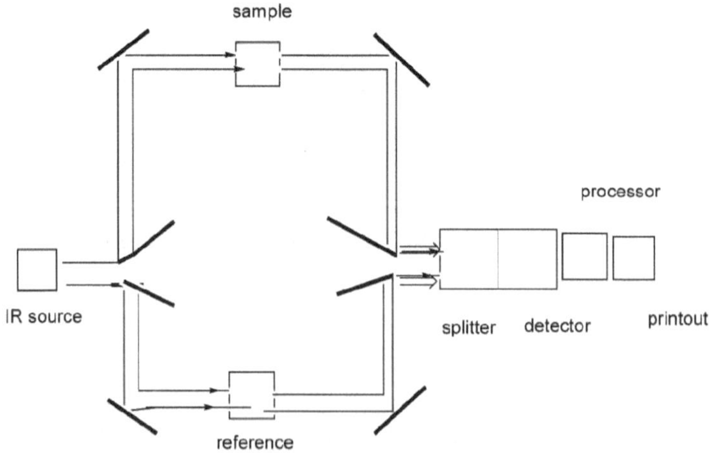

Schematic Diagram of IR Spectrophotometer

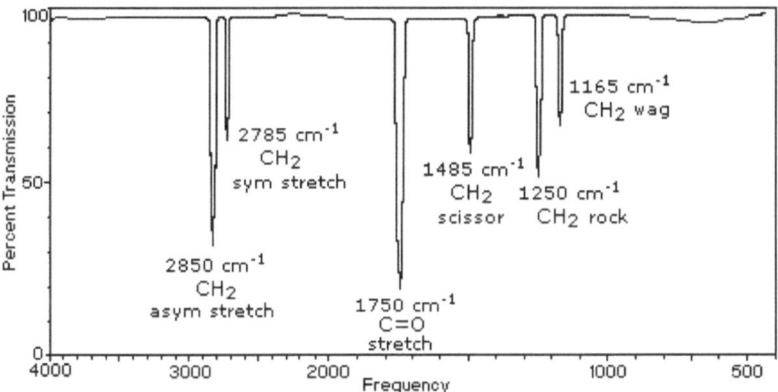

Gas Phase Infrared Spectrum of Formaldehyde, $H_2C=O$

Approximate positions of some I.R. bands

Group	λ (μm)
C-H (aliphatic)	3.33-3.70
C-H (aromatic)	3.23-3.33
S-H	3.85-3.89
N-H	2.97-3.03
C-O	9.52-10.00
C=O (aldehyde)	5.75-5.8
C=O (ketone)	5.85-5.86
C=O (ester)	5.71-5.88
C=O (acid)	6.06
C-N	6.02-8.23
C-C	9.09-13.33
C=C	5.99-6.17
C≡C	4.44-4.76
C≡N	4.44-4.76
C-Cl	12.5-16.67
O-H (phenolic)	2.7
O-H (phenolic–H-bonding)	2.7-3.03
C-F	7.14-10.0
C-Br	16.67-20.0
C-I	20.00

17

FLAME PHOTOMETRY

When an element burns in a flame it emits radiation characteristic of its identity. Flame photometry is the determination of an element in solution by measurement of emission of light by atoms of the element in flame. Initial tests for identification of a few elements by flame tests are well known. Na burns with yellow flame, K with lilac flame, Ca, with red flame, Cu, with green flame etc. Emission of characteristic radiation by an element and correlation of emission intensity with the concentration of that element form the basis for flame photometry.

The sample to be analyzed is prepared in solution and sprayed under controlled conditions into a flame. The light from the flame enters a monochromator to isolate the desired region of the spectrum. A photocell and an amplifier measure the intensity of the isolated radiation. After proper calibration of the photometer with solutions of known composition and concentration, it is possible to correlate the intensity of a given spectral line with the unknown concentration of that element. A calibration curve prepared using known concentrations is useful. Internal standards and addition of known concentration of sample to be quantified are more useful.

Flame photometer

The flame photometer consists of mainly four parts.

1. **Source of flame:** A burner that provides flame and can be maintained in a constant form and at a constant temperature.

2. **Nebuliser and mixing chamber:** Helps to transport the homogeneous solution of the substance into the flame at a steady rate.

3. **Optical system (optical filter):** The optical system comprises three parts: convex mirror, lens and filter. The convex mirror helps to transmit light emitted from the atoms and focus the emissions to the lens. The convex lens helps to focus the light on a point called slit. The reflections from the mirror pass through the slit and reach the filters. This will isolate the wavelength to be measured from that of any other extraneous emissions. Hence it acts as interference type color filter.

4. **Photodetector:** It detects the emitted light and measures the intensity of radiation emitted by the flame. The emitted radiation is converted to an electrical signal with the help of the photodetector. The produced electrical signals are directly proportional to the intensity of light.

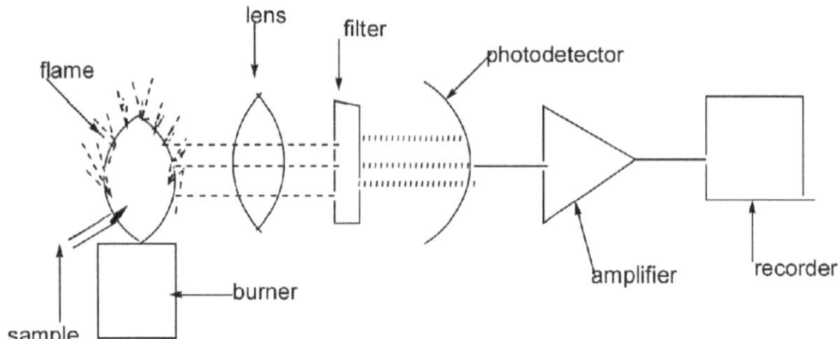

Schematic Diagram of Simple Flame Photometer

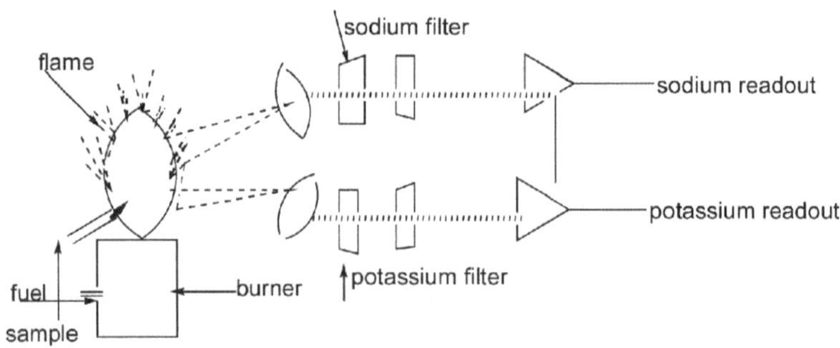

Schematic Diagram of Double Beam Flame Photometer

Events occurring in the flame: Flame photometry employs a variety of fuels mainly air, oxygen or nitrous oxide (N_2O) as oxidant. The temperature of the flame depends on fuel-oxidant ratio. The various processes that take place in the flame are as follows:

1. **Desolvation:** The metal particles in the flame are dehydrated by the flame and hence the solvent is evaporated.
2. **Vapourisation:** The metal particles in the sample are dehydrated and simultaneous evaporation of the solvent.
3. **Atomization:** Reduction of metal ions in the solvent to metal atoms by the flame heat.
4. **Excitation:** The electrostatic force of attraction between the electrons and nucleus of the atom helps them to absorb a particular amount of energy. The atoms then jump to the excited energy state and
5. **Emission process:** Since the higher energy state is unstable the atoms jump back to the stable low energy state with the emission of energy in the form of radiation of characteristic wavelength, which is measured by the photo detector.

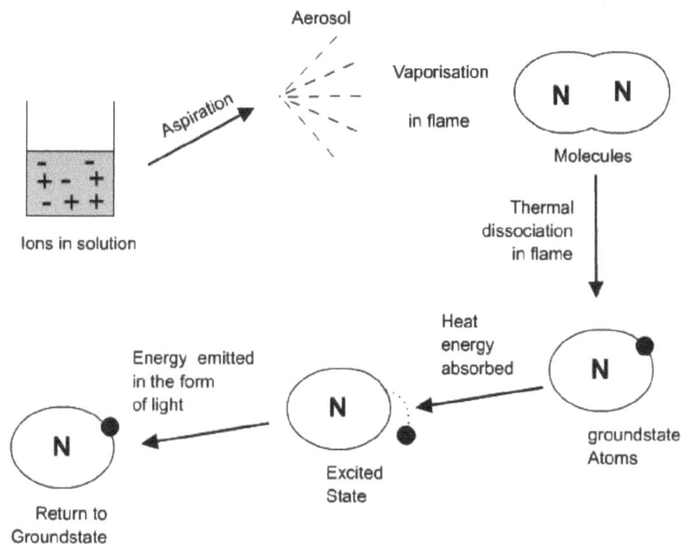

An overview of flame photometric process

Table: Emission spectral details of alkali metals

Some metals and their characteristic wavelengths

Element	Emitted wave Lengrh (nm)	Colour of the flame
Potassium	766	Violet
Lithium	670	Red
Calcium	622	Orange
Sodium	589	Yellow
Barium	554	Lime green

Applications: Flame photometer is useful for both quantitative and qualitative purposes. Flame photometer with monochromators emits radiations of characteristic wavelengths which help to detect the presence of a particular metal in the sample. This helps to determine the availability of alkali and alkaline earth metals which are useful for soil fertility evaluation. In agriculture, the fertilizer requirement of the soil is analyzed by flame test analysis of the soil. In clinical field, Na+ and K+ ions in body fluids, muscles and heart can be determined by diluting the blood serum and aspiration into the flame. Analysis of soft drinks, fruit juices and alcoholic beverages can also be carried out by using flame photometry.

Flame photometer is best for analysis of Na, K and Ca. It is not so useful for measuring and identifying other elements. Sensitivity of measurement depends on the temperature of the flame. For better sensitivity and wider applicability atomic absorption spectrophotometer is useful.

Advantages: It is a simple quantitative analytical test based on the flame analysis. The technique is well understood. It is inexpensive due to low running costs. The determination of elements such as alkali and alkaline earth metals is performed easily with most reliable and convenient methods. It is quite quick, convenient, selective and sensitive ranging from parts per million (ppm) to parts per billion (ppb).

Disadvantages: The flame photometer has some disadvantages. The concentration of the metal ion in the solution cannot be measured accurately. A standard solution with known molarities is required for determining the concentration of the ions which will corresponds to the emission spectra. It is difficult to obtain accurate results of ions with higher concentration. Major disadvantage is related to self absorption due to low population of excited atoms in the outer region of the flame than in the hotter inner region. The information about the molecular structure of the compound present in the sample solution cannot be determined. The elements such as carbon, hydrogen and halides cannot be detected due to their non radiating nature.

18

ATOMIC-ABSORPTION SPECTROSCOPY

Atomic-absorption (AA) spectroscopy uses the absorption of light to measure the concentration of gas-phase atoms. Since samples are usually liquids or solids, the analyte atoms or ions must be vaporized in a flame or graphite furnace. The atoms absorb ultraviolet or visible light and make transitions to higher electronic energy levels. The analyte concentration is determined from the amount of absorption. Applying the Beer-Lambert law directly in AA spectroscopy is difficult due to variations in the atomization efficiency from the sample matrix, and non-uniformity of concentration and path length of analyte atoms (in graphite furnace of AA). Concentration measurements are usually determined from a working curve after calibrating the instrument with standards of known concentration.

Instrumentation

Light source: The light source is usually a hollow-cathode lamp of the element that is being measured. Lasers are also used in research instruments. Since lasers are intense enough to excite atoms to higher energy levels, they allow AA and atomic fluorescence measurements in a single instrument. The disadvantage of these narrow-band light sources is that only one element is measurable at a time.

Atomizer : AA spectroscopy requires that the analyte atoms be in the gas phase. Ions or atoms in a sample must undergo de-solvation and vaporization in a high-temperature source such as a flame or graphite furnace. Flame AA can only analyze solutions, while graphite furnace AA can accept solutions, slurries, or solid samples. Flame AA uses a slot type burner to increase the path length, and therefore to increase the total absorbance. Sample solutions are usually aspirated with the gas flow into a nebulizing/mixing chamber to form small droplets before entering the flame.

The graphite furnace has several advantages over a flame. It is a much more efficient atomizer than a flame and it can directly accept very small absolute quantities of sample. It also provides a reducing environment for easily oxidized elements. Samples are placed directly in the graphite furnace and the furnace is electrically heated in several steps to dry the sample, ash, organic matter, and vaporize the analyte atoms.

Light separation and detection: AA spectrometers use monochromators and detectors for UV and visible light. The main purpose of the monochromator is to isolate the absorption line from background light due to interferences. Simple dedicated AA instruments often replace the monochromator with a band pass interference filter. Photomultiplier tubes are the most common detectors for AA spectroscopy.

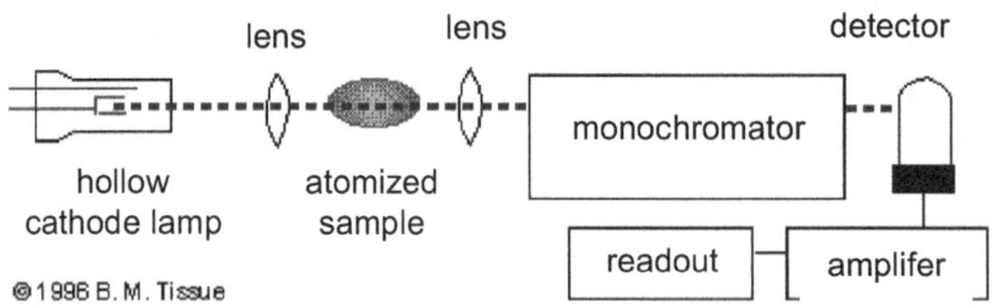

Schematic Diagram of an AA Spectrophotometer

Detection limits: Detection limits are lower and better than that in flame photometry. They are higher by a factor of 100 in AAS. However, detection limits of K, Fe and Sn are about the same in flame and AAS.

Interferences: Incomplete dissociation and formation of refractory compounds give rise to some interferences. Some elements get oxidized in the flame and pose problems in analysis. Ca compounds get dissociated and pose problems. Viscosity of the solution also may create difficulties.

Applications: AA is useful in determination of a large number of metals in trace levels. It is widely used in fields of water pollution, soil analysis, pharmacy and metallurgy.

19

INDUCTIVELY COUPLED PLASMA ATOMIC EMISSION SPECTROSCOPY

Inductively coupled plasma atomic emission spectroscopy (ICP-AES) uses an inductively coupled plasma to produce excited atoms and ions that emit electromagnetic radiation at wavelengths characteristic of a particular element. A plasma is a neutral gas containing a good number of positive and negative ions or free electrons. A plasma can only be created. A combustion flame is one example of a plasma. Advantages of ICP-AES are excellent limit of detection and linear dynamic range, multi-element capability, low chemical interference and a stable and reproducible signal. Disadvantages are spectral interferences (many emission lines), high cost and operating expense and the fact that samples typically must be in solution.

ICP-AES also referred to as inductively coupled plasma optical emission spectrometry (ICP-OES), is an analytical technique used for the detection of trace metals. It is a type of emission spectroscopy that uses the inductively coupled plasma to produce excited atoms and ions that emit electromagnetic radiation at wavelengths characteristic of a particular element. The intensity of this emission is indicative of the concentration of the element within the sample.

The ICP-AES is composed of two parts: the ICP and the optical spectrometer. The ICP torch consists of 3 concentric quartz glass tubes. The output or "work" coil of the radio frequency (RF) generator surrounds part of this quartz torch. Argon gas is typically used to create the plasma.

When the torch is turned on, an intense electromagnetic field is created within the coil by the high power radio frequency signal flowing in the coil. This RF signal is created by the RF generator which is, effectively, a high power radio transmitter driving the "work coil" the same way a typical radio transmitter drives a transmitting antenna. The argon gas flowing through the torch is ignited with a Tesla unit that creates a brief discharge arc through the argon flow to initiate the ionization process. Once the plasma is "ignited", the Tesla unit is turned off.

The argon gas is ionized in the intense electromagnetic field and flows in a particular rotationally symmetrical pattern towards the magnetic field of the RF coil. A stable, high temperature plasma of about 7000 K is then generated as the result of the inelastic collisions created between the neutral argon atoms and the charged particles.

A peristaltic pump delivers an aqueous or organic sample into an analytical nebulizer where it is changed into mist and introduced directly inside the plasma flame. The sample immediately collides with the electrons and charged ions in the plasma and is itself broken down into charged ions. The various molecules break up into their respective atoms which then lose electrons and recombine repeatedly in the plasma, giving off radiation at the characteristic wavelengths of the elements involved.

In some designs, a shear gas, typically nitrogen or dry compressed air is used to 'cut' the plasma at a specific spot. One or two transfer lenses are then used to focus the emitted light on a diffraction grating where it is separated into its component wavelengths in the optical spectrometer. In other designs, the plasma impinges directly upon an optical interface which consists of an orifice from which a constant flow of argon emerges, deflecting the plasma and providing cooling while allowing the emitted light from the plasma to enter the optical chamber. Still other designs use optical fibers to convey some of the light to separate optical chambers.

Within the optical chamber(s), after the light is separated into its different wavelengths (colours), the light intensity is measured with a photomultiplier tube or tubes physically positioned to "view" the specific wavelength(s) for each element line involved, or, in more modern units, the separated colors fall upon an array of semiconductor photodetectors such as charge coupled devices (CCDs). In units using these detector arrays, the intensities of all wavelengths (within the system's range) can be measured simultaneously, allowing the instrument to analyze for every element to which the unit is sensitive all at once. Thus, samples can be analyzed very quickly.

The intensity of each line is then compared to previously measured intensities of known concentrations of the elements, and their concentrations are then computed by interpolation along the calibration lines. In addition, special software generally corrects for interferences caused by the presence of different elements within a given sample matrix.

Applications

1. ICP-OES is used for the determination of metals in wine, arsenic in food, and trace elements bound to proteins.
2. ICP-OES is widely used in minerals processing to provide the data on grades of various streams, for the construction of mass balances.
3. ICP-AES is often used for analysis of trace elements in soil, and it is for that reason it is often used in forensics to ascertain the origin of soil samples found at crime scenes or on victims etc. Taking one sample from a control and determining the metal composition and taking the sample obtained from evidence and determine that metal composition allows a comparison to be made. The soil evidence is used to strengthen other evidences.
4. It is also fast becoming the analytical method of choice for the determination of nutrient levels in agricultural soils. This information is then used to calculate the amount of fertiliser required to maximise crop yield and quality.
5. ICP-AES is used for motor oil analysis. Analyzing used motor oil reveals a great deal about how the engine is operating. Parts that wear in the engine will deposit traces in the oil which can be detected with ICP-AES. This analysis helps to determine the wear and tear of engine parts.

20

NMR SPECTROSCOPY

Nuclear Magnetic Resonance spectroscopy is a powerful and theoretically complex analytical tool. The nuclei of certain isotopes are continuously spinning with an angular momentum, which can give rise to an associated magnetic field. The nuclei of many elemental isotopes have a characteristic spin (I). Some nuclei have integral spins (e.g. I = 1, 2, 3 ....), some have fractional spins (Ex. I = 1/2, 3/2, 5/2 ....), and a few have no spin, I = 0 (Ex. ^{12}C, ^{16}O, ^{32}S,). Isotopes of particular interest and use to organic chemists are ^{1}H, ^{13}C, ^{19}F and ^{31}P, all of which have I = 1/2. If a powerful external magnetic field is applied to the nucleus and made to oscillate in the radiofrequency range, the nucleus resonates between different quantized energy levels at specific frequencies absorbing some of the applied energy. Such a very small change can be detected, amplified and displayed on a chart. The variation in intensity of resonance signal with increasing applied magnetic field is the NMR (nuclear magnetic resonance) spectrum. The isotope of ^{1}H displays this phenomenon while the main isotopes of C, O_2 ,and N_2 do not display.

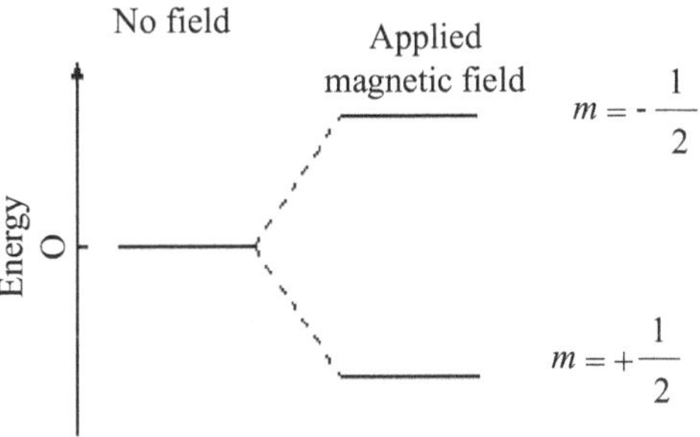

Energy Levels for a Nucleus with Spin

Strong magnetic fields are necessary for NMR spectroscopy. The international unit for magnetic flux is the tesla (T). The earth's magnetic field is not constant, but is approximately 10^{-4} T at ground level. Modern NMR spectrometers use powerful magnets having fields of 1 to 20 T. Even with these high fields, the energy difference between the two spin states is less than 0.1 cal/mole. We know that infrared transitions involve 1 to 10 kcal/mole and electronic transitions are nearly 100 times higher. For NMR purposes, this small energy difference (ΔE) is usually given as a frequency in units of MHz (10^6 Hz), ranging from 20 to 900 MHz, depending on the magnetic field strength and the specific nucleus being studied. Irradiation of a sample with radio frequency (rf) energy corresponding exactly to the spin state separation of a specific set of nuclei will cause excitation of those nuclei in the +1/2 state to the higher $-\frac{1}{2}$ spin state. NMR spectroscopy is therefore the energetically mildest probe used to examine the structure of molecules. The nucleus of a hydrogen atom (the proton) has a magnetic moment $\mu = 2.7927$, and has been studied more than any other nucleus.

The resonance frequencies of H atoms in molecules are often measured and the technique in this instance is referred to as proton magnetic resonance(PMR). PMR is the most valuable form of technique available to organic chemists. The frequency at which any given hydrogen atom in an organic compound resonates is strongly dependent on its precise molecular environment and is subjected to chemical shifts by the presence of other adjacent atoms. The fine structure of each absorption band is determined by the weak magnetic forces of neighboring hydrogen atoms and this feature (known as spin-spin coupling) together with the extent of chemical shift, may be used to deduce the arrangement of hydrogen atoms and often the complete structure of the unknown compound. As the area under the absorption band is proportional to the number of 'H' atoms in an identical environment in the molecule responsible for that signal, additional information in the structures of unknown compounds can be obtained by integrating the signals while recording the spectrum.

There are two types of PMR spectroscopy, namely, broad band and high resolution. The former can be used to obtain valuable information on the physical state and environment of molecules. High resolution PMR spectroscopy is a valued technique for obtaining information on the structure of unknown compounds and has been widely used to solve structural problems. Instruments operating at 60-100MHz are the popular ones in laboratories.

Chemical shift

The magnetic field at the nucleus is not equal to the applied magnetic field; electrons around the nucleus shield it from the applied field. The difference between the applied magnetic field and the field at the nucleus is termed the *nuclear shielding*.

Consider the s-electrons in a molecule. They have spherical symmetry and circulate in the applied field, producing a magnetic field which opposes the applied field. This means that the applied field strength must be increased for the nucleus to absorb at its transition frequency. This *upfield shift* is also termed *diamagnetic shift*.

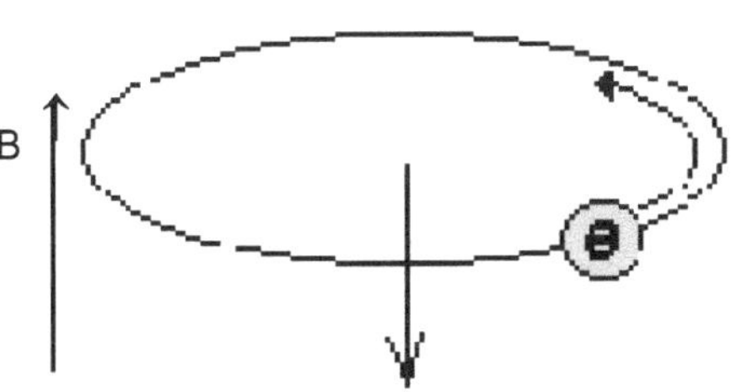

Magnetic Field Produced by Circulating Electron

Electrons in p-orbitals have no spherical symmetry. They produce comparatively large magnetic fields at the nucleus, which give a *low field shift*. This "deshielding" is termed *paramagnetic shift*. In proton (^{1}H) NMR, p-orbitals play no part (there is no p-orbital), which is why only a small range of chemical shift (10 ppm) is observed. We can easily see the effect of s-electrons on the chemical shift by looking at substituted methanes, CH_3X. As X becomes increasingly electronegative, so the electron density around the protons decreases, and they resonate at lower field strengths.

Chemical shift is defined as *nuclear shielding / applied magnetic field*. Chemical shift is a function of the nucleus and its environment. It is measured relative to a reference compound. For ^{1}H NMR, the reference is usually tetramethylsilane, $Si\ (CH_3)_4$.

PMR spectra of aliphatic compounds are not always distinctive as the signals from many of the methylene protons tend to overlap. Some resolution of these can be achieved by the addition of chemical shift agents (generally lanthanide derivatives) which affect a separation of signals from certain protons. Chiral shift agents are used to identify enantiomers.

Compounds must be in solution for analysis and the solvents should preferably not contain the isotope of 'H' . CCl_4 is the most suitable non-polar solvent. But deutero $CHCl_3$ can also be used. Deuterated methanol also is available. Chemical shifts are not measured in absolute units but are recorded as ppm of the resonance magnetic field. Tetramethyl silane is added as internal standard.

The technique has been widely used for organic structure determinations. Identification of double bonds ($=$) and to detect other functional groups. OH, cycopropane rings, triple bonds ($\equiv$), oxirane rings etc. can be detected along with their location in a compound.

Purity determination

NMR is primarily used for structural determination, however it can also be used for purity determination, providing that the structure and molecular weight of the compound is known. This technique requires the use of an internal standard of a known purity. Typically this standard will have a high molecular weight to facilitate accurate weighing, but relatively few protons so as to give a clear peak for later integration Ex. 1,2,3,4-tetrachloro-5-nitrobenzene. Accurately weighed portions of both the standard and sample are combined and analyzed by NMR. Suitable peaks are selected for both compounds and the purity of the sample determined via the following equation.

$$\text{Purity} = \frac{\text{Wt (Std) x } \eta \text{ [H] (Std) x MW (Spl)}}{\text{Wt (Spl) x MW (Std) x [H] (Spl)}} \text{ x P}$$

Where: *Wt(Std)*: Weight of internal standard, *Wt(Spl)*: Weight of sample.

η[H] (*Std*): The integrated area of the peak selected for comparison in the standard, corrected for the number of protons in that functional group.

η[H] (*Spl*): The integrated area of the peak selected for comparison in the sample, corrected for the number of protons in that functional group.

MW(Std): Molecular weight of standard.

MW(Spl): Molecular weight of sample.

P: Purity of internal standard.

21

RADIOCHEMICAL METHODS

In 1896, French physicist Becquerel observed that uranium compounds affected a photographic plate that was wrapped inside a black paper. He found that all uranium compounds gave off penetrating rays. He called the production of radiation by uranium compounds "radioactivity". Madame and Pierre Curie later identified thorium, polonium, and radium as the other radioactive elements.

Rutherford, identified the rays given off by radioactive elements. He named them α, β, γ rays. Alpha (α) rays are charged helium (He) atoms (2p+ 2n with 2+ charge). Alpha rays have a velocity of 10,000 miles/sec. They collide with air particles and stop after 8cm. The collisions by α rays produce heat and ionized particles. Beta particles are nothing but streams of electrons and travel at the speed of light ($3x10^{10}$ cm/sec or 150,000 miles/sec). They are more penetrating than α rays and are of opposite (-) charge. Gamma rays are similar to X-rays and are the most penetrating of these three rays.

Álpha rays (being heavier–helium ions) can be stopped by a thin sheet of paper.

Beta rays (high speed electrons) are 100 times more penetrating and can pass through thin sheets of metal.

Gamma rays are 10,000 times more penetrating than alpha rays. About 5cm of lead or 30 cm of iron can stop them.

Radioactivity: The proportion of neutrons in a nucleus should be sufficient enough to offset the repelling effect of positively charged protons in any nucleus.

As the atomic weight of elements increase the proportion gets imbalanced around that of the element radium. Elements with atomic weight 83 and above are unstable and undergo change and are radioactive. Such elements undergo decay and try to return back to stability or change to stable atoms. When an atom disintegrates, it releases particles or electromagnetic radiation. This is what we call radioactivity.

The identity of an element is determined by the number of protons in its nucleus. Atoms having the same number of protons but different neutrons are called isotopes. For example, hydrogen is an element with one proton (and zero/no neutrons) in its nucleus. It has two isotopes which have one (deuterium) and two (tritium) neutrons in the nucleus. Of the three isotopes, the one with two neutrons, namely, tritium, is radioactive due to imbalance in the proportion of protons and neutrons in the nucleus. The element uranium with 92 protons has an atomic weight of 238 and 146 neutrons. The number of neutrons are insufficient to offset positive repulsions of 92 protons in uranium. The element disintegrates to return back to stability. Radioactive elements like uranium, thorium and radium undergo disintegration sequentially till they return to stable position like that of the element lead. Tritium is an artificial radioisotope, while uranium, thorium and radium are naturally occurring radioisotopes.

Natural radioactive decay usually proceeds by loss of an α- or β- particle from the unstable nucleus and is always accompanied by γ – ray emission.

$^{238}U_{92} \rightarrow {}^{234}Th_{90} + {}^{4}He_{2}$ half life = 4.49x 10^9 years

$^{234}Th_{90} \rightarrow {}^{234}Pa_{91} + {}^{0}e_{-1}$ half life = 24.1 days

The rate at which various unstable nuclei undergo radioactive decay varies as a function of the nuclear stability, but always proceeds as a first order reaction. The first order process depends only on the quantity of radioactive matter. For such purposes, the time required for half of the nuclei of a given isotope to decay is a constant and is called the half-life.

Half life $t_{1/2} = 0.693/\lambda$

λ= dissociation constant,

$t_{1/2}$ is constant for each radioactive nuclide

Detection of radioactivity

The detection of radiation is based on its conversion to an electrical signal in a radiation detector. The signal is amplified, modulated and then transmitted to an identification device or to an automatic control system.

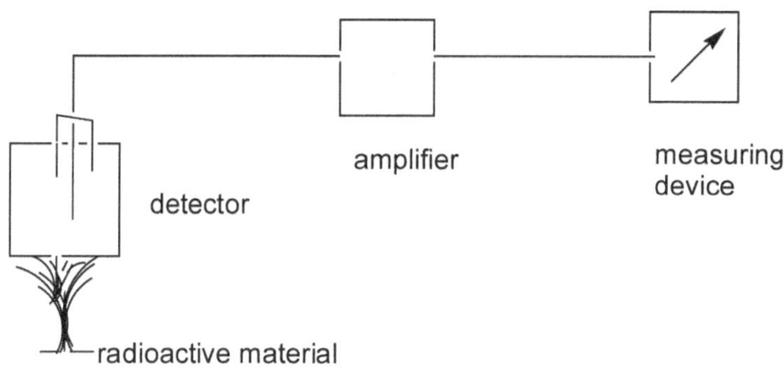

A schematic Diagram of Radiation Detection

Three types of radiation measurement devices are available. The simplest one uses the fact that the radiations affect the X- ray film. This technique is known as autoradiography. In the other method, the ionization caused by the radiations are utilized and measured quantitatively. In the third method, the light emitted by phosphors (scintillations), when they are exposed to radiations are utilized. This is called scintillation counting.

G.M.Counter: The Geiger- Mueller (G.M.) counter is used to quantitatively determine the α- and β- radiations. The main component, the G.M. tube is a sealed tube consisting of two electrodes and is filled with an ionizing gas (alcohol vapour or argon). A potential difference is maintained between the two electrodes. One end of the tube is a thin mica window. When a radioactive sample is brought near this window, the α- or β- particles penetrate the window and enter the chamber. The radiation collides with the gas molecules causing ionization. Primary and secondary ionization takes place. The positive ions and the electrons move towards the cathode and the anode respectively. A potential difference of 1000-1500 V is initially maintained between the electrodes. Due to ionization by the radioactive particles, a gas discharge occurs. It is measured as a pulse of electricity or "count " by suitable circuitry.

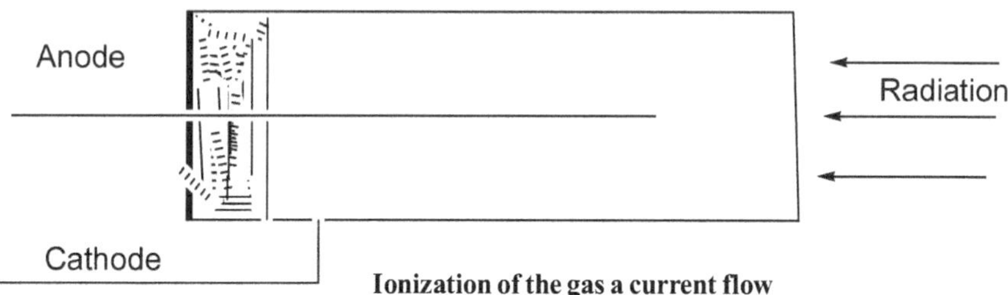

Ionization of the gas a current flow

G.M. Tube: The tube is filled with neon or argon and a voltage is applied between the electrodes.

Scintillation counter: Low energy β- radiation and γ- ray activity can be measured with the scintillation counter. Radiation from the radioactive nuclides excites the shell electrons of many atoms to emit light. This primarily applies to substances where the electrons shifted to a higher level immediately return to their original position emitting light. For example in a NaCl or calcium tungstate crystal, any impact by a particle is observed as a flash of light. This is known as "scintillation" and is the guiding principle of construction of scintillation counters. In these counters, the photons of the flashes of light from a photocathode, consisting for example of CaO, releases electrons which are accelerated in an electric field, bound against a secondary emission electrode (dynode) where they release other electrons. The electron current can be multiplied more than a million fold and the flash of light electronically is recorded in the scintillator. Background sound correction is possible and the amount of radioactivity is measured. The device used to multiply electrons is called photoelectric electron multiplier.

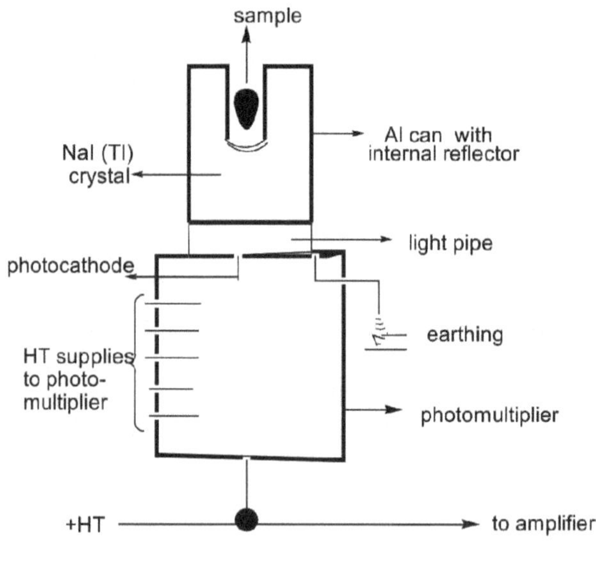

Scintillation Counter

Semiconductor detectors: They depend on the interaction of high energy radiation with crystals of semiconductors like silicon or germanium. The silicon detectors are useful for low energy gamma and X- radiation. The germanium semiconductors are useful for medium and high energy gamma radiation. When the energy of the radiation is absorbed in the crystal, free electrons and positive holes are produced in numbers proportional to the energies of the incident photons. A thin gold layer is plated on to each end of the crystal and electrodes are attached to this so that a potential of 3-5 K volts may be applied across the crystal. The electrons and the positive holes move to the electrodes and produce signals proportional to the incident radiation.

These semiconductor devices are complicated. However, they furnish excellent, high resolution gamma and X-ray spectra. Computer aided spectral analysis is much in vogue these days. Álpha particle spectrometry can also be carried out using semiconductor detectors.

Autoradiography: It is a broad based technique used to determine where a particular radioisotope is located. Autoradiography takes advantage of the ability of radiation emitted from a radioisotope to activate a photographic emulsion. The sample containing radioactivity, namely, a chromatogram or a tissue slice or a section of a whole animal or plant is kept in close contact with ordinary X-ray film. The emanating radiations affect the chemicals in the photographic plate in much the same way as ordinary light or X-rays does. After suitable time of exposure, the X-ray film is developed. Areas which have been in contact with radiation would appear as dark spots.

The intensity of the spot can be measured using a densitometer. It works on the principle similar to the colorimeter. A chart recorder attached will do the necessary job and the radiation can be quantified. For this purpose known quantities of standards can be used.

The main use of autoradiography is in locating the position of a radiolabel in a specimen, before its determination. For example the different amino acids may be located on a single autoradiograph from a TLC plate. Locations of the spots can easily be made out initially followed by densitometry to relate density to the amount of analyte.

Radioisotope dilution analysis: A known amount w_1 of radio analyte of known specific activity a_1 is added to the sample to be analyzed. After thorough mixing, sufficient quantity of the analyte is separated (no need for complete separation). The new specific activity a_2 is measured. Then w_2 of the analyte is calculated.

Total activity added = $a_1 w_1 = a_2(w_2 + w_1)$

This method has been used to quantify the vitamin B_{12} in a biological sample using ^{60}Co label. Isotope dilution does not require quantitative separations. However, determination of chemical yield is the limitation in this method. Isotope dilution analysis is a definitive method and is quantitative. The work can be carried out only in specialized radioisotope laboratory.

Radioactivation analysis: Nuclear bombardment with neutrons is used to produce radioactive elements, called irradiation of the sample. The product namely, radioisotope serves to measure and relate to the quantity of the original element. Neutron activation analysis is an extremely sensitive, selective and accurate technique. It is common to irradiate a standard alongside the sample. The formula $a_1/w_1 = a_2/w_2$ is used to calculate the unknown quantity of the sample, w_2 . a_1 and a_2 are the activities of the standard and the sample, while w_1 is the weight of the standard.

Activation analysis requires the use of a powerful source of neutrons as the activator. It is suitable for elements which form an isotope which has a suitable half life. It is generally not suitable for biological materials. It is used for measurement of lead in hairs and nails.

Units of radioactivity: Radioactivity is measured as disintegrations per second (d/s). One unit of Curie = 3.7×10^{10} d/s or 2.22×10^{12} dpm (disintegrations per minute). This is equal to disintegrations in 1g of radium. With the introduction of SI units the accepted unit of activity is the Becquerel (Bq) which is equal to one nuclear transformation per second. 1 curie = $3{,}7 \times 10^{10}$ Becquerels.

Units of radioactivity

Unit	Abbreviation	Details
Disintegrations/min	d.p.m.	The rate of decay
Disintegrations/sec	d.p.s.	The rate of decay
Curie	Ci	The no. of d.p.s. in 1g of Ra (3.7×10^{10} d.p.s.)
Millicurie	mCi	Ci x 10^{-3} or 2.22×10^{9} d.p.m.
Microcurie	µCi	Ci x 10^{-6} or 2.22×10^{6} d.p.m.
Becquerel (SI unit)	Bq	1 d.p.s.
Mega Becquerel	MBq	10^{6} Bq 27.027 µCi
Electron volt	eV	1.6×10^{-19} J
Roentgen	R	Radiation that produces 1.61×10^{15} ion pairs per kg

Utility of radionuclides: Radio-nuclides are used to label normal isotopes and their activity is measured to understand metabolism. ^{32}P, ^{131}I, ^{3}H ^{36}Cl, etc. are used to trace movement, absorption, efficiency etc. in agriculture, biology and medicine.

Some examples of radionuclide utilization are presented below:

1 ^{14}C acetate can be used to study the tricarboxylic acid cycle involved in plant metabolism.

2. Glucose metabolism in plants/ animals can be studied using ^{14}C or tritium isotopes

3. Turn over of proteins in rats – Rats can be injected labelled amino acids. Rats after killing can be analyzed for protein in different organs of the rat.

4. Absorption, accumulation and translocation of nutrients (^{32}P, ^{36}Cl) by plants/animals etc.
5. Labelled drugs can be used to study drug accumulation (site, rate, degradation products etc.)

Radio- isotopes commonly used

Radionuclide	Half life	radiation
^{3}H	12.3 yr	Beta
^{14}C	5700yr	Beta
^{24}Na	15.1hr	Beta
^{32}P	14.3d	Beta
^{35}S	87.1d	Beta
^{42}K	12.4hr	Beta, gamma
^{45}Ca	152d	Beta
^{59}Fe	45d	Beta, gamma
^{60}Co	5.3yr	Beta, gamma
^{64}Cu	12.8hr	Beta, gamma
^{65}Zn	250d	Beta, gamma
^{131}I	8.0d	Beta, gamma
^{36}Cl	$3x10^{5}y$	Beta
^{28}Mg	21.4hr	Beta

Safety

Handling of radioisotopes need great care. Due to their ionizing and penetrating power they cause a huge level of harm to the personnel handling them. Even the disposal of used radioisotope material needs special attention. Laboratory benches should be water proof. Personnel should wear protective clothing and use disposable gloves etc. Suitable detectors to monitor personal radiation hazard should be available in the laboratories.

22

MASS SPECTROMETRY

Mass spectrometry (MS) is a useful technique which produces spectra of the masses of the atoms or molecules in a sample. The spectra are used to determine the elemental or isotopic details of a sample. The masses of particles and of molecules, are used to elucidate the chemical structures of molecules, such as peptides and other chemical compounds. Mass spectrometry works by ionizing chemical compounds to generate charged molecules and measuring their mass-to-charge ratios.

A sample, which may be solid, liquid, or gas, is ionized, by bombarding with electrons. This may cause some of the sample's molecules to break into charged fragments. These ions are then separated according to their mass-to-charge ratio, typically by accelerating them and subjecting them to an electric or magnetic field. Ions of the same mass-to-charge ratio will undergo the same amount of deflection. The ions are detected by a mechanism capable of detecting charged particles, such as an electron multiplier. Results are displayed as spectra of the relative abundance of detected ions as a function of the mass-to-charge ratio. The atoms or molecules in the sample can be identified by correlating known masses to the identified masses or through a characteristic fragmentation pattern.

The mass spectrometer

In order to measure the characteristics of individual molecules, a mass spectrometer converts them to ions so that they can be moved about and manipulated by external electric and magnetic fields. The three essential functions of a mass spectrometer, and the associated components, are:

1. A small sample is ionized, usually to cations by loss of an electron - The Ion Source
2. The ions are sorted and separated according to their mass and charge- The Mass Analyzer
3. The separated ions are then measured, and the results displayed on a chart - The Detector

Because ions are very reactive and short-lived, their formation and manipulation must be conducted in a vacuum. Atmospheric pressure is around 760 mm of mercury. The pressure under which ions may be handled is roughly 10^{-5} to 10^{-8} mm (less than a billionth of an atmosphere). Each of the three tasks listed above may be accomplished in different ways. In one common procedure, ionization is effected by a high energy beam of electrons, and ion separation is achieved by accelerating and focusing the ions in a beam, which is then bent by an external magnetic field. The ions are then detected electronically and the resulting information is stored and analyzed in a computer.

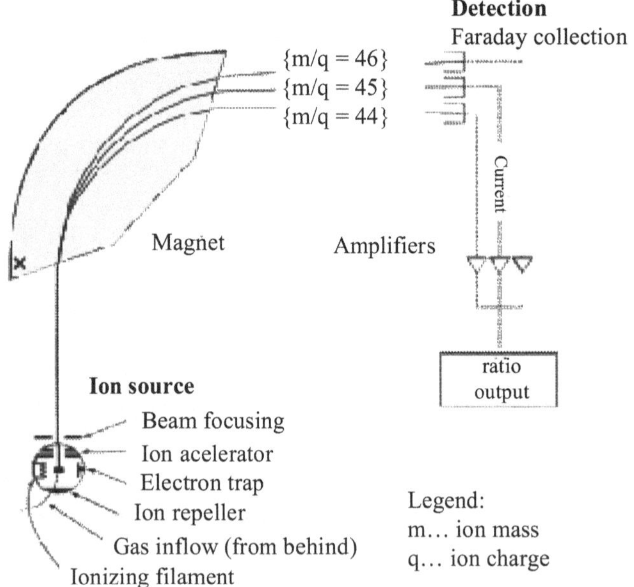

Schematic Diagram of a Mass Spectrometer

The heart of the spectrometer is the ion source. Here molecules of the sample are bombarded by electrons emanating from a heated filament. This is called an EI (electron-impact) source. Gases and volatile liquid samples are allowed to leak into the ion source from a reservoir. Non-volatile solids and liquids may be introduced directly. Cations formed by the electron bombardment are pushed away by a charged repeller plate (anions are attracted to it), and accelerated toward other electrodes, having slits through which the ions pass as a beam. Some of these ions fragment into smaller cations and neutral fragments. A perpendicular magnetic field deflects the ion beam in an arc whose radius is inversely proportional to the mass of each ion. Lighter ions are deflected more than heavier ions. By varying the strength of the magnetic field, ions of different masses can be focused progressively on a detector fixed at the end of a curved tube (under a high vacuum).

When a high energy electron collides with a molecule it often ionizes it by knocking away one of the molecular electrons (either bonding or non-bonding). This leaves behind a molecular ion. Residual energy from the collision may cause the molecular ion to fragment into neutral pieces and smaller fragment ions. The molecular ion is a radical cation, but the fragment ions may either be radical cations or carbo-cations, depending on the nature of the neutral fragment.

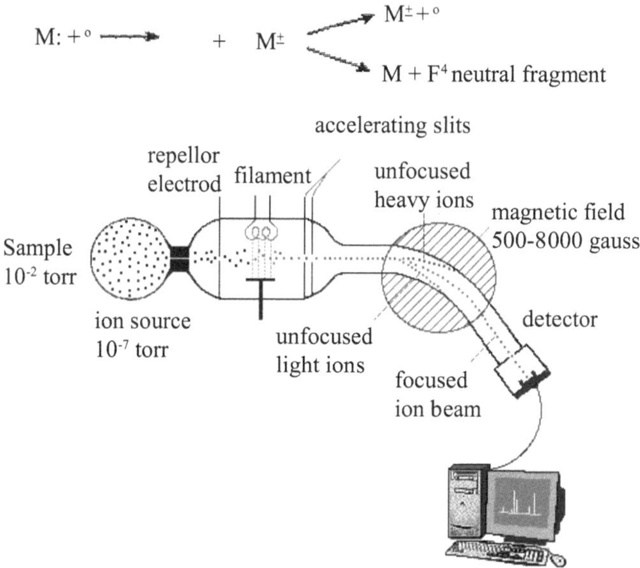

A mass spectrophotometer

The mass spectra

A mass spectrum is normally presented as a vertical bar graph, in which each bar represents an ion having a specific mass-to-charge ratio (m/z) and the length of the bar indicates the relative abundance of the ion. The most intense ion is assigned an abundance of 100, and it is referred to as the base peak. Most of the ions formed in a mass spectrometer have a single charge, so the m/z value is equivalent to mass itself. Modern mass spectrometers easily distinguish (resolve) ions differing by only a single atomic mass unit (amu), and thus provide completely accurate values for the molecular mass of a compound. The highest-mass ion in a spectrum is normally considered to be the molecular ion, and lower-mass ions are fragments from the molecular ion, assuming the sample is a pure compound.

The following diagram displays the mass spectra of three simple gaseous compounds, carbon dioxide, propane and cyclopropane. The molecules of these compounds are similar in size, CO_2 and C_3H_8 both have a nominal mass of 44 amu, and C_3H_6 has a mass of 42 amu. The molecular ion is the strongest ion in the spectra of CO_2 and C_3H_6, and it is moderately strong in propane. The unit mass resolution is readily apparent in these spectra (note the separation of ions having m/z=39, 40, 41 and 42 in the cyclopropane spectrum). Even though these compounds are very similar in size, it is a simple matter to identify them from their individual mass spectra. By clicking on each spectrum in turn, a partial fragmentation analysis and peak assignment will be displayed. Even with simple compounds like these, it should be noted that it is rarely possible to explain the origin of all the fragment ions in a spectrum. Also, the structure of most fragment ions is seldom known with certainty.

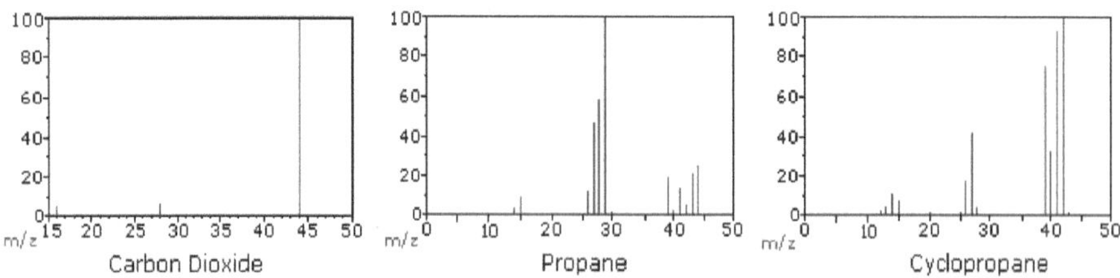

Since a molecule of carbon dioxide is composed of only three atoms, its mass spectrum is very simple. The molecular ion is also the base peak, and the only fragment ions are CO (m/z=28) and O (m/z=16). The molecular ion of propane also has m/z=44, but it is not the most abundant ion in the spectrum. Cleavage of a carbon-carbon bond gives methyl and ethyl fragments, one of which is a carbo-cation and the other a radical. Both distributions are observed, but the larger ethyl cation (m/z=29) is the most abundant, possibly because its size affords greater charge dispersal. A similar bond cleavage in cyclopropane does not give two fragments, so the molecular ion is stronger than in propane, and is in fact responsible for the base peak. Loss of a hydrogen atom, either before or after ring opening, produces the stable allyl cation (m/z=41). The third strongest ion in the spectrum has m/z=39 (C_3H_3). The small m/z=39 ion in propane and the absence of a m/z=29 ion in cyclopropane are particularly significant in distinguishing these hydrocarbons.

Applications

Mass spectrometry has both qualitative and quantitative uses. These include identifying unknown compounds, determining the isotopic composition of elements in a molecule, and determining the structure of a compound by observing its fragmentation. Other uses include quantifying the amount of a compound in a sample or studying the fundamentals of gas phase ion chemistry (the chemistry of ions and neutrals in a vacuum). MS is now in very common use in analytical laboratories that study physical, chemical, or biological properties of a great variety of compounds.

As an analytical technique it possesses distinct advantages such as: 1. Increased sensitivity over most other analytical techniques because the analyzer, as a mass-charge filter, reduces background interference 2. Excellent specificity from characteristic fragmentation patterns to identify unknowns or confirm the presence of suspected compounds. 3. Information about molecular weight. 4. Information about the isotopic abundance of elements and 5. Temporally resolved chemical data.

The disadvantages are that it often fails to distinguish between optical and geometrical isomers and the positions of substituent in o-, m- and p- positions in an aromatic ring. Also, its scope is limited in identifying hydrocarbons that produce similar fragmented ions.

23

POLARIMETRY

Compounds having one or more asymmetric carbon atoms can rotate the plane of polarized light. Such compounds are called optically active. Measurement of this property is called polarimetry.

The magnitude and direction in which the plane of polarized light is rotated depend on 1. The nature of the compound, 2. The concentration of the solution, 3. The path length of light, 4. The solvent, 5. The wave length of light and 6. The temperature. These variables are related and can be presented in the formula

$$[\alpha]_{\lambda}^{t} = a / \text{lxc}$$

(α) = specific rotation, α = observed rotation, t = temperature λ = wavelength of light, l = path length (decimeters- dm) and c = concentration of solution in g/ml.

If the light is rotated to the right , the compound is dextro-rotatory and the direction is denoted by (+); and if the light is rotated to the left, the compound is laevorotatory and the direction is denoted by (-).

The specific rotation of a compound at a specific wavelength of light and temperature is used as a standard physical constant. It can be used in the identification or in measurement of concentration of a compound.

Specific Rotation

Optical rotation measurements can be used to determine concentration and/or purity of a substance or simply to detect the presence of an optically active chemical in a mixture. The *specific rotation* of a chemical is simply an angular rotation obtained under standard measuring conditions: concentration, tube length, temperature and wavelength.

Most specific rotations refer to the sodium wavelength of 589 nm. Specific rotation is a unique characteristic of a chemical and can of course be any angle and often has a magnitude greater than ± 90°. The sodium wavelength of 589 nm is by far the most common light source used in polarimetry. Another popular source is the Mercury 546 nm source and there is an increasing interest in the near infrared wavelength of 880 nm because of its ability to penetrate dark, highly coloured, light absorbing samples.

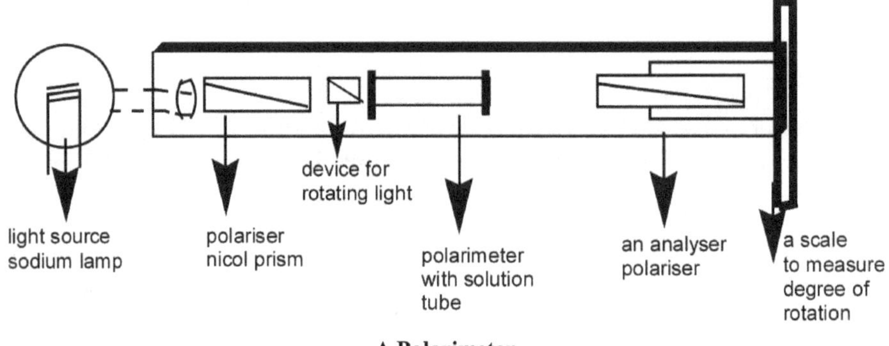

A Polarimeter

Polarimetry measures the rotation of polarized light as it passes through an optically active fluid. The measured rotation can be used to calculate the value of solution concentrations; especially substances such as sugars, peptides and volatile oils. A polarimeter consists of a polarized light source, an analyzer, a graduated circle to measure the rotation angle, and sample tubes.

The polarized light passes through the sample tube and exhibits angular rotation to the left (-) or right (+). On the side opposite the polarizer is the analyzer. Using optics, visual fields are manually adjusted by the user to measure the optical rotation angle.

Polarimeters offer high accuracies where precision is critical in determining the concentration of samples. In manual polarimeters we look through a viewing scope to read values on a vernier scale. Semiautomatic polarimeters have a digital display. Polarimeters can measure the angle of rotation (¡) or International Sugar Scale (°Z), or both.

Sugar Industry, Pharmaceutical Industry, Chemical Industry, Flavours, Fragrances and Essential Oils, Starch, Food and Drink, Agriculture, Amino acids, and Monosodium Glutamate industries utilize polarimetry.

24

CONDUCTIMETRY

When a fixed voltage is applied to two electrodes dipping into a solution, a current will flow, depending on the conductivity of the solution. Conductivity measurements do not give information about the nature of the ions in solution. However, they are useful in the quantitative measurement of the ions.

The current passing between the two electrodes is carried by the ions in solution and is related to the number of ions present which is a result of the molar concentration and the extent of ionization. The anions donate electrons to the anode while the cations accept electrons from the cathode. It is this flow of electrons that determine the amount of current flowing. The contribution of ionic species is determined by the ionic mobility (velocity).

The current flowing through a conductor is defined by the Ohm's law:

$$\frac{\text{voltage applied}\,(V)}{\text{resistance}\,(R)} = \text{current}\,(I)$$

In terms of conductance $(C = 1/R)$, $I = C \times V$, ohm^{-1} or mho (reverse of ohm) is the conventional unit of conductance.

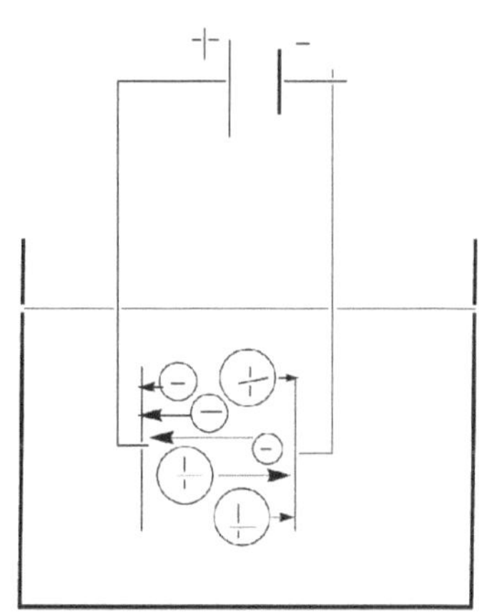

Conductivity of a Solution

The specific conductance (k) of a solution is defined as the conductance per centimeter of a solution which has a cross sectional area of 1cm^2 and is measured in Siemens/ cm or as ohms^{-1}cm^{-1}. The molar conductance (μ is the specific conductance of a solution, μ = k x volume of solution which contains 1gram mole. Using conductimetric measurements one can determine the ionic mobilities of ions and calculate the molar conductance of an electrolyte in infinite dilution.

Electrolytic solutions have constantly moving ions. Hence conductance is a measure used for solutions. Electrolytic conductivity is the conductivity of a solution in between two electrodes of unit area of cross section, separated by a unit distance. Electrolytic conductivity of electrolytes remains constant at a constant temperature.

Conductimetric titrations, particularly when they are automated, are useful if either of the reactants is deeply coloured, which prevents visual monitoring of the titration using indicators. This is also useful if the reactants are very dilute.

In the titration of a strong acid against a strong base the hydrogen ions are replaced by the cation from the base which has a small ionic mobility and hence a lower conductivity. A gradual reduction in conductance of the solution results. At the end point an increase in hydroxyl ions results in an increase in conductance. An extrapolation of the lines before and after the end point gives an intercept at the end point.

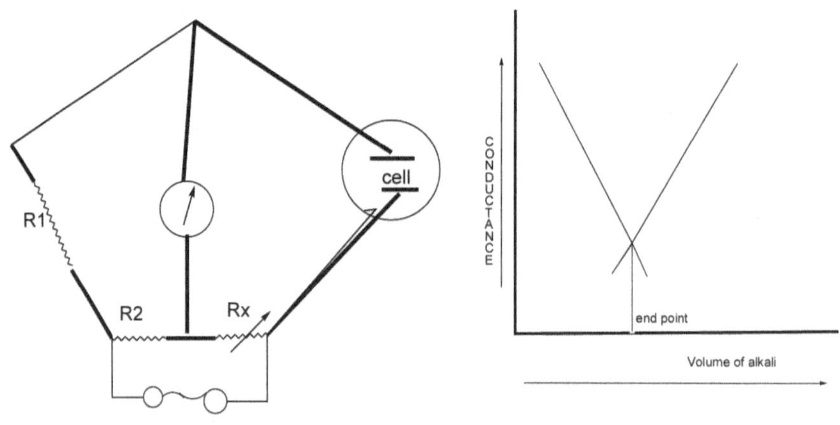

Wheat Bridge Circuit

Conductimetric Tridition Strong Acid vs Strong base

25

POTENTIOMETRY

When a metal M is immersed in a solution containing its own ion M^{n+}, then an electrode potential develops. The value of the electrode potential is given by the Nernst equation.

$$E = E_0 - 2.303\ RT\ x \frac{\log_a}{nF}$$

Where E = the electrode potential at the specified concentration

E_0 = the standard electrode potential, R = the gas constant

T = the absolute temperature, n = the no of electrons involved

F = the Faraday constant, a = is the activity of the ion

for measurements made at 25^0C the equation simplifies to

$$E = E_0 - 0.059 \log_{a/n}$$

The standard electrode potential of an element is its electrical potential when it is in contact with a molar solution of its ions. It is impossible to measure the potential of half a cell. Hence standard electrode potentials are presented as the difference in potential between the test electrode and a hydrogen electrode (arbitrary electrode potential = 0 at all temperatures, accepted internationally). Any half-cell will reduce another cell which has a lower potential.

It is difficult to prepare a hydrogen electrode and inconvenient to use it. Hence the saturated calomel electrode is most frequently used. It has a standard potential of +0.242V relative to the hydrogen electrode. It consists of a mercury electrode in equilibrium with Hg ions in the slightly soluble salt mercuric chloride.

In direct potentiometry a galvanic cell is set up, whose voltage depends on the activity of the analyte. The cell always has two electrodes, a reference electrode whose potential remains constant and an indicator electrode whose potential responds to the activity changes in the test solution. In a potentiometric titration, the progress towards the equivalence point is assessed by monitoring the cell voltage which depends on the activity of one of the reactants.

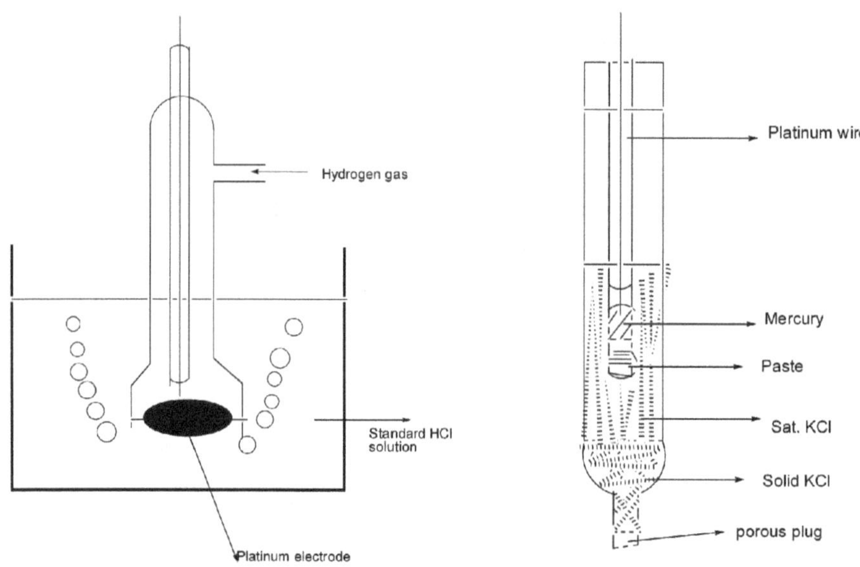

The Hydrogen Electrode **The Saturated Calomel Electrode**

Potentiometric titrations

Titrations can be conveniently followed potentiometrically. The potential of the solution varies, p^H and redox measurements are well suited for measurement with the help of a sensitive galvanometer. The end point is indicated by a significant change in potential. For redox reactions, the indicator electrode is platinum and ion selective electrodes (ISE) are used in other cases. The end point can be determined from the slope of the curve. Hence much better precision is possible in potentiometry.

Measurement of p^H

Acidity is due to an excess H^+ ions over OH^- ions in solution. H^+ ion concentration is a measure of acidity and alkalinity. But this is usually very small. Hence the acidity or alkalinity is expressed as the hydrogen ion exponent (p^H). p^H is defined as the negative logarithm to the base 10 of the hydrogen ion concentration [pH = - log10 (H+)]. A neutral solution has p^H of 7

For measurement of p^H, glass electrode is commonly used. The membrane is of special glass (hydrated aluminosilicate with Na or Ca ions. It is selectively permeable to H^+ ions and the potential developing depends on the H^+ ion concentration of the test solution inside the electrode. This potential can be measured against a reference calomel electrode using a high impedance voltmeter. For ease of operation the glass electrode is combined with a reference calomel electrode in a single glass probe.

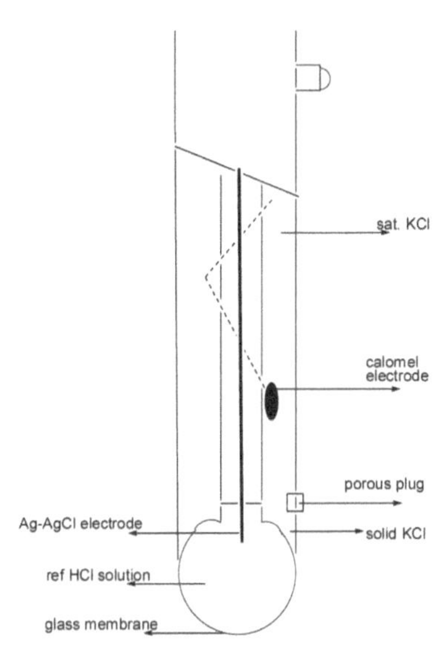

A Combined pH electrode

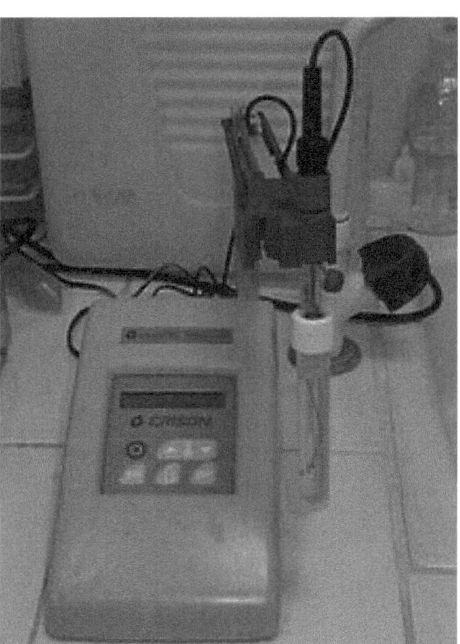

A p^H meter

26

ELISA AND RADIOIMMUNOASSAY (RIA)

Immunoassays rely on the ability of an antibody to recognize and bind a specific macromolecule in what might be a complex mixture of macromolecules. In immunology the particular macromolecule bound by an antibody is referred to as an antigen and the area on an antigen to which the antibody binds is called an epitope.

In some cases an immunoassay may use an antigen to detect for the presence of antibodies, which recognize that antigen, in a solution. In other words, in some immunoassays, the analyte may be an antibody rather than an antigen.

In addition to the binding of an antibody to its antigen, the other key feature of all immunoassays is a means to produce a measurable signal in response to the binding. Most immunoassays involve chemically linking antibodies or antigens with some kind of detectable label. A large number of labels exist in modern immunoassays, and they allow for detection through different means. Many labels are detectable because they either emit radiation, produce a color change in a solution, fluoresce under light, or because they can be induced to emit light.

One of the most popular labels used in immunoassays is enzymes. Immunoassays which employ enzymes are referred to as enzyme-linked immunosorbent assays (ELISAs), or sometimes enzyme immunoassays (EIAs). Enzymes used in ELISAs include horseradish peroxidase(HRP), alkaline phosphatase (AP) or glucose oxidase. These enzymes allow for detection often because they produce an observable color change in the presence of certain reagents.

ELISA involves detection of an "analyte" in a liquid sample by a method that continues to use liquid reagents during the "analysis" that stays liquid and remains inside a reaction chamber or well needed to keep the reactants contained. ELISA separates some component of the analytical reaction mixture by adsorbing certain components onto a solid phase which is physically immobilized. In ELISA, a liquid sample is added onto a stationary solid phase with special binding properties and is followed by multiple liquid reagents that are sequentially added, incubated and washed followed by some optical change (e.g. color development) in the final liquid in the well from which the quantity of the analyte is measured. Since enzyme reactions are very well known amplification processes, the signal is generated by enzymes which are linked to the detection reagents in fixed proportions to allow accurate quantification – thus the name "enzyme linked".

The analyte is also called the ligand because it will specifically bind or ligate to a detection reagent, thus ELISA falls under the bigger category of ligand binding assays. The ligand-specific binding reagent is "immobilized", i.e., usually coated and dried onto the transparent bottom and sometimes also side wall of a well which is usually constructed as a multiple-well plate known as the "ELISA plate". Conventionally, like other forms of immunoassays, the specificity of antigen-antibody type reaction is used because it is easy to raise an antibody specifically against an antigen in bulk as a reagent. Alternatively, if the analyte itself is an antibody, its target antigen can be used as the binding reagent.

The steps of direct ELISA are many and are as follows:

- A buffered solution of the antigen to be tested for is added to each well of a microtiter plate, where it is given time to adhere to the plastic through charge interactions.
- A solution of non-reacting protein, such as bovine serum albumin or casein, is added to well (usually 96-well plates) in order to cover any plastic surface in the well which remains uncoated by the antigen.
- The primary antibody is added, which binds specifically to the test antigen coating the well. This primary antibody could also be in the serum of a donor to be tested for reactivity towards the antigen.
- A secondary antibody is added, which will bind the primary antibody. This secondary antibody often has an enzyme attached to it, which has a

negligible effect on the binding properties of the antibody. In other cases, the primary antibody itself is conjugated to the enzyme.

- A substrate for this enzyme is then added. Often, this substrate changes color upon reaction with the enzyme. The color change shows the secondary antibody has bound to primary antibody, which strongly implies the donor has had an immune reaction to the test antigen. This can be helpful in a clinical setting, and in research.
- The higher the concentration of the primary antibody present in the serum, the stronger the color change. Often, a spectrometer is used to give quantitative values for color strength.

The enzyme acts as an amplifier; even if only few enzyme-linked antibodies remain bound, the enzyme molecules will produce many signal molecules. Within common-sense limitations, the enzyme can go on producing color indefinitely, but the more primary antibody is present in the donor serum, the more secondary antibody + enzyme will bind, and the faster the color will develop.

ELISA may be run in a qualitative or quantitative format. Qualitative results provide a simple positive or negative result(yes or no) for a sample. The cut off between positive and negative is determined by the analyst. In quantitative ELISA, the optical density (OD) of the sample is compared to a standard curve, which is typically a serial dilution of a known-concentration solution of the target molecule. For example, if a test sample returns an OD of 1.0, the point on the standard curve that gave OD = 1.0 must be of the same analyte concentration as the sample.

Radioimmunoassay

Radioimmunoassay is useful in quantification of hormones, drugs, vitamins and other compounds at very low concentrations. This method is highly specific and sensitive and can detect concentrations as low as 10^{-12}g.

Proteins are species specific and injection of a protein from one animal to another species elicits an immune response. The second animal identifies the new protein or antigen (Ag) as a foreign compound and responds with the production of an antibody (Ab). It is a protein molecule which circulates in the blood and forms a complex with the antigen.

Small molecules such as drugs, poisons, etc., do not elicit immune response. But if the small molecule is covalently linked to a protein (Ex. bovine serum albumin), then antibodies that react selectively with the small molecule may be produced. The small molecule is called hapten.

The requirement for the RIA assay is a sample of antigen or hapten analyte that has been labeled with a radioactive isotope. With insulin, it is easy to introduce an iodine isotope (I^{125} $t_{1/2}$ =60days) by iodination of aromatic amino acids. For hapten analytes like drugs, steroids and vitamins, tritium (H^3 $t_{1/2}$ = 5700years) is the preferred label. In general, for RIA, commercial kits containing antibody, radiolabelled analyte and analyte standards for calibration are available.

The Procedure

The principle of RIA is very simple. The analyte in a sample or standard is allowed to compete with a measured quantity of radiolabelled analyte for complexing with a limited quantity of antibody. The more the analyte in the sample, the less is the radioactivity in the Ag-Ab (or hapten-Ab) complex. The working curve obtained with standards inspite of being generally non-linear is useful the estimation.

The radioactive isotope decay does not depend on the chemical state. The radioactivity measurement does not differentiate the complexed from the uncomplexed radioactive material. This means that the Ag-Ab complex must be separated from unbound Ag before the radioactive measurement step. There are many ways to accomplish this through chromatographic and electrophoretic procedures. The Ag-Ab complex is normally stable and does not dissociate while moving away from excess Ag.

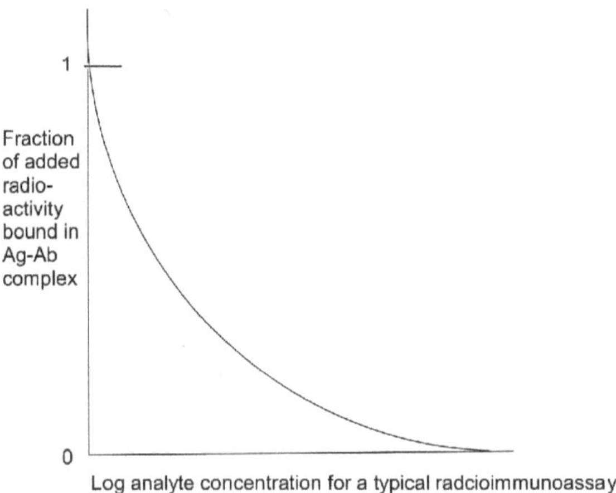

Calibration Curve for a Radioimmunoassay

27

GRAVIMETRIC ANALYSIS

Gravimetric analysis describes a set of methods in analytical chemistry for the quantitative determination of an analyte based on the mass of a solid. A simple example is the measurement of solids suspended in a water sample. A known volume of water is filtered, and the collected solids are weighed.

In gravimetry, the analyte needs to be first converted to a solid by precipitation with an appropriate reagent. The precipitate can then be collected by filtration, washed, dried to remove traces of moisture from the solution, and weighed. The amount of analyte in the original sample can then be calculated from the mass of the precipitate and its chemical composition.

In some cases, it is easier to remove the analyte by evaporation. The analyte might be collected—perhaps in a cryogenic trap or on some adsorbent material such as activated carbon—and measured directly. Or, the sample can be weighed before and after it is dried; the difference between the two masses gives the mass of analyte lost. This is especially useful in determining the water content of complex materials such as foodstuffs.

Procedure

After appropriate dissolution of the sample the following steps should be followed for successful gravimetric procedure:

1. The sample is dissolved, if it is not already in solution. This may involve several steps including adjustment of the pH of the solution in order for

the precipitate to occur quantitatively and get a precipitate of desired properties, removing interferences, adjusting the volume of the sample to suit the amount of precipitating agent to be added.

The precipitating reagent is added at a concentration that favors the formation of a "good" precipitate. This requires addition of a precipitating agent in solution to the sample solution. Upon addition of the first drops of the precipitating agent, super-saturation occurs, then nucleation starts to occur where every few molecules of precipitate aggregate together forming a nucleus. At this point, addition of extra precipitating agent will either form new nuclei or will build up on existing nuclei to give a precipitate. The optimum conditions for precipitation which make the super-saturation low are: **a.** Precipitation using dilute solutions, **b.** Slow addition of precipitating agent, **c.** Stirring the solution during addition of precipitating agent to avoid concentration sites, **d.** Increase solubility by precipitation from hot solution, **e.** Adjust the pH in order not to loose precipitate by dissolution and **f.** Usually add a little excess of the precipitating agent for quantitative precipitation and check for completeness of the precipitation.

2. After the precipitate has formed and has been allowed to "digest", the solution is carefully filtered. This may require low concentration, extensive heating (also described as "digestion"), or careful control of the p^H. Digestion can help reduce the amount of co-precipitation. The filter is chosen to trap the precipitate; smaller particles are more difficult to filter.

   - Depending on the procedure followed, the filter might be a piece of ashless filter paper in a fluted funnel, or a filter crucible. Filter paper is convenient because it does not typically require cleaning before use; however, filter paper can be chemically attacked by some solutions (such as concentrated acid or base), and may tear during the filtration of large volumes of solution.

   - An alternative is a crucible whose bottom is made of some porous material, such as sintered glass, porcelain or sometimes metal. The porous materials are chemically inert and mechanically stable, even at elevated temperatures. Crucibles are often used with a mat of glass or asbestos fibers to trap small particles.

   - After the solution has been filtered, it should be tested to make sure that the analyte has been completely precipitated. This is easily done by adding a few drops of the precipitating reagent; if a precipitate is observed, the precipitation is incomplete.

3. After filtration, the precipitate – including the filter paper or crucible – is heated. This achieves three purposes:
   - The remaining moisture is removed (drying).
   - Secondly, the precipitate is converted to a more chemically stable form. For instance, calcium ion might be precipitated using oxalate ion, to produce calcium oxalate (CaC_2O_4); it might then be heated to convert it into the oxide (CaO). It is vital that the empirical formula of the weighed precipitate is known, and that the precipitate is pure.
   - The precipitate cannot be weighed with the necessary accuracy in place on the filter paper; nor can the precipitate be completely removed from the filter paper in order to weigh it. The precipitate can be carefully heated in a crucible until the filter paper has burned away; this leaves only the precipitate. (As the name suggests, "ashless" paper is used so that the precipitate is not contaminated with ash.).
4. After the precipitate is allowed to cool (preferably in a desiccator to keep it from absorbing moisture), it is weighed (in the crucible). The mass of the crucible is subtracted from the combined mass, giving the mass of the precipitated analyte. Since the composition of the precipitate is known, it is simple to calculate the mass of analyte in the original sample.
5. Precipitation from Homogeneous Solution: In some situations, it is better to generate the precipitating agent in the precipitation medium rather than adding it. For example, in order to precipitate iron as the hydroxide, we dissolve urea in the sample. Heating of the solution generates hydroxide ions from the hydrolysis of urea. Hydroxide ions are generated at all points in solution and thus there are no sites of concentration. We can also adjust the rate of urea hydrolysis and thus control the hydroxide generation rate. This type of procedure can be very advantageous in case of colloidal precipitates.

Example: A chunk of ore is to be analyzed for sulfur content. It is treated with concentrated nitric acid and potassium chlorate to convert all of the sulfur to sulfate (SO_4^{2-}). The nitrate and chlorate are removed by treating the solution with concentrated HCl. The sulfate is precipitated with barium (Ba^{2+}) and weighed as $BaSO_4$ following some of the steps mentioned above.

Advantages: Gravimetric analysis, provides for exceedingly precise analysis. In fact, gravimetric analysis has been used to determine the atomic masses of many elements to six figure accuracy. Gravimetry provides very little room for instrumental error and does not require a series of standards for calculation of an unknown. Also, methods often do not require expensive equipment. Gravimetric analysis, due to its high degree of accuracy, when performed correctly, can also be used to calibrate other instruments in lieu of reference standards.

Disadvantages: Gravimetric analysis usually only provides for the analysis of a single element, or a limited group of elements, at a time. Comparing modern dynamic flash combustion coupled with gas chromatography with traditional combustion analysis will show that the former is both faster and allows for simultaneous determination of multiple elements while traditional determination allows only for the determination of carbon and hydrogen. Methods are often convoluted and a slight mis-step in a procedure can often mean disaster for the analysis (for Ex: colloid formation in precipitation gravimetry).

28

VOLUMETRIC ANALYSIS

Titration, also known as titrimetry, is a common laboratory method of quantitative chemical analysis that is used to determine the unknown concentration of an identified analyte. Since volume measurements play a key role in titration, it is also known as volumetric analysis. A reagent, called the *titrant* or *titrator* is prepared as a standard solution. A known concentration and volume of titrant reacts with a solution of *analyte* or *titrand* to determine concentration. The volume of titrant reacted is called *titre*. Volumetric analysis originated in late 18th-century in France.

Analysis of samples by titration

A typical titration begins with a beaker or Erlenmeyer flask containing a known volume of the analyte and a small amount of indicator placed underneath a calibrated burette containing the titrand. Small volumes of the titrant are then added to the analyte and indicator until the indicator changes, reflecting arrival at the endpoint of the titration. Depending on the endpoint desired, single drops or less than a single drop of the titrant can make the difference between a permanent and temporary change in the indicator. When the endpoint of the reaction is reached, the volume of reactant consumed is measured and used to calculate the concentration of analyte by:

$$C_a = \frac{C_t V_t M}{V_a}$$

where C_a is the concentration of the analyte, in molarity (normality); C_t is the concentration of the titrant, also in molarity (normality); V_t is the volume of the titrant used, in ml or liters; M is the mole ratio of the analyte and reactant from the balanced chemical equation; and V_a is the volume of the analyte used, typically in ml or liters.

Procedure

Titrations require titrant and analyte in a solution form. Though solids are usually dissolved into an aqueous solution, other solvents such as glacial acetic acid or ethanol are used for special purposes (as in petrochemistry). Concentrated analytes are often diluted to improve accuracy. Many non-acid-base titrations require a constant pH throughout the reaction. Therefore a buffer solution may be added to the titration chamber to maintain the pH. Where two reactants in a sample may react with the titrant and only one is the desired analyte, a separate masking solution may be added to the reaction chamber which masks the unwanted ion. Some redox reactions may require heating the sample solution and titrating while the solution is still hot to increase the reaction rate. For instance, the oxidation of some oxalate solutions requires heating to 60 °C to maintain a reasonable rate of reaction.

Titration curves

A titration curve is a curve in the plane whose X-coordinate is the volume of titrant added since the beginning of the titration, and whose Y-coordinate is the concentration of the analyte at the corresponding stage of the titration (in an acid-base titration, the Y-coordinate is usually the pH of the solution). The equivalence point (endpoint) is clearly visible.

In an acid-base titration, the titration curve reflects the strength of the corresponding acid and base. For a strong acid and a strong base, the curve will be relatively smooth and very steep near the equivalence point. Because of this, a small change in titrant volume near the equivalence point results in a large pH change. Indicators like litmus, phenolphthalein or bromothymol blue can be used.

If one reagent is a weak acid or base and the other is a strong acid or base, the titration curve is irregular and the pH shifts less with small additions of titrant near the equivalence point. For example, the titration curve for the titration between oxalic acid (a weak acid) and sodium hydroxide (a strong base) is shown in the picture. The equivalence point occurs between pH 8-10, indicating the solution is basic at the equivalence point and an indicator such as phenolphthalein would be appropriate. Titration curves corresponding to weak bases and strong acids are similarly behaved, with the solution being acidic at the equivalence point and indicators such as methyl orange and bromothymol blue being most appropriate.

Titrations between a weak acid and a weak base have titration curves which are highly irregular. Because of this, no definite indicator may be appropriate and a pH meter is often used to monitor the reaction. The type of function that can be used to describe the curve is called a sigmoid function.

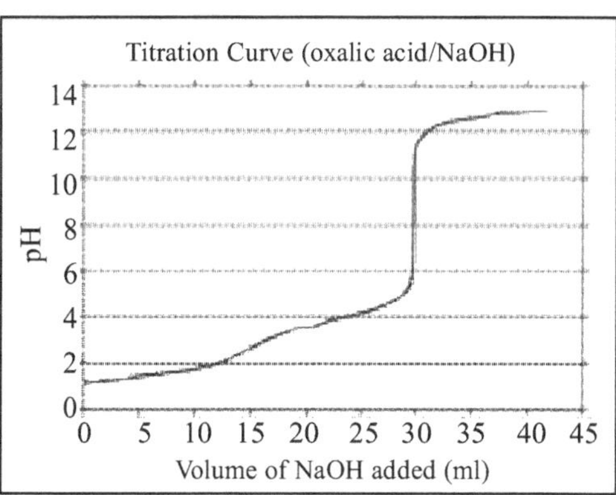

Types of titrations

There are many types of titrations with different procedures and goals. The most common types of qualitative titration are acid-base titrations and redox titrations.

Acid–base titration: Acid-base titrations depend on the neutralization between an acid and a base when mixed in solution. In addition to the sample, an appropriate indicator is added to the titration chamber, reflecting the pH range of the equivalence point. The acid-base indicator indicates the endpoint of the titration by changing color. The endpoint and the equivalence point are not exactly the same because the equivalence point is determined by the stoichiometry of the reaction while the endpoint is just the color change from the indicator. Thus, a careful selection of the indicator will reduce the indicator error. For example, if the equivalence point is at a pH of 8.4, then the phenolphthalein indicator should be used. Common indicators, their colors, and the pH range in which they change color are given in the Table (page 121). When more precise results are required, or when the reagents are a weak acid and a weak base, a pH meter or a conductance meter are used.

Redox titration: Redox titrations are based on a reduction-oxidation reaction between an oxidizing agent and a reducing agent. A potentiometer or a redox indicator is usually used to determine the endpoint of the titration, as when one of the constituents is the oxidizing agent like potassium dichromate. The color change of the solution from orange to green is not definite, therefore an indicator such as sodium diphenylamine is used. Analysis of wines for sulfur dioxide requires iodine as an oxidizing agent. In this case, starch is used as an indicator; a blue starch-iodine complex is formed in the presence of excess iodine, signalling the endpoint.

Some redox titrations do not require an indicator, due to the intense color of the constituents. For instance, in permanganometry a slight persisting pink color signals the endpoint of the titration because of the color of the excess oxidizing agent, potassium permanganate.

Complexometric titration: Complexometric titrations rely on the formation of a complex between the analyte and the titrant. In general, they require specialized indicators that form weak complexes with the analyte. Common examples are Eriochrome Black T for the titration of calcium and magnesium ions, and the chelating agent EDTA used to titrate metal ions in solution.

Measuring the endpoint of a titration

Different chemicals like indicators and instruments like potentiometer (p^H meter), conductivity meter etc. and reaction properties like precipitation and colour change (self indicator) are utilized to determine the endpoint.

1. **Indicator:** A substance that changes color in response to a chemical change is called an indicator. An acid-base indicator (Ex. phenolphthalein) changes color depending on the pH. Redox indicators are also used. A drop of indicator solution is added to the titration at the beginning. The endpoint is reached when there is change in color .
2. **Potentiometer:** An instrument that measures the electrode potential of the solution. These are used for redox titrations; the potential of the working electrode will suddenly change as the endpoint is reached.
3. **pH meter:** A potentiometer with an electrode whose potential depends on the amount of H^+ ion present in the solution. The pH of the solution is measured throughout the titration, more accurately than with an indicator; at the endpoint there will be a sudden change in the measured pH.
4. **Conductivity:** Ion concentration can change significantly in a titration, which changes the conductivity (For instance, during an acid-base titration, the H^+ and OH^- ions react to form neutral H_2O). As total conductance depends on all ions present in the solution and not all ions contribute equally (due to mobility and ionic strength), predicting the change in conductivity is more difficult than measuring it.

5. **Color change:** In some reactions, the solution changes color without any added indicator. This is often seen in redox titrations when the different oxidation states of the product and reactant produce different colors.

6. **Precipitation:** If a reaction produces a solid, a precipitate will form during the titration. A classic example is the reaction between Ag^+ and Cl^- to form the insoluble salt AgCl. Cloudy precipitates usually make it difficult to determine the endpoint precisely. To compensate, precipitation titrations often have to be done as "back" titrations

Endpoint and equivalence point

Though equivalence point and endpoint are used interchangeably, they are different terms. *Equivalence point* is the theoretical completion of the reaction: the volume of added titrant at which the number of moles of titrant is equal to the number of moles of analyte. *Endpoint* is what is actually measured, a physical change in the solution as determined by an indicator or a specific instrument. There is a slight difference between the endpoint and the equivalence point of the titration. This error is referred to as an indicator error, and it is indeterminate.

Back titration

Back titration is a titration done in reverse. Instead of titrating the original sample, a known excess of standard reagent is added to the solution, and the excess is titrated. A back titration is useful if the endpoint of the reverse titration is easier to identify than the endpoint of the normal titration, as with precipitation reactions. Back titrations are also useful if the reaction between the analyte and the titrant is very slow, or when the analyte is in a non-soluble solid.

Indicators and their colour changes

Indicator	Color on acidic side	Range of color change	Color on basic side
Methyl violet	Yellow	0.0–1.6	Violet
Bromophenol blue	Yellow	3.0–4.6	Blue
Methyl orange	Red	3.1–4.4	Yellow
Methyl red	Red	4.4–6.3	Yellow
Litmus	Red	5.0–8.0	Blue
Bromothymol blue	Yellow	6.0–7.6	Blue
Phenolphthalein	Colorless	8.3–10.0	Pink
Alizarin yellow	Yellow	10.1–12.0	Red

Some specific examples of titrations

- Kjeldahl method: It is used to measure the nitrogen content in a sample. Organic nitrogen is digested into ammonia with sulfuric acid and potassium sulfate. Finally, ammonia (distilled and collected) is back titrated with boric acid and then sodium carbonate.
- Acid value: The mass in milligrams of potassium hydroxide (KOH) required to neutralize carboxylic acid in one gram of sample. An example is the determination of free fatty acid content. These titrations are achieved at low temperatures.
- Saponification value: The mass in milligrams of KOH required to saponify carboxylic acid in one gram of sample. Saponification is used to determine average chain length of fatty acids in fat. These titrations are achieved at high temperatures.
- Ester value (or ester index): A calculated index. Ester value = Saponification value – Acid value.
- Amine value: The mass in milligrams of KOH equal to the amine content in one gram of sample.
- Hydroxyl value: The mass in milligrams of KOH corresponding to hydroxyl groups in one gram of sample. The analyte is acetylated using acetic anhydride and then titrated with KOH.
- Winkler test for dissolved oxygen is used to determine oxygen concentration in water. Oxygen in water samples is reduced using manganese (II) sulfate, which reacts with potassium iodide to produce iodine. The iodine is released in proportion to the oxygen in the sample, thus the oxygen concentration is determined with a redox titration of iodine with thiosulphate using a starch indicator.
- Vitamin C: Also known as ascorbic acid, vitamin C is a powerful reducing agent. Its concentration can easily be identified when titrated with the blue dye Dichlorophenol-indophenol (DCPIP) which turns colorless when reduced by the vitamin.
- Benedict's reagent: Excess glucose in urine may indicate diabetes in the patient. Benedict's method is the conventional method to quantify glucose in urine using a prepared reagent. In this titration, glucose reduces cupric ions to cuprous ions which react with potassium thiocyanate to produce a white precipitate, indicating the endpoint.
- Bromine number/ Iodine number: A measure of unsaturation in an analyte, expressed in milligrams of bromine / iodine absorbed by 100 grams of sample.

Karl Fischer titration: A potentiometric method to analyze trace amounts of water in a substance. A sample is dissolved in methanol, and titrated with Karl Fischer reagent. The reagent contains iodine, which reacts proportionally with water. Thus, the water content can be determined by monitoring the potential of excess iodine.

29

AMPLIFICATION AND SEQUENCING OF NUCLEIC ACIDS

DNA and RNA are the carriers of genetic information in all the living organisms. The study and analysis of the genetic code can provide information that can be used in clinical diagnostics, drug discovery, genetic engineering and forensic sciences. Hence there is need to understand the techniques in isolation of DNA and RNA followed by their amplification and sequencing.

Extraction and isolation of nucleic acids: The isolation process consists of three steps:

1. **Cell lysis**: To release the contents of the cell, the cellular walls are ruptured by treatment with surfactants like sodium dodecyl sulphonate(SDS) and Triton-X100. Alternatively, the cells are incubated with an enzyme having cell lytic properties like lysozyme.
2. **Isolation:** The nucleic acids are isolated by removing cell debris, separating the nucleic acids from other cell contents and breaking down any complexes between proteins and nucleic acids. Methods such as centrifugation, precipitation, size exclusion chromatography, liquid chromatography, treatment with protein digesting enzymes and CsCl (cesium chloride) density gradient centrifugation are used.
3. **Amplification:** The DNA and RNA of the resulting solution is amplified or concentrated.

CsCl density gradient centrifugation

CsCl density gradient centrifugation can be used to isolate DNA. Due to difference in buoyant densities DNA is separated from RNA or proteins. Upon centrifugation, the different molecules in the sample form distinct bands depending on their buoyant densities. RNA has higher density and sinks to the bottom. Proteins are lighter and float on top. The buoyant density is about 1.7g/ml (equal to the density of CsCl - at the centre).

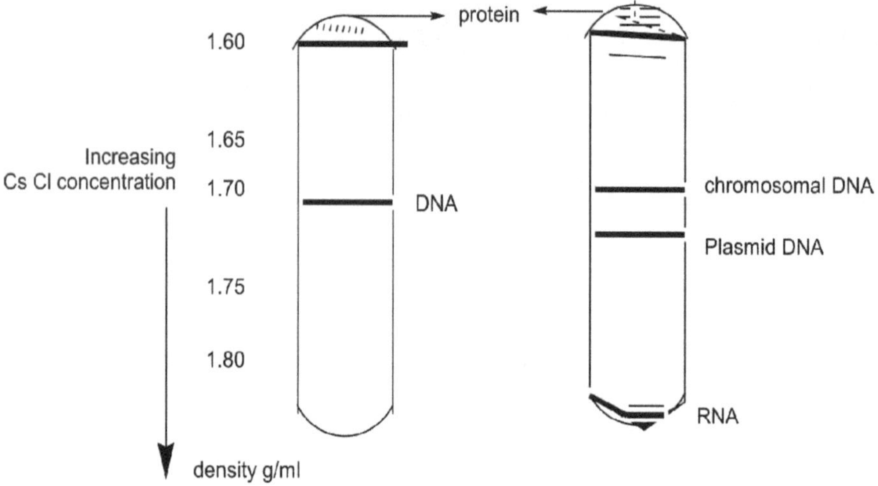

CsCl density Gradient Centrifugation

Nucleic acid amplification- PCR reaction

The polymerase chain reaction(PCR) is an *in vitro* technique that allows the enzymatic – catalyzed amplification of specific nucleic acid sequences. With PCR, minute quantities of nucleic acids can be analysed. Even a single DNA molecule is enough for amplification.

The principle of PCR: The amplification is carried out in a single sample tube, which is placed into a thermocycler. The sample tube contains: 1). An excess of two primes, that is, oligonucleotides, that are complimentary to the ends of targeted nucleic acid region. 2). The enzyme DNA polymerase. 3). The template DNA sample to be amplified. 4). A buffer to maintain the correct p^H and to supply the ions such as Mg^{++} that are necessary for the reaction. The raction takes place in three steps.

1. Denaturation: DNA strands are separated at 95^{0}C.
2. Annealing of primers: Primers are hybridised to their complimentary sequences at 50-60^{0}C.

3. Primer extension at 72^{0}C: the polymerase catalyzes the synthesis of complimentary single stranded DNA by extending the 3' end of hybridized polymer.

The three stages constitute " one cycle ". a typical PCR consists of 20-35 such cycles. This procedure of repeated heating and cooling of reaction mixture is referred to as ***thermal cycling.*** The three steps together are carried out for 30-120 s each. Only the very first denaturation step takes about 5min. A PCR with 30cycles can be completed within 1-2hr.

In theory, the number of DNA copies are doubled during each cycle, resulting in an exponential amplification. In practice, the doubling is not achieved. For example, the theoretical yield is 1x 10^6, but the real yield is 4x10^5 amplifications.

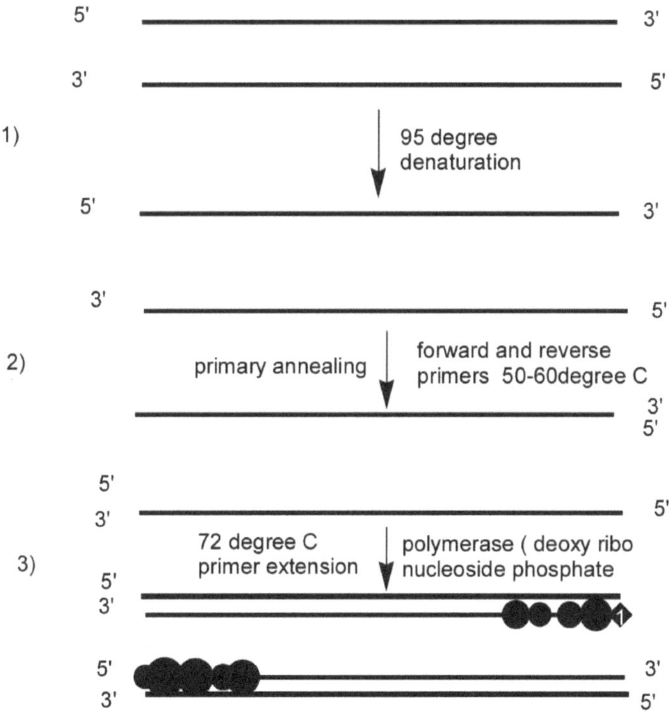

Nucleic acid amplification – PCR reaction

Nucleic acid sequencing

As the function of nucleic acids is determined by the sequence of the bases within the molecule, the sequencing plays an important role in nucleic acid analysis.

Restriction enzymes in sequencing

DNA extracts from cells or tissues are generally too long to be sequenced directly. They must be broken down in an orderly manner into smaller fragments of upto 800 base pairs. For this purpose, restriction endonucleases or simply restriction enzymes, are employed. These enzymes recognize a specific base sequence of four to eight bases within a double stranded DNA (ds DNA) molecule and cleave both strands at a specific point close to this recognition site.

Digestion with restriction enzymes produces a series of precisely defined fragments, which can be separated according to their size by gel electrophoresis. The fragments then need to be denatured into single strands. This can be achieved by melting the ds DNA. The single strands are then separated by gel electrophoresis.

Part II: Analytical Methods and Procedures

1

METHODS OF ANALYSES

Organic material especially, seeds, grains, leaves, fruits and nuts contain prominently three groups of compounds. They are carbohydrates, proteins and fats/oils, mostly in that order. They also contain 3-5% minerals and smaller quantities of vitamins and other minor constituents. Cereal grains contain more of carbohydrates while oilseeds and nuts contain more oil. Tree nuts and even oilseeds may contain 30-50% protein while cereal grains contain 5-8% protein only. The starch and sugar contents of oilseeds may be less than 10% while the oil content of cereal grains ranges around 5%. The actual content and the composition of these groups of compounds really account for their quality as well as commercial value.

The composition and quality of grains and other plant products vary with respect to their genetic background, location or the soils under which they are grown, season or the environment, the maturity of the grain/product or its harvest stage and the post harvest processes including storage etc. Labeling of the consumer product has become mandatory in recent years. Crop Scientists are regularly bringing out made to order or designer seeds, whose quality and composition is very much different from the traditional produce. Hence, one needs to examine the composition to satisfy the needs of the quality control and consumer requirements.

This book deals with methods of oil, protein and sugar analysis in organic materials like grains, nuts, leaves and their products. Analytical methods of carbohydrate components like reducing and non reducing sugars along with methods of individual sugar (glucose, fructose, sucrose, starch, cellulose etc.) estimation are presented. Similarly methods for various proteins, peptides and amino acids also have been included. Fatty acid, and amino acid content and their composition evaluation find sufficient place. Antinutrients especially aflatoxins, glucosinolates, ricin, oxalate, phytic acid, oligosaccharides, etc. also are included. Analytical methods of some vitamins and minerals have also been included. Pesticide residues which have gained importance in recent years also find a place with respect to their estimation. Some aspects of structural elucidation,based on individual component composition etc. have also been included.

Biochemical constituents like enzymes, nucleic acids, steroids, alkaloids, etc. play an important role in in the overall quality of the grains and their products. Hence techniques for their assay have also been presented. Wherever possible, more than one procedure is included. The analyst can select a method based on the facilities available in his/her laboratory. Only a few important and most essential analytical procedures of the above organic components have been presented. The analyst can find more details and methods of evaluation of other constituents in standard laboratory analytical manuals.

2

CARBOHYDRATES AND THEIR METHODS OF ANALYSIS

Carbohydrates are the most abundant biomolecules on earth. They occur in nature as glucose, starch and cellulose etc. Each year, photosynthesis converts more than 100 billion metric tonnes of carbon dioxide and water into starch, cellulose and other plant products. Carbohydrates are polyhydroxy aldehydes or ketones and their derivatives. They provide 4K calories of energy per gram. They promote utilization of lipids and reduce wastage of proteins. They are readily available in roots, tubers, cereal grains, sugarcane and sugar beet.

Carbohydrates have the general formula $C_n(H_2O)_n$, indicting that they are the hydrates of carbon. Some carbohydrates also contain nitrogen, phosphorous or sulfur. They are characterized by the number of carbon atoms they contain. If n=3, they are trioses, n=4 tetroses, n=5 pentoses, n=6 hexoses and n=7, heptoses. The most common sugars in plants are the hexoses. Two or more simple carbohydrates condense together to form polymer sugars like sucrose, starch and cellulose etc. The basic condensing bond is glycosidic bond.

A carbohydrate is also called saccharide, meaning sugar. Sugars are synthesized by plants through a series of reactions from carbon dioxide in the presence of sunlight and a green pigment called chlorophyll by a process named photosynthesis. Sugars serve as sources of energy to all the living beings. Sugars are broken down through a process called respiration to release energy.

$$6CO_2 + 6H_2O \underset{\text{respiration}}{\overset{\text{photosynthesis}}{\rightleftharpoons}} C_6H_{12}O_6 + 6O_2$$

Carbohydrates vary greatly in size ranging from the simple glyceraldehyde molecule to the plymeric starch molecule. The carbohydrates have been classified based on the degree of polymerization into mono, di, oligo, and polysaccharides. Saccharide in Greek means sugar.

Classification of carbohydrates

Major types	1. Simple carbohydrates Ex. monosaccharides
	2. Complex carbohydrates. Ex. di- oligo- and polysaccharides
Carbon number	1. Tetrose (C_4)
	2. pentose (C_5)
	3. hexose (C_6)
	4. heptose (C_7)
Functional group	1. Aldoses : sugars with –CHO group
	2. Ketoses : sugars with -C=O group
Reactivity	1. Reducing sugars: sugars oxidized by Tollens' reagent
	2. Non-reducing sugars: sugars not oxidized by Tollens' reagent

Analysis of carbohydrates

A large number of analytical techniques have been developed to measure the total concentration and type of carbohydrates present in plants or foods. The carbohydrate content can easily be determined by calculating the percent remaining after all the other components have been measured: % Carbohydrates = 100 – (% moisture + % protein + % lipid + % mineral). This method can lead to erroneous results due to experimental errors in any of the other assay methods, and so it is usually better to directly measure the carbohydrate content for accurate measurements.

The amount of preparation needed to prepare a sample for carbohydrate analysis depends on the nature of the sample being analyzed. Aqueous solutions, such as fruit juices, syrups and honey, usually require very little preparation prior to analysis. On the other hand, many samples contain carbohydrates that are physically associated or chemically bound to other components, *e.g.,* nuts, cereals, fruit, breads and vegetables. In such cases it is necessary to isolate the carbohydrates from the rest of the food before they can be analyzed. The precise method of carbohydrate isolation depends on the carbohydrate type, the sample matrix type and the purpose of analysis. However, there are some procedures that are common to many isolation techniques.

One of the most commonly used methods of extracting low molecular weight carbohydrates is to boil a defatted sample with an 80% alcohol solution. Monosaccharides and oligosaccharides are soluble in alcoholic solutions, whereas, proteins, polysaccharides and dietary fiber are insoluble. The soluble components can be separated from the insoluble components by filtering the boiled solution and collecting the filtrate and the residue (the part retained by the filter). These two fractions can then be dried and weighed to determine their concentrations. In addition to monosaccharides and oligosaccharides various other small molecules may also be present in the alcoholic extract that could interfere with the subsequent analysis Ex. amino acids, organic acids, pigments, vitamins, minerals etc. It is usually necessary to remove these components prior to carrying out a carbohydrate analysis. This is commonly achieved by treating the solution with clarifying agents or by passing it through one or more ion-exchange resins.

Clarifying agents. Water extracts of many foods contain substances that are colored or produce turbidity, and thus interfere with spectroscopic analysis or endpoint determinations. For this reason solutions are usually *clarified* prior to analysis. The most commonly used clarifying agents are heavy metal salts (such as lead acetate) which form insoluble complexes with interfering substances that can be removed by filtration or centrifugation. However, it is important that the clarifying agent does not precipitate any of the carbohydrates from solution as this would cause an underestimation of the carbohydrate content.

Ion-exchange. Many monosaccharides and oligosaccharides are polar non-charged molecules and can therefore be separated from charged molecules by passing the samples through ion-exchange columns. By using a combination of a positively and a negatively charged column it is possible to remove most charged contaminants. Non-polar molecules can be removed by passing a solution through a column with a non-polar stationary phase. Thus proteins, amino acids, organic acids, minerals and hydrophobic compounds can be separated from the carbohydrates prior to analysis.

Prior to analysis, the alcohol can be removed from the solutions by evaporation under vacuum so that an aqueous solution of sugars remains.

Chromatographic and Electrophoretic methods

Chromatographic methods are the most powerful analytical techniques for the analysis of the type and concentration of monosaccharides and oligosaccharides in foods. Thin layer chromatography (TLC), Gas chromatography (GC) and High Performance Liquid chromatography (HPLC) are commonly used to separate and identify carbohydrates. Carbohydrates are separated on the basis of their differential adsorption characteristics by passing the solution to be analyzed through a column. Carbohydrates can be separated on the basis of their partition coefficients, polarities or sizes, depending on the type of column used.

HPLC is currently the most important chromatographic method for analyzing carbohydrates because it is capable of rapid, specific, sensitive and precise measurements. In addition, GC requires that the samples be volatile, which usually requires that they be derivitized, whereas in HPLC samples can often be analyzed directly. HPLC and GC are commonly used in conjunction with NMR or mass spectrometry so that the chemical structure of the molecules that make up the peaks can also be identified.

Carbohydrates can also be separated by electrophoresis after they have been derivitized to make them electrically charged, *e.g.,* by reaction with borates. A solution of the derivitized carbohydrates is applied to a gel and then a voltage is applied across it. The carbohydrates are then separated on the basis of their size. The smaller the size of a carbohydrate molecule, the faster it moves in an electrical field.

Chemical methods

A number of chemical methods used to determine monosaccharides and oligosaccharides are based on the fact that many of these substances are *reducing agents* that can react with other components to yield precipitates or colored complexes which can be quantified. The concentration of carbohydrate can be determined gravimetrically, spectrophotometrically or by titration. Non-reducing carbohydrates can be determined using the same methods if they are first hydrolyzed to make them reducing. It is possible to determine the concentration of both non-reducing and reducing sugars by carrying out an analysis for reducing sugars before and after hydrolyzation. Many different chemical methods are available for quantifying carbohydrates. Most of these can be divided into three catagories: titration, gravimetric and colorimetric. Some specific methods are furnished in this text.

Sugars by volumetric method

Principle

The method utilizes the reducing property of sugars. The reducing sugars convert the copper in Fehling's solution to red and insoluble Cu_2O in alkaline medium. The unknown solution is titrated with standardized Fehling's solution using methylene blue indicator. The volume of sugar solution required for reduction is noted. The sugar solution is clarified with zinc ferricyanide or lead acetate

Reagents and equipment

1. Reagent A: 69.278g $CuSO_4.H_2O$ is dissolved in one litre of water. Filter and store the solution.
2. Reagent B: Dissolve 346g of sodium potassium tartarate and 100g of NaOH in one litre of water. Filter the solution.
3. Equal volumes of the above two solutions are pipetted and mixed thoroughly. For titration 10-25 ml of the solution is taken.
4. Methylene blue indicator: 0.5% in 80% ethanol.
5. Standard invert solution: Weigh 2.375g of sucrose and dissolve in 100ml water. Add 15ml of 1NHCl and boil for 2min. Neutralize with 10% NaOH and adjust the volume to 500ml in a volumetric flask.

   1ml std invert sugar = 0.00475g of sucrose or 0.005g invert sugar

Procedure

1. **Standardization of Fehling's solution**: Take 50ml each of reagents A and B and mix well. Take 10 ml into a conical flask and add 50ml water. Take standard sugar solution into a burette. Titrate the Fehling's solution with sugar solution while heating the conical flask. Add methylene blue towards the end and continue the titration till the blue colour disappears. Note the burette reading.
2. **Sample solution:** Weigh accurately 10-15g sample and dissolve in cold water. Dilute to 250 ml with water, shake well and filter. If necessary the solution needs to be clarified with lead acetate solution.
3. Titrate an aliquot (10-25ml) of the sample solution against Fehling's solution and calculate the reducing sugar content.
4. **Total sugars:** Take 50ml of prepared sample solution. Add 5g of citric acid and 50ml water. Boil for 10min. Neutralize with dilute NaOH and adjust the volume to 250ml. Carry out titration against 10 ml Fehling's solution.

   Calculate reducing sugars and total sugars as invert sugars.

   Percent non-reducing sugars = Difference between the above two values.

Sugars by polarimetric method

Principle

The sugars have optical acitivity. The extent of rotation which varies from sugar to sugar is a function of the amount of sugar. The specific rotation of some sugars are: dextrose = +52.7, fructose = - 92.7, lactose = +52.6, maltose = 138.4, sucrose =+66.5, invert sugars = - 20.0.

Reagents and equipment

1. Sample containing single sugar: The specific rotation of clarified sample is measured. The amount of sugar is calculated using the relationship below:

$$\text{Sugar \%} = \frac{\text{Angular momentum}}{\text{Specific rotation}} \times \frac{100}{\text{tube length}} \times \frac{100}{\text{\% conc}}$$

2. Sample containing a mixture with sucrose: When sucrose alone is present, its quantity can be calculated

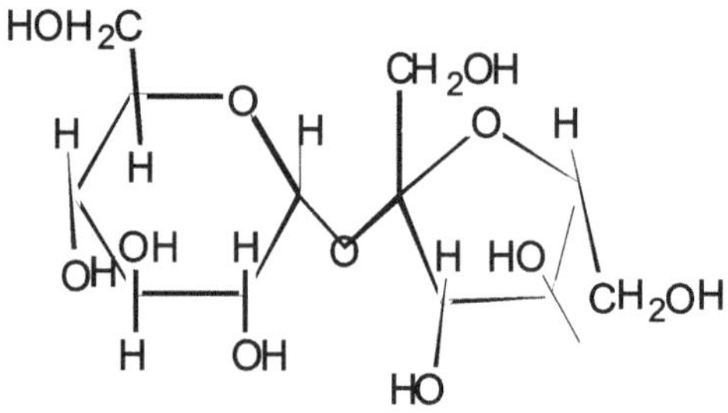

$$\text{Sucrose \%} = \frac{100 \times \text{angular rotation}}{66.5 \times \text{length of tube}}$$

   When other sugars or optically active substances are present then this formula cannot be used. The optical rotation of the solution is measured. The solution is allowed to undergo inversion. The change after inversion in rotation is taken as a measure of sucrose (from + 66.5 to – 22.15)

3. Polarimeter

Procedure

1. Clarified sample solution: Take 50ml aliquot in 100 ml volumetric flask. Add 25ml water and 7ml 5N HCl. Keep the flask at 60^0 C for 12min. Cool and make upto 100ml.

2. Measure the rotation of solution in a polarimeter before and after inversion. Calculate % sucrose.

$$\% \text{ sucrose} = \frac{\text{Rotation of original solution} - \text{rotation after inversion}}{0.8865}$$

3. The difference between the angular rotation of the original solution and (% sucrose X 0.665) represents rotation due to the second sugar.

   % of the other sugar

$$= \frac{\text{Rotation of original solution} - 100\ (0.665\ x\ \text{Sucrose\%})}{\text{Specific rotation of second sugar}}$$

Total carbohydrates (Phenol - Sulphuric acid method)

Principle

The soluble sugars in a sample are extracted with water at high temperature and pressure. They are made to react with phenol- sulphuric acid reagent and the amber color developed is measured to estimate the sugar content.

Reagents and equipment

1. Phenol reagent: 5 g of reagent grade phenol in 100 ml of distilled water.
2. Reagent-grade concentrated sulfuric acid.
3. Thick-walled Pyrex boiling tubes: 6 inches to 1 inch diameter.
4. Spectrophotometer or colorimeter with (quartz) cuvettes.
5. *Glucose standard* : Dissolve 100mg of glucose in 100 ml of 0.15% (wt/vol) benzoic acid (which is added as a preservative). Store at 5^0C. This solution is stable for several months. Dilute 1:10 in distilled water just before use to give a solution containing 100 µg/ml. Prepare standards, 10 to 100 µg/ml from the diluted solution.
6. *Mannose and galactose standard* : 20 mg mannose and 10 mg galactose are taken in a 100ml volumetric flask and made upto volume (300ppm solution). 10 ml of this solution is diluted to 100ml to obtain 30 ppm (30 µg/ ml) solution. The latter standard solution is more representative and preferable.
7. Autoclave.
8. Centifuge.

Procedure

1. Weigh 0.5 g plant (seed/defatted cake or leaf powder etc.) sample in a 100 ml flask, add 40 ml of distilled water, autoclave at 15 lb pressure for 4 hr. Cool and filter the supernatant. Make up the filtrate to 50 ml, in a volumetric flask.
2. For testing take 0.1 ml of extract in a test tube, add 2.9 ml distilled water, add 0.5 ml of saturated aqueous phenol and shake thoroughly. Add 5.0 ml of concentrated sulphuric acid carefully. Amber golden colour develops.
3. Keep it for 30 min and measure the absorbance at 490 nm in a colorimeter or spectrophotometer.
4. For blank take 3 ml of distilled water in a test tube, add 0.5 ml saturated aqueous phenol, shake thoroughly and add 5.0 ml of sulphuric acid, keep for 30 min.
5. Measure the intensity of colour at 490 nm in a colorimeter.
6. Extrapolate the concentration from the standard curve and calculate the percent sugars in the sample.

Standard curve

1. Take 7 test tubes and add 0.0, 0.5, 1.0, 1.5, 2.0, 2.5 and 3.0 ml of 30 ppm standard, add 3.0, 2.5, 2.0, 1.5, 1.0, 0.5, and 0.0 ml distilled water.
2. Shake well and add 0.5 ml of saturated aqueous phenol. Shake and add 5.0 ml concentrated sulphuric acid, shake, plug and keep for 30 min.
3. Measure the intensity of colour at 490 nm. Plot the readings against concentrations in ppm.

Calculation

$$\mu g \text{ sugar} = \frac{\text{absorbance of sugar x } \mu g \text{ standard.}}{\text{Absorbance of standard}}$$

Extraction of sugars

1. Sugars can also be extracted from plant material using the following procedure. Take 0.5 g of the defatted seed/ plant sample in a mortar. Add 10 ml of 60:25:15 methanol : chloroform: water. Grind and centrifuge. Separate the supernatant. Repeat extraction of the residue with 5ml each of the above solvent mixture two more times. Centrifuge and pool the supernatants. Make up the supernatants to 25 or 50 ml.

Caution

Phenol causes severe skin and eye burns. Protect by wearing goggles, gloves, etc. when using this reagent. Do not pipette by mouth.

Ref: Dubois, M.K., Giles, J.K., Hamilton, P.A., Rebers and Smith, F. 1956. A Colorimetric Method for the Determination of Sugars and Related Substances. Analytical Chemistry. 28:350-356

Carbohydrate Content (Anthrone Method)

Principle

The carbohydrate content of the sample is estimated by Anthrone method. The aqueous extract is hydrolyzed and made to react with anthrone reagent and the green colour developed is measured at 630nm to estimate the total sugars in the sample.

Reagents and equipment

1. HCl (2.5N).
2. Anthrone reagent : Dissolve 200 mg anthrone in 100 ml of ice cold 95% H_2SO_4. The reagent is prepared fresh before use.
3. Standard glucose(stock): Dissolve 100 mg of glucose in 100 ml water.
4. Working standard: Dilute 10 ml of stock glucose solution to 100ml with distilled water. A few drops of toluene are added and stored in refrigerator.
5. Spectrophotometer with quartz cuvettes.

Procedure

1. Hundred mg of the sample is weighed into a boiling tube. It is then hydrolyzed by keeping it in a boiling water bath for 3hr with 5ml of 2.5N HCl and cooled to room temperature.
2. It is neutralized with solid sodium carbonate until the effervescence ceases. The volume is made upto 100ml and centrifuged.
3. The supernatant is collected and from this 0.5 or 1ml of the aliquots are taken for analysis.
4. The standard curve is prepared by taking 0, 0.2,0.4,0.6,0.8, and 1 ml of the working standard. '0' serves as the blank. The volume is made upto 1ml in all the tubes including the sample tubes by adding water.
5. Then 4ml of anthrone reagent is added, heated for 8 mins in a boiling water bath, cooled rapidly and the intensity of green to dark green colour is read at 630 nm.

6. Standard graph is plotted by taking concentration of the standards on the X axis and the absorbence on the Y axis.
7. From the graph the amount of carbohydrate present in the sample tube is read and later the % carbohydrate is calculated.

Calculations:

Amount of carbohydrate present in 100 mg of the sample =

$$\frac{\text{mg of glucose x 100}}{\text{volume of the test sample}}$$

Note: The contents of all the tubes are cooled in ice before the addition of the ice-cold anthrone reagent.

Ref: Hodge, J.E. (1962). In Methods in carbohydrates(eds. Whistler,R.L. and BeMiller,J.N.), Academic Press,New York.

Total water soluble sugars - T.S.S. (Anthrone reagent method)

Principle

The carbohydrates are hydrolysed by concentrated H_2SO_4 to form furfural and other degradation products. The furfural condenses with anthrone to form blue coloured complex which is subjected to quantification (Fong, *et al.*, 1953, A.O.A.C. 1965). This method is also based on Molische test.

Reagents and equipment

1. Saturated lead acetate (neutral).
2. Sodium oxalate.
3. Anthrone reagent: 0.2 or 0.1 percent (w/v) in 70 percent H_2SO_4 (A.R.). The reagent is freshly prepared each day and allowed to stand for 30-40 min before use.
4. Glucose solution (10-100 μg/ml) for standard curve.
5. Water bath.
6. Spectrophotometer with cuvettes.

Extraction

1. One to two grams of the dried sample material is taken into thick walled pyrex test tubes. About 25 ml of distilled water is added in each tube.

The contents are heated in a boiling water bath (for 15 minutes) and the supernatant, after decantation for 2 minutes, is filltered through whatman No.1 filter paper.

2. The material which settles in the tubes is heated again with distilled water (about 20 ml) for 10 minutes and the supernatant is filtered.
3. This process of extraction is repeated ten times after which no sugar components can be detected in the residue.
4. All the filtrates are collected in a standard volumetric flask (250 ml). The filtrate is then clarified.

Clarifications

1. Before making the volume of the aforesaid volumetric flasks upto mark, enough saturated neutral lead acetate solution is added to produce flocculent precipitates. After thorough shaking, the flasks are allowed to stand for 15 minutes.
2. The supernatant is tested with a few drops of lead acetate solution. When no further precipitate is formed the contents are diluted to 250 ml mark with distilled water.
3. Mix thoroughly and filter through a dry filter paper in dried conical flask. The excess of lead is precipitated with sodium oxalate and the clear solution obtained is used for analysis.

Procedure

1. Ten ml of anthrone reagent is pipetted into each of the thick walled pyrex test-tubes (150 x 250 mm) and diluted in ice cold water. The solution under test (exactly 1 ml) is layered on the acidic reagent, cooled for a further 3-5 minute period and then thoroughly mixed while still immersed in ice cold water.
2. The tubes containing reaction mixture is fitted with small capillary tubes and heated in a vigorously bioling water bath for 7.5 min (in case of 0.2 percent anthrone) and 10 minutes (in case of 0.1 percent anthrone) and then immediately cooled in ice cold water.
3. The percent transmission of the solution is read in a spectrophotometer at 625 nm wave length against a blank.
4. A standard curve is obtained by plotting the spectrophotometric readings of standard sugar solutions (10-100 μg/ml).

5. The concentration of the test solution is calculated from the standard curve.

Calculations

$$\text{T.S.S. (\%)} = \frac{\text{O.D. for test solution x dilution factor}}{\text{O.D for standard x 100 x mg of glucose in the standared}}$$

References

A.O.A.C, 1965. Association of official Analytical Chemists. 1965. Official Methods of Analysis 10th Ed. Washington D.C.

Fong, J., Scheffer, P.L. and Kirk, P.K. 1953. The Ultramicro Determination of Glycogen in Liver. A Comparison of Anthrone and Reducing Sugar Methods. Archives of Biochemistry and Biophysics. 45: 319.

Total reducing sugars (Potassium ferricyanide method)

Principle

Reducing sugars are oxidized by alkaline potassium ferricyanide. This reduced potassium ferrocyanide is precipitated by $ZnSO_4$ as Zinc salts. Excess of potassium ferricyanide oxidises KI to I_2 which is titrated with sodium thiosulphate solution in the presence of starch indicator.

Reagents and equipment

1. Potassium ferricyanide: Take 8.25 g of potassium ferricyanide and 10.6g of anhydrous sodium carbonate and make up the solution to one litre in water. This solution is kept in a dark bottle and is not used until 2 or 3 days after preparation.
2. Glacial acetic acid (5%).
3. Iodine solution (KI+ $ZnSO_4$ + NaCl). Dissolve 12.5 g KI, 25.0 g $ZnSO_4$ and 125 g NaCI in 500 ml of water. This solution is filtered through a double filter paper. Keep in the dark.
4. 0.01 N sodium thiosulphate.
5. 1 percent starch as indicator.
6. Water bath

Procedure

1. 5 ml each of the test sugar solutions are mixed with 5 ml. of potassium ferricyanide solution in each test tube. The test tubes are closed with rubber corks (fixed with capillary tubes bent at right angle). Heat on a boiling water bath for 15 minutes and then cool immediately in ice cold water.
2. To each tube, 5 ml of iodine solution is added followed by 3 ml of 5% glacial acetic acid. The liberated iodine is immediately titrated against 0.01 N sodium thiosulphate using 1 per cent starch as an indicator near the end point.
3. Blank experiment is also run alongside.

Calculations

mg reducing sugars = b (X + 0.05)

b= a factor for invert sugar which is 0.338

X = volume of 0.01 N sodium thiosulphate (Difference of the two titre values i.e. blank minus titer value of test ample)

Sugar content (Alkaline ferricyacide – Colorimetric method)

Principle

Sugars above pH 10.5 reduce ferricyanide to ferrocyanide which react with ferrous ions to produce the Prussian blue colour (Whistler, and Wolfrom, 1962) .

Reagents and equipment

1. Alkaline cyanide: 0.53% w/v Na_2CO_3 containing 0.065% KCN.
2. Potassium ferricyacide : 0.05% w/v in water
3. Ferric ammonium sulphate: 1.5 g dissolved in 1 litre of 0.05 N H_2SO_4
4. Spectrophotometer with quartz cuvettes
5. Water bath

Procedure

1. Mix 1-3ml of sugar solution containing 1 to 9 μg reducing sugars. Add 1 ml alkaline cyanide + 1 ml potassium ferricyanide.

2. Heat the mixture for 15 min on a water bath. A blue colour develops.
3. Measure at 700 nm in a spectrophotometer.
4. A standard curve is obtained by plotting the spectrophotometric readings of standard sugar solutions (10-100 μg/ml) against their absorbence.
5. The concentration of the test solution is read from the standard curve and further calculations are made based on the sample taken and extract dilutions.

Reference

Whistler R.L. and Wolfrom. M.L. 1962. Methods in Carbohydrate Chemistry. Academic Press, London.

Determination of total and reducing sugars (Somogyi, 1940)

Principle

The aqueous extract of the sugars is titrated with Somogyi reagent (phosphate – copper) as per iodimetric procedure and the content of sugars is calculated.

Reagents and equipment

1. **Solution A:-** (Phosphate – copper reagent): 14g of anhydrous disodium phosphate (Na_2HPO_4) and 20 g of Rochelle salt are dissolved in 350 ml of distilled water. 50 ml of 1N NaOH solution and 40 ml of 10% $CuSO_4$ solution are slowly added to it while stirring. Finally 90 g of anhydrous sodium sulphate and 12.5 ml of 1N potassium iodate (3.567 g KIO_3/100 ml) solution are added. The entire solution is then made up to 500 ml and allowed to stand for two days during which time impurities separate out. The clear top part of the solution is decanted and the remainder filtered through a good grade filter paper. If kept in a stoppered Pyrex bottle and protected from strong light, the solution remains stable for months.
2. **Solution B:-** 2.5% KI solution: Alkalinised by a knife tip of Na_2CO_3. Prepare solution afresh each week and store in a dark coloured bottle.
3. **Solution C:-** Approximately 2N H_2SO_4 made by diluting concentrated H_2SO_4.
4. **Solution D:-** 0.005 N Hypo (sodium thiosulphate) solution. This is prepared by diluting 50 ml of 0.1 N hypo solution to a litre, incorporating 2ml of 10% NaOH solution in it for protection against atmospheric carbon dioxide and for increasing the stability of the solution.

5. Hot plate and water bath.

Procedure

A. Extraction

1. 2.5g of powdered sample is placed in a 250 ml Pyrex beaker. Add 0.5 g of $CaCO_3$ and 100ml of distilled water and boil gently for 15 minutes on a hot plate.
2. The beaker is cooled to room temperature and the contents are quantitatively transferred to a 250 ml volumetric flask and 5 ml of saturated neutral lead acetate solution are added and the contents of the flask are thoroughly mixed and allowed to stand for 15 minutes. Volume is made up to the mark with D.
3. The contents are shaken well. 30 to 50 ml portion is decanted off. Sufficient anhydrous disodium phosphate (Na_2HPO_4) is added to the decanted solution to precipitate all the lead.
4. The contents are shaken well once again and filtered through dry Whatman no. 42 fluted filter paper. A few drops of the filtrate are tested with disodium phosphate to make sure that all the lead is precipitated.
5. Filtrate is re-filtered if necessary using a fresh dry filter paper. The filtrate thus obtained is called "clarified extract" and is stored in stoppered flask in a refrigerator and can be used for the determination of total and reducing sugars.

B. Determination of reducing sugars

1. 10 ml of clarified extract is placed in a 50 ml volumetric flask, diluted to volume and mixed. 5 ml of this solution is used for determination of reducing sugars using Somogyi's phosphate copper reagent on the same day as per the Iodimotric procedure given below.

C. Determination of total sugars

1. 10ml of clarified extract is placed in 50 ml volumetric flask. 5ml of (1:4v/v) HCl are added. The contents are mixed well and allowed to stand for 24 hours at room temperature to invert the sugars.
2. The solution is neutralized with normal NaOH solution using a piece of red litmus paper as indicator and keeping the flask in ice-cold water bath while NaOH solution is added slowly. Solution is made upto volume and mixed.

3. 5 ml of this solution are used for the determination of total sugars using Somogyi's phosphate-copper reagent as per the Iodimetric procedure given below.

D. Iodimetric procedure

1. 5 ml of solution A (phosphate-copper reagent) and 5ml of sugar solution are mixed in a 25 x 200 mm Pyrex test-tube, covered with a cotton plug, and heated by immersion in a vigorously boiling water bath for exactly fifteen minutes. They are firmly held in metal racks during the heating to avoid undue agitation.
2. At the end of the heating period the tubes are removed to a container with running tap water and cooled for three minutes to about 30^{0}C. After cooling, 3 ml of solution B are added by running it from a pipette down the wall of the test tube, without stirring or agitation.
3. Following this, about 1.5 ml of solution C is rapidly dropped (rather than permitted to flow into the test tube) with simultaneous agitation, so that the entire contents of the tube are mixed and acidified at once. A graduated pipette with a cracked off tip serves the purpose.
4. The cotton plug is replaced and the solution is allowed to stand for five minutes. It is then titrated with solution D using starch solution (1 ml of 1%) as indicator at end stages.
5. A pair of blanks (5ml of water with 5 ml of phosphate-copper reagent) is run with each set of ten determinations.

Calculations

$$\% \text{ Reducing sugars (as glucose)} = \frac{338 \times (B-T)}{A.D.M. \times (100-M)}$$

where B= Titre value for blank, T = titre value for the sample.

A.D.M = Air dry material of sample, M = % moisture in it

The above calculation is based on the assumption that 1 mg of pure glucose solution gives a (B-T) value of 7.4 ml of 0.005 N thiosulphate.

N.B:

1. With these reagents it is possible to determine reducing sugars equivalent to 2.0 to 3.0 mg of glucose in 5 ml of sugar solution.
2. The reagent is quite suitable when the (B-T) value falls between 15 and 20 ml of 0.005 N hypo.

3. For 2-5% reducing sugars- add 5 ml of N KIO_3 per 500 ml, for 1-2 % or less than 1% - Add 2.5 ml of N KIO_3 per 500 ml, for 6-32% — Add 12.5 ml of N KIO_3 per 500 ml

Reference

Somogyi, M.I. 1940. A New Reagent for the Determination of Sugars. Journal of Biological Chemistry. 160:61-8.

Lactose in milk

Principle

Lactose is a major carbohydrate constituent in milk (4-6%). It is a reducing sugar and hence the method of reducing sugars can be used for its determination.

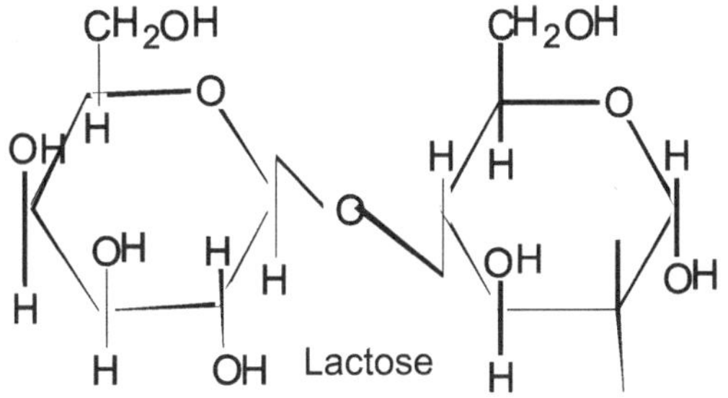

Lactose

Reagents equipment

1. 10% sodium tungstate
2. 1/3 M H_2SO_4
3. Alkaline copper reagent: Dissolve 40g anhydrous Na_2CO_3 in 400mlwater. Add 7.5g tartaric acid and after dissolution add 4.5g $CuSO_4 . 5H_2O$ to the carbonate solution. Shake well and dilute to one litre.
4. Phosphomolybdic acid reagent: Dissolve 70g molybdic acid and 10g sodium tungstate in 700ml of 5% NaOH. Boil for 40min to remove NH_3 in the molybdic acid. Cool and add 85% phosphoric acid. Dilute the reagent to one litre.
5. Lactose stock solution: 1g lactose in 100ml 0.2% benzoic acid.
6. Folin – Wu tubes
7. Spectrophotometer

Procedure

1. To 1ml fat free milk in a 100ml volumetric flask, add 2ml of 10% sodium tungstate. Later add slowly with shaking 2ml 1/3M H_2SO_4 . Make up to mark and allow to stand for 5min.
2. Filter the mixture using no. 42 filter paper.
3. Into three sets of Folin-Wu tubes pipet, in duplicate, a)1ml filtrate and 1ml distilled water, b) 2ml lactose standard, 0.6g lactose and c) 2ml distilled water as a reagent blank.
4. To each tube add 2ml alkaline copper reagent and heat in a water bath for 8min.
5. Cool and while shaking add 4ml phosphomolybdic acid.
6. Allow the tubes to stand for 1min. Dilute the mixture to 25ml mark with 1:4 phosphomolybdic acid.
7. Read absorbance against reagent blank at 630nm.

Calculation

$$\text{Lactose conc.} \frac{\text{g}}{\text{100ml}} = \frac{\text{A unknown x 0.6 x 100}}{\text{A standard x 1000 x 0.01}}$$

A standard = absobance of lactose standard,

A unknown = absorbance of milk sample.

Sugar content of fruits

Principle

The sugar content in fruits (aqueous extract-mainly invert sugar) is determined titrimetrically by oxidation with iodine in alkaline conditions. An aliquot is hydrolysed to estimate sucrose content.

Reagents and equipment

1. Fruits: Ex. orange.
2. 0.1M sodium thiosulphate solution
3. 0.05 M iodine solution
4. 0.05M NaOH

5. 1% soluble starch solution
6. 2M HCl
7. 2M NaOH
8. Methyl red indicator solution
9. Homogenizer

Procedure

1. Weigh 50g skinned and de-pipped (skin removed) orange (or any other fruit) and homogenize. Centrifuge and transfer supernatants quantitatively to 100ml flask and make up.
2. Pipet in triplicate 5ml extracts into conical flasks. Add iodine solution from a burette: 5ml each time. Wait for a min and add 0.05M NaOH thrice the volume of iodine. Continue addition till an excess is reached (appearance of pale yellow colour) . Record the exact volume of iodine added.
3. Add 2M HCl to the flask until the mixture is just acid to methyl red.
4. Titrate against 0.1M sodium thiosulphate solution using starch solution as indicator during final titration stages
5. Pipet 25ml sugar extract into a 250ml flask, add 5ml 2M HCl and heat in a boiling water bath for 20min. Cool and adjust p^H to 6.0 with 2M NaOH.
6. Transfer the hydrolysate to 50ml volumetric flask and make up.
7. Pipet 5ml and follow steps 2 to 4 to determine glucose content.

Calculation

Glucose content in 5ml sugar extract (2.5g fruit) = (total volume of iodine-excess volume of iodine) X molarity of iodine X 180g glucose

Calculate the value for 100g fruit. Invert sugar conc. = 2 X glucose conc.

Total glucose in hydrolyzed extract (5ml or 1.25g fruit) = glucose content from invert sugar + glucose content from sucrose;

Invert sugar x 0.95 = sucrose

Express values as percentage

Estimation of cellulose

Cellulose is a major polysaccharide in plants. It is the most abundant organic compound in nature. Cellulose is composed of glucose units joined together in the form of repeating units of the disaccharide cellobiose with many cross linkages. It is also the major component of farm wastes.

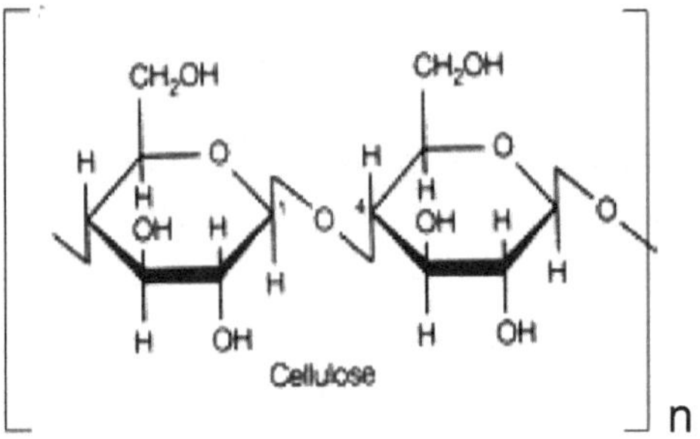

Cellulose

Principle

The cellulose content of the sample can be determined after it's hydrolysis to simple sugars and their estimation carried out by anthrone reagent method.

Reagents and equipment

1. Acetic/nitric reagent: Mix 150 ml of 80% acetic acid and 15ml of concentrated nitric acid.
2. Anthrone reagent: Dissolve 200mg anthrone in 100 ml concentrated H_2SO_4. It is prepared freshly and cooled for 2 hrs before use.
3. H_2SO_4 (67%).
4. Water bath
5. Vortex mixer
6. Spectrophotometer with quartz cavettes.
7. Centrifuge.

Procedure

1. To 500mg of the sample in a test tube, 3ml of acetic/nitric reagent is added and mixed in a vortex mixer.
2. Then the test tubes are placced in a boiling water bath for 30 min, cooled and then the contents are centrifuged for 15-20min.

3. The supernatant is discarded and the residue is washed with distilled water. To this add 10ml of 67% H_2SO_4 and allowed to stand for 1 hr.
4. One ml of this solution is diluted to 100 ml. To 1ml of this diluted solution, 10 ml of anthrone reagent is added and mixed well.
5. The test tubes are then heated in a boiling water bath for 10min, cooled and the colour developed is read at 630nm.
6. A blank is also set up with anthrone reagent and distilled water.

Standard curve

1. To 100mg of cellulose, 10ml of 67% H_2SO_4 is added and allowed to stand for 1hr.
2. For standard curve, a series of volumes of 0.4 to 2ml corresponding to 40-200 μg of cellulose are taken, and 10ml of anthrone reagent is added and mixed well.
3. It is then heated for 10min in a boiling water bath, cooled and the intensity of the colour developed is read at 630nm.
4. A standard graph is prepared by plotting concentration versus O.D. of standard cellulose.

Calculations

The content of cellulose of the sample is read from the standard graph and further calculations are made to obtain the percent cellulose in the sample based on weight of sample taken and dilutions made.

Determination of crude fibre

Principle

All soluble/digestible material are removed in a fat free sample using strong acids and alkalies. Insoluble fibre with minerals remain which is weighed. Minerals are separately determined by burning and deducted to obtain the crude fibre content.

Reagents

1. Sulphuric acid (0.255 N or 1.25% w/v) : To 198. 58 ml of water, 1.42ml of concentrated H_2SO_4 is added to obtain 200ml of 0.255 N H_2SO_4
2. NaOH (0.313N) : Dissolve 2.504 g of NaOH in 200 ml of D.W.

3. Ethyl alcohol 95%.
4. Petroleum ether/ hexane(65°C).
5. Hot plate.
6. Muffle furnace.

Procedure

1. Two grams of moisture and fat free sample is weighed into a 500 ml beaker. To this 200ml of boiling 0.255N H_2SO_4 is added and the mixture is boiled for 30 min keeping the volume constant by the addition of water at frequent intervals.
2. At the end of this period, the mixture is filtered through a muslin cloth and the residue is washed with hot water till free from acid.
3. The material is then transferred to the same beaker and 200 ml of boiling 0.313 N NaOH is added.
4. After boiling for 30min (keeping the volume constant as before) the mixture is filtered through a muslin cloth.
5. The residue is washed with hot water till free from alkali followed by washing with some alcohol and ether.
6. The residue is then transferred to a crucible, dried overnight at 80-100°C and weighed (M_1).
7. The crucible is heated in a muffle furnace at 600°C for 2 to 3 hrs, cooled and weighed again (M_2).
8. The difference in the weights (M_1-M_2) represents the weight of crude fiber.

Calculation

$$\text{Crude fiber (on dry basis) \% by mass} = \frac{100\,(M_1 - M_2)}{M}$$

Where M_1 = mass in g of the crucible contents before ashing.

M_2 = mass in g of the crucible contents after ashing.

M = mass in g of the dry fat free sample taken for the test.

Estimation of Lignins

Lignins are phenolic polymers which are present (upto 30%) in cell walls of plants. Along with cellulose, lignins are the structural components responsible for stiffness and rigidity. Lignins act as physical barriers against pathogens.

Principle

Refluxing the sample material with acid detergent solution removes the water soluble and other non fibrous material. The left out material is weighed after filtration, dried and ashed. The loss in weight on ignition gives the the acid detergent fibre.

Reagents and equipment

1. Acid detergent solution: Dissolve 20g of cetyl ammonium bromide in one liter of 1N H_2SO_4.
2. 72% H_2SO_4.
3. Acetone.
4. Round bottom (RB) flask and refluxing set.
5. Muffle furnace.
6. Sintered glass crucible- G_2.

Procedure

A. Acid detergent fibre (ADF)

1. Place 1g powdered sample in a RB Flask and 100ml of acid detergent solution. Heat to boil for 5-10 min. Reduce heat to avoid foaming as boiling begins. Reflux for 1hr after the onset of boiling. Make the boiling slow and steady.
2. Remove container, swirl and filter the contents through a pre-weighed sintered crucible G_2 by suction and wash with hot water twice.
3. Wash with acetone and break up the lumps. Repeat acetone washing until filtrate is colorless.
4. Dry at 100^{0}C overnight.
5. Weigh after cooling in a desiccator (wt of fibre).
6. Express ADF content in percentage as

   W/S X 100 , where W = wt of fibre, S = wt of sample

B. Acid detergent lignin (ADL)

1. Tranfer ADF to a 100ml beaker with 25 – 50 ml of 72% H_2SO_4. Add 1g asbestos. Stand for 3hr with stirring intermittently.
2. Dilute the acid with distilled water and filter with pre-weighed Whatman no.1 paper. Wash glass rod and residue to get rid of acid.
3. Dry the filter paper at 100°C. Record the weight after cooling in a desiccator.
4. Transfer the filter paper to a pre- weighed silica crucible and ash at 550°C for 3hr.
5. Cool in a desiccastor. Weigh and calculate ash content.
6. For blank determination take 1g asbestos, add 72% H_2SO_4 and follow steps 2 to 5.

Calculation

$$\% \text{ ADL} = \frac{\left[\begin{array}{l}\% \text{ ADL} = \text{wt of } 72\% \text{ } H_2SO_4 \\ \text{washed fibre (test value-} \\ \text{asbestos)}\end{array}\right] - \left[\text{ash(testvalue- asbestos)}\right] \text{X } 100}{\text{wt of sample}}$$

Estimation of hemicelluloses

Hemicelluloses are non-cellulosic, non-pectic cell wall polysaccharides. Xylans, mannans, glucomannans, galactans and arabinogalactans are the main components of hemicelluloses. They are not digestible by humans and are called unavailable carbohydrates.

Principle

The sample material is refluxed with neutral detergent solution to remove the water soluble and non-fibrous constituents. The residue is weighed and expressed as neutral detergent fibre (NDF).

Reagents and equipment

1. Neutral detergent solution: Weigh 18.61g of ethylenediamine tetraacetate and 6.81g sodium borate decahydrate. Put them in a beaker and dissolve in about 200ml of distilled water by heating. Add (about 100-200 ml)

solution containing 30g of sodium lauryl sulphate and 10ml 2-ethoxy ethanol. To this add 100ml solution containing 4.5g of disodium hydrogen phosphate. Make up the volume to one litre and adjust the p^H to 7.0.

2. Decahydronaphthalene
3. Sodium sulphite
4. Acetone

Procedure

1. Take 1g of powdered sample in a refluxing flask. Add 10ml of cold neutral detergent solution.
2. Add 2ml of decahydronaphthalene and 0.5g sodium sulphite.
3. Heat to boiling and reflux for 60min.
4. Filter the contents through a sintered glass crucible (G_2) by suction and wash with hot water.
5. Finally give two washings with acetone.
6. Transfer the residue to a crucible.
7. Cool the crucible in a desiccator and weigh.
8. This is the neutral detergent fibre (NDF).
9. ADF determined separately can be used to calculate the content of hemicelluloses.

Calculation

Hemicelluloses = NDF – ADF

Determination of amylose

Starch is a polysaccharide which is composed of amylose and amylopectin. Amylose is a linear or non-branched polymer of glucose while amylopectin is a branched polymer. The glucose units are joined by α-1-4 glycosidic linkages. Amylose exists in coiled form with each coil having six glucose residues.

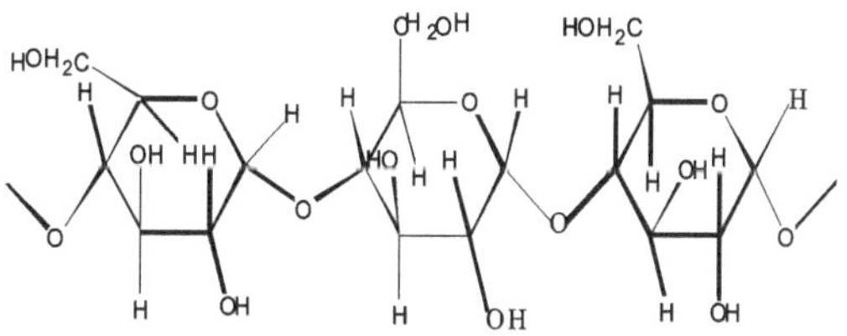

AMYLOSE

Principle

The iodine is adsorbed within the helical coils of amylase to produce a blue-coloured complex. The same is measured colorimetrically to quantify amylose.

Reagents and equipment

1. Distilled ethanol.
2. 1N NaOH.
3. 0.1% phenolphthalene.
4. Iodine reagent: Dissolve 1g iodine and 10g KI in water and make up to 500ml.
5. Amylose standard: Dissolve 100mg amylose in 10ml 1N NaOH and make up to 100ml with distilled water.
6. Water bath.
7. Spectrophotometer.

Procedure

1. Weigh 100mg of powdered plant sample, add 1ml ethanol. Also add 10ml of 1N NaOH and leave it overnight (the sample suspension can be heated for 10min in a water bath instead of overnight suspension).
2. Make up the volume to 100ml.
3. Take 2.5ml of the extract. Add about 20ml distilled water and two drops of 0.1% phenolphthalene.
4. Add 0.1N HCl drop by drop till the pink colour disappears.
5. Add 1ml of iodine reagent and make up to 50ml, read the colour at 590 nm.
6. Read the amount of amylose from the graph prepared using the standards.
7. Dilute 1ml iodine reagent to 50ml with distilled water for the blank.

Calculation

Absorbance value read from the spectrophotometer corresponds to 2.5ml of the sample solution.

The amount read from the standard graph = x mg amylose

Hence 100ml contains = x/ 2.5 X 100mg amylose = % amylose *(amylopectin = total starch- amylase)*

Starch content (As glucose by anthrone method)

Principle

Starch is hydrolysed to simple sugars, which are estimated by the anthrone method.

Reagents and equipment

1. Anthrone : Dissolve 100mg of anthrone in 100ml of ice cold 95% sulphuric acid.
2. Ethanol (80%).
3. Perchloric acid (52%).
4. Glucose standard (Stock solution): Dissolve 100mg of glucose in 100ml of water.
5. Working standard: Dilute 10 ml of the stock solution to 100ml with water.
6. Centrifuge.
7. Water bath.
8. Spectrophotometer with quartz cuvettes.

Procedure

1. Hundred mg of the sample is homogenized in hot 80% ethanol to remove sugars. This is then centrifuged and the residue is retained.
2. The residue is washed repeatedly with hot 80% ethanol till the washings give no colour with anthrone reagent.
3. The residue is then dried over a water bath. To the dried residue is added 5ml of water and 6.5ml of 52% perchoric acid.
4. It is then extracted at $0^{0}C$ for 20 min, centrifuged and the supernatant is saved.
5. The extraction is repeated using fresh perchloric acid. Again it is centrifuged and the supernatants are pooled and made upto 100ml.
6. Pipet 0.1 or 0.2 ml of the supernatant and make upto 1ml with water.
7. Four ml of anthrone reagent is added to each tube followed by heating for 8 min in a boiling water bath. It is then cooled rapidly and the intensity of green to dark green colour that developes is read at 630 nm.

8. Standard curve is prepared by taking 0.2,0.4,0.6,0.8, and 1ml of the working standard and volume is made upto 1ml in each test tube with water. Four ml of anthrone reagent is added to each tube followed by heating for 8 min in a boiling water bath. It is then cooled rapidly and the intensity of green to dark green colour that developes is read at 630 nm.

Calculation

The glucose content in the sample is calculated from the standard graph. The value is multiplied by a factor 0.9 to arrive at the starch content.

Starch content (Enzymatic method)

Principle

Starch can also be estimated after it's enzymatic hydrolysis by the phenol – sulphuric acid method (Southgate, 1976 and Dubois *et al.,* 1956). Starch comprises of amylose and amylopectin. On hydrolysis by glucosidase enzyme, starch is converted into maltose and finally to glucose. The sugar thus obtained is analyzed quantitatively using phenol - sulphuric acid reagent.

Reagents and equipment

1. Ethyl alcohol
2. 2M sodium acetate buffer, pH 4.8: Dissolve 164 g sodium acetate in water and add 120 ml glacial acetic acid, adjusting the pH to 4.8 and make up to 1 litre.
3. Enzyme- Amyloglucosidase.
4. Nitrogen free paper.
5. Balance.
6. Autoclave.
7. Water bath.
8. Spectrophotometer.
9. pH meter.
10. Water bath shaker.

Procedure

1. Weigh 75mg defatted flour into a 50 ml conical flask.
2. Add a few drops of ethyl alcohol to disperse the flour, if necessary.
3. Add 10 ml of distilled water.
4. Cover the flask with nitrogen free paper.
5. Autoclave the contents for 90 minutes at 19 1b (126^{0}C) to gelatinize the starch.
6. Cool and add 1 ml of 2M sodium acetate buffer.
7. And 25mg amyloglucosidase and approximately 15 ml distilled water.
8. Incubate the contents after covering with parafilm at 55^{0}C for 2hr in a water bath- shaker (for hydrolysis)
9. After incubation, make up the volume to 250ml.
10. Pipet 10 ml aliquot and dilute to 100 ml.
11. Pipet 1 ml of aliquot of the diluted extract and add 1 ml 5% phenol and 5ml 96% sulphuric acid, and mix well.
12. Cool and read the absorbance at 490 nm against the blank.
13. Also develop color with standard glucose (10, 20, 30, 40, 50 μg/ml) and read the absorbence.
14. Develop the color with water, phenol and sulphuric acid (for blank).
15. Measure absorbances against a blank in a spectrophotometer at 490 nm.

Calculations

Distribution factor =

$$\frac{\text{Concentration of standard } (\mu g)}{\text{Absorbence of standared}} \text{ X } \frac{1 \text{ (conversion for g)}}{10{,}00{,}000} \text{ X }$$

$$\frac{250 \text{ ml (sample dilution)}}{1 \text{ ml (taken for color development)}} \text{ X } \frac{100 \text{ (per cent)X } 0.9}{0.075 \text{ g (sample weight)}}$$

Distribution factor x Absorbance of sample = % of starch. 0.9 is the conversion factor for sugars.

References

Dubois, M.K., Giles, J.K., Hamilton, P.A., Rebers and Smith, F. 1956. A Colorimetric Method for the Determination of Sugars and Related Substances. Analytical Chemistry. 28:350-356

Southgate, D.A.T. 1976. On Determination of Food Carbohydrates. Applied Science Publishers Ltd. pp 52-55.

Fructose content

Principle

Fructose is a keto-hexose. It is called fruit sugar. Honey is a rich source of fructose. The hydroxyl methyl furfural formed from fructose reacts with resorcinol to give a red colour. It is measured to estimate the fructose content.

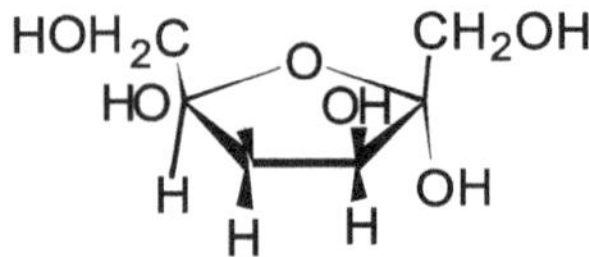

Reagents and equipment

1. Resorcinol reagent: Dissolve 1g resorcinol and 0.25g thiourea in 100ml glacial acetic acid. This solution is stable in the dark.
2. Dilute HCl: 1:5 conc. HCl : H_2O
3. Standard fructose solution : Dissolve 50mg fructose in 50ml water. Dilute 5ml to 50ml to obtain a working standard.
4. Spectrophotometer.
5. Water bath.

Procedure

1. To 2ml fructose extract (20-80µg fructose) in a test tube add 1ml resorcinol reagent.
2. Then add 7ml dil. HCl
3. Pipet 0.2, 0.4, 0.6, 0.8 and 1.0ml working standards into separate test tubes. Make up to 2ml with water. Add 1ml resorcinol reagent and 7ml dil. HCl.
4. Prepare a blank simultaneously.

5. Heat all tubes in a water bath at 80^{0}C for 10 min.
6. Remove all test tubes and cool by immersing in tap water for 5 min.
7. Read the colour at 520nm in a spectrophotometer within 30 min.
8. Draw a standard graph. From the graph read the value of the unknown sample.
9. Calculate the concentration of fructose taking into dilutions etc. and express as percentage.

Inulin content

Principle

Inulin is a polymer made of fructose units with 2,1 linkage. It is found in onion, garlic and in many other plant parts.The method depends on the formation of a blue colour with Folin and Wu reagent, which is discharged with permanganate on titration.

Reagents and equipment

1. Folin and Wu reagent: A). alkaline copper reagent: Take 40g anhydrous Na_2CO_3 and dissolve in 400ml water in a 1litre volumetric flask. Dissolve 7.5g tartaric acid in the solution. Add 100ml of 4.5% $CuSO_4$ to the volumetric flask and make up the contents. B). phosphomolybdic acid reagent: Add 5g sodium tungstate to 25g phosphomolybdic acid. Add 20ml 10% NaOH and 50ml distilled water. Boil for 40min to drive off NH_3 in molybdic acid . Cool and dilute to 350ml. Add 125ml o-phosphoric acid and make up to 500ml.
2. Lead acetate solution (1N).
3. Disodium phosphate solution.
4. 0.01N $KMnO_4$ solution.
5. Water bath.

Procedure

1. Sample extraction and clarification: Add ethanol to 1g of finely powdered sample to remove glycosides.
2. Then treat the residue with 20-25ml of water thrice at 40-45^0 C. Filter.
3. Add 5ml each normal lead acetate solution and disodium phosphate solution to the filtrate.

4. Dilute the filtrate to 100ml in a volumetric flask.
5. Pipet 2ml of filtrate into a test tube. Add 2ml of alkaline copper reagent and heat in a water bath for ten min. Cool and add 2ml phosphomolybdic acid reagent. Mix the contents and dilute to about 25ml.
6. Cool and titrate with 0.01N $KMnO_4$ solution from a burette. The end point is a faint rose colour. If the inulin content is high the end point is faint yellow.

Calculation

1mg inulin = 1.85ml 0.01N $KMnO_4$

Based on this background the calculations are made taking into account weight of sample and related dilutions.

Reference

Ashwell, G (1957) In: Methods in Enzymol 3 (Eds Colowick, S. J. and Kaplan, N. O.) Academic Press New York p 75

Pectin content (Gravimetric method)

Principle

Pectic substances occur as constituents of cell walls. They are mixtures of polysaccharides formed from galactose, arabinose and galacturonic acid. The pectin in the sample is hydrolyzed with alkali and then precipitated as calcium pectate in an acid medium. The precipitate is washed free from chlorides, dried and weighed.

Reagents and equipment

1. Dilute NaOH solution.
2. 1N $CaCl_2$.
3. Acetic acid.
4. Water bath.

Procedure

1. Accurately weigh 30-50g sample in a beaker. Add 200ml water and heat on a water bath. Dissolve by stirring with a glass rod.
2. Dilute the solution to 250ml with distilled water in a volumetric flask. Then filter the contents.

3. Take two 50ml aliquots in separate beakers and dilute with equal volumes of water.
4. Neutralize with NaOH and then add 10ml alkali. Keep for 12hr.
5. Add 50ml of acetic acid and after 5min 20ml 1N $CaCl_2$ solution is added with stirring.
6. Keep for 1hr. and boil for 2min. Filter the precipitate through a filter paper and wash with boiling water. Dry the precipitate at 100°C, cool and weigh.

Calculation

Express the calcium pectate as percentage based on the weight of the precipitate and the sample weight.

Pectin content (Colorimetric method)

Principle

EDTA is added to a sugar free sample to precipitate the metallic constituents. The pectin esters are hydrolyzed with alkali and pectinase enzymes are added. Galacturonic acid formed is converted to a colored complex with carbazole in the presence of H_2SO_4. The color is measured at 520nm.

Reagents and equipment

1. 0.5% aqueous EDTA solution.
2. Pectinase enzyme.
3. 0.15% carbazole in alcohol.
4. Galacturonic acid standard solution (100ppm in water).
5. Ice bath.
6. Water bath.
7. Spectrophotometer.

Procedure

1. The sample is extracted with 100% alcohol to remove the sugars.
2. Weigh exactly 1g sugar free sample.
3. Adjust p^H to 11.5 with 1NNaOH and keep at 25°C for 30 min.

4. Acidify the solution to p^H 5.0 - 5.5 with acetic acid.
5. Add 0.1g pectinase , stir and dilute to 250ml
6. Dilute 2 ml to 100 ml.
7. Take 12 ml conc. H_2SO_4 in a big test tube and cool to 3^0C in an ice bath.
8. Add 2 ml solution (6 above). Mix well. Re-cool to 5^0C.
9. Heat for 10 min on a boiling water bath and cool to 20^0C.
10. Add 1ml of carbazole reagent. Mix well and keep for 30min.
11. Measure O.D. of the solution at 520nm.
12. Standard curve: Prepare a calibration curve using galacturonic acid (5-80ppm).
13. From the standard curve read the concentration.
14. Make calculations based on the sample weight, dilutions and the concentration read from the standard graph.

Sugar content of glycoproteins (Phenol sulphuric acid method)

Principle

Seeds of lentil and guar contain glycoproteins which undergo changes during post harvest operation. The glycoprotein is hydrolysed and the liberated sugar is estimated by the phenol- sulphuric acid method.

Reagents

1. 0.2 M boric acid (12.4g/litre).
2. 0.05 M borax (19.05g/litre): 0.2M in terms of sodium borate.
3. 50ml of solution1 + 59ml of solution 2 mixed and diluted to 200ml.
4. Conc. H_2SO_4.
5. 5% phenol in water.
6. Centrifuge.
7. Water bath.
8. Spectrophotometer with cuvettes.

Procedure

1. Finely powdered sample (0.5-1.0 g) is mixed with ten volumes of 0.02M boric acid buffer p^H 9.0 and shaken for 30 min.
2. Keep overnight at 4^{0}C and centrifuge at 10,000g.
3. The supernatant is diluted with the same buffer in cold till free from sugar.
4. Dialyze and bring the supernatant to p^H 4.5 by mixing with dil HCl.
5. The precipitate formed is separated.
6. 2ml of glycoprotein solution (10-70μg sugar) is pipetted into a boiling test tube.

9. Add 1 ml of 5% phenol in water and 5 ml Conc.H_2SO_4

7. Allow tubes to stand for 10min. Shake for 10-20 min and place in a water bath at 25-30^{0}C.
8. Record the absorbance at 490nm against a blank.
9. Read the concentration of the sample from a standard graph (70-210μg).

3

FATS / OILS / LIPOPHILIC CONSTITUENTS

Lipids are a diverse group of chemical molecules. They are generally insoluble in water but soluble in organic solvents. Lipids include long-chain hydrocarbons, alcohols, aldehydes, fatty acids, and their derivatives, especially, glycerides, phospholipids, glycolipids and sulfolipids. Other compounds like sterols, fatty acid esters of sterols, vitamins A, D, E, and K, carotenoids, and other poly-isoprenoids also are part of the lipids. Oils and fats are mainly triglycerides and neutral molecules. Those in the liquid state at ambient temperatures are called oils while those in the solid state are called fats. Lipids and lipid components have varied uses in the food and non food sectors. Vegetable and animal matter contain a variety of these constituents in different proportions. It is neccessry to evaluate the composition and quality of these constituents from the point of view of their utility.

Extraction of crude total lipids

The samples are first ground in a hand operated/ electric grinder or wiley mill. Take 1 gm of ground sample immediately after grinding in 250 ml flask. Add 50 ml of petroleum ether to it. The mixture is stirred in an electric shaker for one hour and kept overnight. Then the mixture is filtered through sintered funnel. Two more washings are given to the residue with 5-10 ml petroleum ether each time. The filtrate is collected and petroleum ether is removed completely on a boiling water bath. The weight of the residue indicates the content of lipids.

There are many variations of this procedure. The organic materials are mixed with chloroform and methanol in amounts that yield a monophasic solution when combined with water in the tissue. Upon dilution with water or chloroform, a biphasic solution results. The lipids are retained in the chloroform phase, and non-lipid materials are retained in the methanol-water phase. The chloroform layer is isolated, and the lipids can be purified and saparated into classes as desired. The recovery of total crude lipids after separation from the nonlipid contaminants should exceed 95%.

Plant tissues must be extracted first with a solvent that inhibits the action of lipases; isopropanol is used most frequently for the purpose. The plant tissues are macerated with 100 parts by weight of isopropanol. The mixture is filtered, the solids extracted again in a similar manner and finally is shaken overnight with 199 parts of chloroform-isopropanol (1:1v/v). The combined filtrates are taken almost to dryness, the residue is taken up in chloroform-methanol (2:1 v/v) and given water wash. The purified lipids are recovered from the lower layer.

Removal of non-lipid contaminants: Most polar organic solvents used to extract lipids from tissues also extract significant amounts of non-lipid contaminants such as sugars, urea, amino acids and salts. When the polar solvents have been removed by evaporation or distillation, the lipids may be taken up in a small volume of a relatively non-polar solvent such as hexane: chloroform (3:2 v/v) leaving many of the extraneous non-lipid substancs behind. It has been demonstrated that a pre-extraction of tissues with 0.25% acetic acid removes all potential contaminants of lipid extracts, while simultaneously deactivating the lipolytic enzymes.

Most of the contaminating compounds can removed from chloroform-methanol (2:1 v/v) mixtures simply by shaking the combined solvents with one quarter of their total volume of water, or even better with a quarter of the total volume of a dilute salt solution (e.g. 0.88 per cent potassium chloride solution). The lower phase, which composes about 60 per cent of the total volume, contains the purified lipid and the upper phase contains the non-lipid contaminants. With any washing procedure, it should be noted that centrifugation may be necessary.

Artefacts of extraction procedures: If chloroform-methanol, or in general, alcoholic extracts which contain lipids, are refluxed or stored for long periods in the presence of very small amounts of tissue sodium carbonate or bicarbonate, trans-esterification of many of the lipids occur and large amounts of methyl esters are found in the extracts.

Some practical considerations

Polyunsaturated fatty acids auto-oxidize very rapidly if left unprotected in air. Although natural tissue antioxidants such as tocopherols may afford some protection, it is advisable to add an additional antioxidant such as BHT ('butylated hydroxy toluene') or 2,6-di-tert-butyl-p-cresol at a level of 50-100 mg per litre to the solvents. Whenever possible, extraction procedures should be carried out in an atmosphere of nitrogen, and both tissues and tissue extracts should be stored at –20° under nitrogen. It is helpful to deaerate solvents by flushing them with nitrogen before use.

Solvent should be removed from lipid extracts under vacuum in a rotary film evaporator at or near room temperature. When a large amount of solvent must be evaporated, it should be concentrated to a small volume and then transferred to as small a flask as is convenient so that the lipids do not dry out as a thin film over a large area of glass. There is no need to bleed in nitrogen continuously during the eveporation process as the solvent vapours effectively displace any air, but the vacuum should eventually be broken with nitrogen. Lipids should not be left for any time in the dry state but should be taken up and stored in an inert non-alcoholic solvent such as chloroform.

Sephadex G-25 columns for removing non-lipid contaminants

Chloroform, methanol and water in the ratio 8:4:3 (v/v/v) are mixed and partitioned. The upper phase (UP) and lower phase (LP) are separated and retained. 25 g of Sephadex G-25 are soaked overnight in 100ml of UP. Then the Sephadex G-25 is washed with 4 x100ml of UP. Columns (1cm diameter x 10 cm height) are packed with a slurry of this material. The column is rinsed with 10ml of UP followed by 10ml of LP. The crude unwashed lipid extract (200mg), is taken up in 2-5 ml of LP, filtered to remove any precipitated non-lipid and is applied to the column, which is eluted with 25-30 ml of LP at a flow rate of up to 1ml per minute. The pure lipid is eluted in this solvent and is then recovered. The column can be regenerated for further use by washing with 20ml of LP.

Determination of total phosphlipids

The quantity of phospholipids in the lipid sample is determined by estimating the lipid phosphorous. A suitable aliquot of the lipid extract is evaporated in a test tube at 7°C on a water bath. After drying, 0.4 ml of acid mixture (H_2SO_4; $HClO_4$ 9:1 v/v) is added and digestion is carried out in a hot air even at 180°C for 45 min, After cooling, 9.6ml of ANSA-ammonium molydbate mixture is added and the tubes placed in a water bath at 9°C for 20 min. The tubes are then cooled and absorabance read at 820 nm. A standard curve (0.5 – 2.5 μg) is prepared using potassium dihydrogen orthophosphate and the phosphorous content in the sample is extrapolated from the standard curve.

1 **Amino-naphthol sulfuric acid (ANSA) reagent:** 0.25 g ANSA and 0.50g sodium suphite are added in 20ml of water. 12.72 g of sodium matabisulphite is then added and the volume made up to 100ml. Just before use, ANSA reagent is mixed with 0.26% ammonium molybdate in the ratio of 11:250 (v/v), respectively.

Determination of total galactolipids

The galactolipids are estimated by determining the amount of hexose present in the lipid extract. For this galactolipids are hydrolysed to give free hexoses which are measured colorimetrically. A suitable aliquot of the chloroform extract is evaportated to dryness on a water bath at 70°C. Three ml of dilute H_2SO_4 (1:12) is added to it and galactolipids are hydrolysed by heating on a boiling water bath for 1 hr. One ml of the hydrolysate is layered on 5 ml of anthrone reagent (0.2 g of anthrone in 100 ml of dilute suphuric acid (5:2))in chilled test tubes. The contents are mixed and tubes kept in a water bath at 90°C for 20 min. After cooling the absorbance is read at 620nm. Hexose content of the samples is read from a standard curve for galactose (10-50 μg).

Lipid analysis

Lipid analysis is highly sensitive and sophisticated. It involves various steps. The initial step is extraction of lipids followed by elimination of non-lipid components. The pure lipids are then fractionated into individual classes using mainly thin layer chromatography (TLC). These individual lipids can be further analyzed by high performance liquid chromatography (HPLC) to identify the molecular species and structural elucidation. Triacylglycerols after saponification yield fatty acids. The fatty acids are converted to their methyl esters, which can be separated and identified with the help of gas liquid chromatography(GLC). GC-MS (gas chromatography- mass spectrometry) is more useful in structural elucidation. The reader may refer to advanced and specialized books for more information on lipid analysis.

References

Bligh, E.G. and Dyer, W.J. 1959. Canadian Journal of Biochemistry and Physiology. 37: 911-917.

Folch, J., Less, M. and Stanley, G.H.S. 1957.Journal of Biological Chemistry. 226: 497.

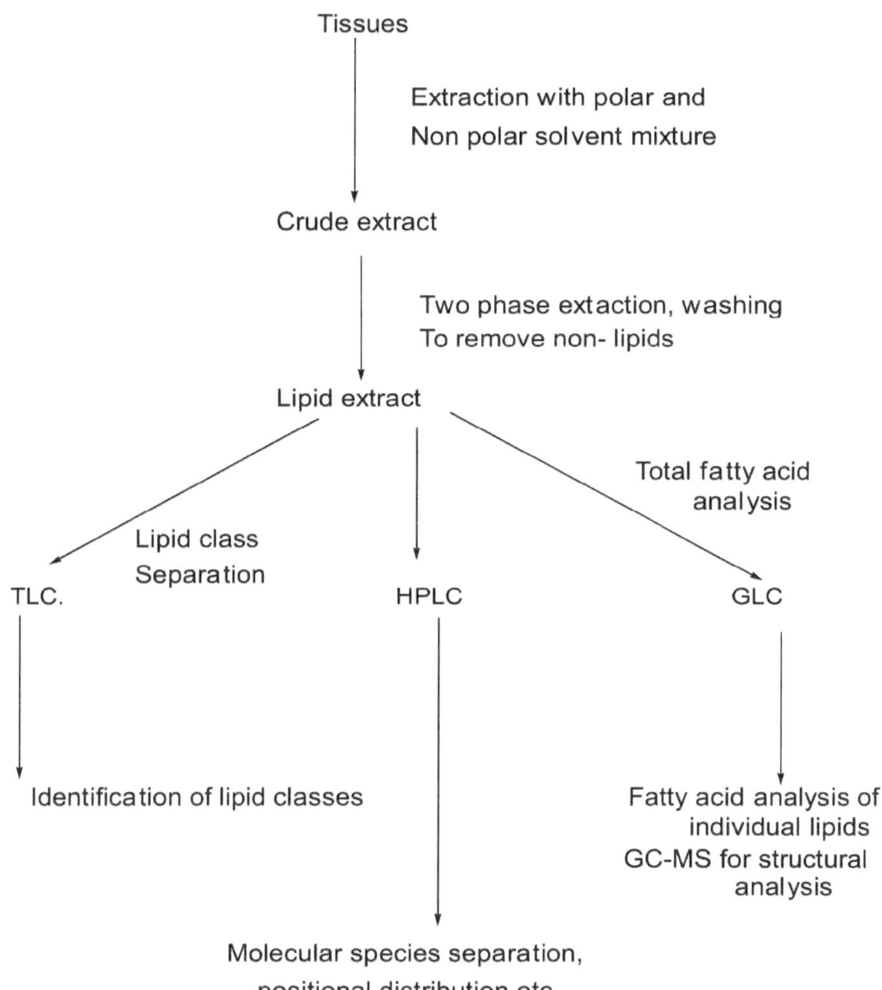

Extraction and purification of lipids

Principle

The lipids are extracted with organic solvents like chloroform and methanol. Water can be used to remove water solubles. Traces of moisture can be removed by a zeotropic distillation with benzene.

Reagents and equipment

1. Absolute methanol and chloroform (Analytical grade, freshly distilled reagents).

2. Filter Paper: Whatman no. 1.
3. Porcelain Buchner funnel: 43mm.
4. Suction flask: glass 125 ml.
5. Rotary evaporator.
6. Mortar and pestle.

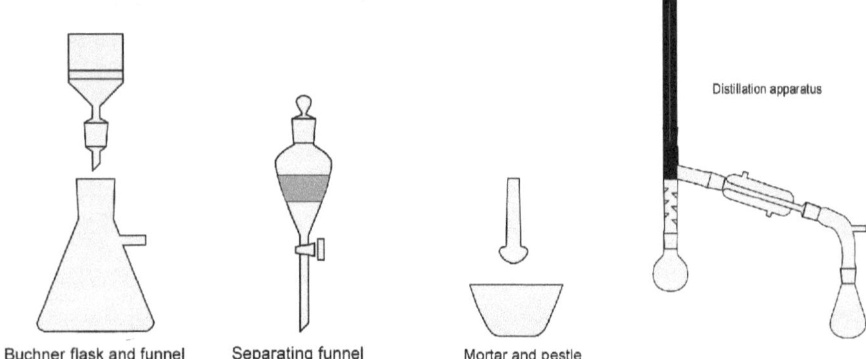

Procedure

1. The following modified Bligh and Dyer procedure may be used, for extaction of lipids:
2. 100g of plant material is cut into small pieces or finely powdered and immediately added to 300 ml of methanol-chloroform (2:1 v/v) and 50 ml H_2O.
3. The mixture is blended for 2 min in a mixer or waring blender. The homogenate is filtered with suction, and the residue is reblended with a mixture of 300 ml of methanol-chloroform (2:1 v/v) and 50 ml H_2O, washed with 150 ml of methanol: chloroform (2:1 v/v).
4. The homogenate is again filtered with suction, and the residue is washed with 150 ml of methanol-chloroform (2:1 v/v).
5. To the combined filtrates in a separatory funnel are added 250 ml of $CHCl_3$ and 290 ml H_2O and the phases are allowed to separate.
6. The $CHCl_3$ layer is withdrawn, diuted with benzene and concentrated in vacuum.
7. The residual lipids are immediately dissolved in 250 ml $CHCl_3$ – Methanol (1:1) and the solution cleared by centrifugation (if necessary) and stored at -10^{0}C

Extraction of seeds

1. Powder 2-5g seeds in a Wiley mill or a mortar and pestle. Transfer the powdered sample to 50 ml centrifuge tube and extract with 20ml boiling isopropanol with vortexing for several min.
2. Centrifuge and remove the supernatant. Repeat with 20 ml isopropanol. Combine the extracts. Evaporate on a rotary evaporator.
3. Re-extract seed with 38ml $CHCl_3$:MeOH:H_2O (1: 2: 0.8 v/v). Add extract to isopropanol residue.
4. Add 20 ml each of $CHCl_3$ and H_2O mixture and centrifuge.
5. Remove the lower $CHCl_3$ phase. Dilute with benzene, evaporate and dissolve the residue in $CHCl_3$ MeOH (1:1), centrifuge if neccessary. Store at -10°C .

Quantification

1. If neccessry, take an aliquot of the extract into a tared glass or porcelain container, and evaporate to dryness under a stream of nitrogen in a water bath at 35 to 40°C to give a dry crude total lipid preparation. Work in a fume hood.
2. Weigh the amount of crude total lipid on an analytical balance.

Storage of lipid extracts

1. Due to general lability of lipids towards peroxidation and hydrolysis they should never be stored for extended periods of time (i.e. more than 1-2 years).
2. Remove water by azeotropic distillation with benzene by dissolving in $CHCl_3$ containing 10-50% methanol and storing at –10 to -15°C.
3. For storage periods of upto 2 years dissolve in benzene containing 10-15% methanol or 99% ethanol(N_2 flushed).
4. An antioxidant such as tocopherol or BHT (2,6 di-tert-butyl-4 methyl phenol) at 0.05% should be added. Store at -40°C or –80°C.

Purification (Removal of water soluble contaminants)

1. To the lipids in 10 ml of $CHCl_3$: MeOH (2:1), add 4.5ml of H_2O, mix and centrifuge.
2. Remove upper phase and wash again with MeOH: H_2O (10:9) . Dilute with benzene, bring to dryness under N_2.

3. Dissolve in $CHCl_3$ and use the lipid extract for further analysis wherever required.

Chromatographic purification of lipids

Principle

The crude lipids are purified on sephadex G-25 column by elution with chloroform: methanol solvent.

Reagents and equipment

1. Solvents: Mix redistilled chloroform (200 parts) and methanol (100 parts) with 75 parts by volume of water, shake vigorously in a separatory funnel, and allow the mixture to separate into an upper phase (solvent I) and a lower phase(solvent II). Collect the phases separately and store them at room temperature.

2. Sephadex G-25, Fine bead form.

3. Chromatography tube: 1 cm by 15cm with sintered-glass bottom and Teflon stopcock.

Procedure

1. Soak 25 g of Sephadex G-25 beads overnight in 100 ml of solvent I. Allow the beads to settle, decant the solvent, and rinse the beads four times with solvent I.

2. Prepare a thick slurry of beads (50-ml total volume) in solvent I. De-gas the slurry by placing it in a vacuum flask, and apply vacuum with an aspirator for 10 to 15 min. Swirl occasionally. Pour the slurry of beads into a column containing a few ml of solvent I. Continue adding resin slurry until a resin bed, 10 cm high is obtained.

3. Apply pressure of 1 to 2 lb/in^2 (50-100mmHg or 0.07-0.14 atmospheres) from a nitrogen tank to settle the resin bed, and place a glass-fiber filter disk on top of the resin bed. Rinse the column with 10ml of solvent 1 and 10 ml of solvent II.

4. Dissolve 150 to 200 mg of the dry crude total lipids in 2 to 5ml of solvent II. Filter the dissolved lipid fraction through a medium-porosity sintered-glass funnel directly onto the Sephadex column. Rinse the funnel with 2 ml of solvent II.

5. Apply pressure of 2 to3 lb/in^2(100-150 mm Hg) with a nitrogen tank to achieve a flow rate of 0.5 to 1.0 ml/min. Add solvent II and elute the column until 25ml of solvent is collected.
6. Concentrate the eluate under reduced pressure at room temperature until a cloudy aqueous emulsion forms. Lyophilize this emulsion to dryness. Dissolve the dried lipid in chloroform:methanol (2:1 v/v).
7. Transfer it to tared glass stoppered vials, and evaporate the solvents under a stream of nitrogen at 35 to 40^{0}C. Complete the drying in an evacuated vacuum desiccater containing paraffin shavings.
8. After drying, weigh the vial containing the dried lipid extract.
9. Recovery of lipids can be checked by using known amounts of purified total lipids in the extraction and purification steps.

Soxhlet extraction of oil

Principle

Lipids are extracted into non-polar organic sovents like hexane or petroleum ether and the quantity determined gravimetrically. Waxes, resins organic acids, colouring matter, and other lipophilic constituents along with fat get extracted.

Reagents and equipments

1. Soxhlet extraction apparatus with flasks, condensers and heaters etc.
2. Thimbles.
3. Analytical balance.
4. Water bath.
5. Oven (preferably vacuum oven).
6. Separating funnels.
7. Petroleum ether (40-60^{0}C).
8. Methanol.

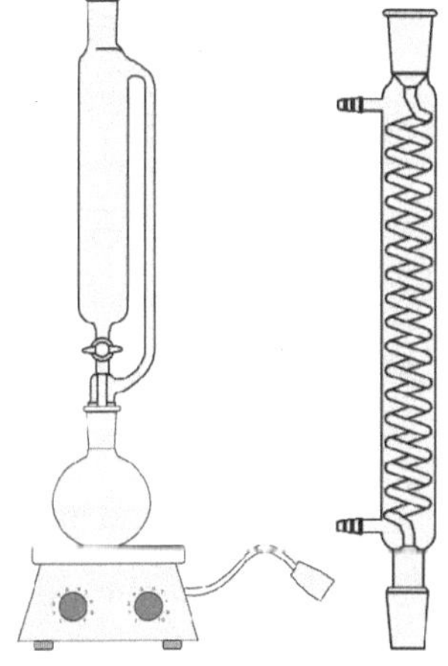

Lipid extraction apparatus with hot plate and condenser

Procedure

Part I: Total lipids

1. 10-15 g of plant or oil bearing sample is weighed into a thimble and the same is put inside the Soxhlet condenser.
2. Add solvent either petroleum ether (40-60^{0}C) or hexane approximately 150 ml. Run the Soxhlet for 4-6 hr.
3. The extraction is complete when a drop of the solvent taken from the drippings of the extractor leaves no greasy stain on the filter paper.
4. Remove the thimble and heat so that the extractor is filled up to about 2/3rd with petroleum ether and only a small quantity is left in boiling flask, disconnect the flask, filter the residual petroleum ether (containing fat in solution), through sintered funnel and collect the filtrate in a weighed 100 ml beaker or a glass dish.
5. Wash the funnel and flask, four times with redistilled petroleum ether till a drop of the filtrate leaves no greasy spot. Evaporate petroleum ether in the dish and find the weight of the residue. Keep in an oven for 3 to 4 hrs for evaporation of the solvent. If necessary use vacuum to remove all solvent.
6. Dry the dish till the consecutive weights are constant. Weight of the dish with the total lipid is recorded.

Calculation

Per cent total lipids =

$$\frac{\text{weight of the dish with extract-empty weight of dish x100}}{\text{weight of sample x}\left(100\text{-M}\right)}$$

M = % moisture

Part II: Neutral lipids

1. After weighing, the total lipids are dissolved in 10 ml petroleum ether.
2. Pour the same into a separating funnel.
3. The flask should be washed two more times (5ml each time) with petroleum ether and the washings transferred to the separating funnel.
4. Add 10 ml methanol (95% aqueous).

5. Shake and separate the lower layer.
6. Again shake two times with 10 ml portions of methanol (95% aqueous).
7. Separate and pool the methanolic lower layers which contain polar lipids.
8. Methanolic layer fraction (pooled) is evaporated (in a rotary flask) and weighed to obtain the polar lipid content.
9. The upper layer (petroleum ether) contains non polar lipids, mainly triglycerides/oil.
10. Collect upper layer into an appropriate flask.
11. Keep it in an oven for 3 or 4 hr and weigh to obtain the neutral lipid content.

Calculation

The polar and neutral lipids can be obtained from the weights above or can be calculated from the total lipid content also.

Percent neutral lipids = % total lipids – % polar lipids

Percent polar lipids = % total lipids - % neutral lipids

Oil content (Cold percolation method)

Principle

Oil/fat is extracted in the cold by repeated washing with petroleum spirit and the same estimated after removal of the solvent. This method is very simple and is useful when a Soxhlet extraction apparatus is not available.

Reagents and equipment

1. Anhydrous sodium sulphate.
2. Percolation tube or a funnel of suitable size.
3. Oven or water bath or a hot plate.
4. Porcelain cups.
5. Petroleum spirit (40-60°C)

Procedure

1. Weigh the oileagenous sample (about 10 g). Crush and powder the sample with one spoonful of anhydrous sodium sulphate.

2. Transfer into a percolation tube with cotton or a funnel closed with cotton plug.
3. Give 3 washings at one hr intervals with 15ml petroleum spirit each time.
4. Collect the solvent which gets decanted to a cup whose empty weight is known.
5. Evaporate the solvent slowly in the cup and dry in an oven for 5 to 10 min to remove all the solvent.
6. Weigh the cup with the oil.

Calculation

Per cent tota lipids =

$$\frac{\text{weight of the cup with extract-empty weight of cup x100}}{\text{weight of sample x}\left(100\text{-M}\right)}$$

Note

1. *Check for any fine particles/ sediment in the extract and filter or centrifuge before evaporation of the sovent*
2. *Completion of extraction can be checked by the absence of greasiness on a filter paper at the time of third washing.*

Oil content (NMR method)

Principle

The nuclei of certain materials when placed in a magnetic field absorb at a specific radio- frequency. This phenomenon is known as nuclear magnetic resonance (NMR). In the 630 gauss field provided by the permanent magnet of the analyzer, hydrogen nuclear resonance occurs at 2.7 MHz. By means of an electronic gate the narrow resonance from hydrogen in mobile liquid like moisture or oil is distinguished from the broad resonance from hydrogen in the solids.

Reagents and equipment

1. Balance.
2. Oven.

3. MK III A New Port analyzer. Standard settings: R.F. level, 100A; A.F. gain, 3000; Intergration, 32 S; Gate width, 5 gauss (Instruction manual of the instrument should be used to set up the equipment).
4. NMR tubes.
5. Pure oil sample (Oil from the sample to be analysed).

Procedure

1. Fill known quantity of pure oil up to the mark in the NMR tube.
2. Place the NMR tube in the sample compartment of the Newport analyzer that is set to the standardized instrument conditions.
3. Record the NMR signal of reference oil.
4. Now fill the groundnut/ any other oil seed sample in another NMR tube up to the atched mark.
5. Note the weight of the sample filled.
6. Dry the sample in an oven at 105^{0}C for 16 h.
7. After drying, stopper and cool the NMR tube.
8. Note the weight of dried sample and calculate the moisture content.
9. Reset and record the signal of dry seed sample, in the NMR insturment.

Calculations

NMR oil equivalent (OE) is obtained by the equation

(In dry seeds % O.E. = % Oil):

$$\text{OE \%} = \frac{\text{weight of oil x NMR reading of sample}}{\text{NMR reading of oil x wet weight of sample}} 100$$

$$\text{Oil \% (NMR) at 5\% moisture level} = \frac{(100\text{-}5)\text{ x NMR Oil \%}}{(100\text{ - sample moisture \%})}$$

Precautions

1. Warm the instrument for at least 30 min before use.
2. Do not keep any iron objects near the instrument lest it would permanently damage the magnet.

Ref: Jambunathan, R., Madhusudhana Raju, S. and Subhadra P. Barde. 1985. Analysis of Oil Content of Groundnuts by Nuclear Magnetic Resonance Spectrometry. Journal of the Science of Food and Agriculture. 36:162-166.

Oil and moisture contents (Pulsed NMR method)

Principle

When a test sample is inserted into the magnetic field of a pulsed NMR spectrometer, an alternating electromagnetic field in the form of an intense 90° radio frequency (RF) pulse excites all the hydrogen nuclei. The free induction decay (FID) following the 90° pulse signal is proportional to the total number of protons from the water and oil phases of the sample. Application of a second RF pulse, a so-called 180° pulse, produces a spin-echo signal when only the signal from the oil phase contributes to the FID. The maximum amplitude of this echo signal is proportional to the oil content. The difference between the two amplitudes is proportional to the moisture content. Automatic conversion of the measured signals, after suitable calibration of the apparatus is made to obtain the percentages of oil or moisture.

Samples for moisture/ oil content calibration

To obtain a reliable calibration curve, the moisture contents of the calibration samples should be 6-9% for oilseeds, in general, and 6 -12% for soya beans. Samples shall be of the same species (same variety or hybrid etc.) as the test samples and of similar fatty acid compositions (especially for the analysis of rapeseeds which are rich in erucic acid, or sunflower seeds etc. which are rich in oleic acid or linoleic acid or castor seeds which are rich in ricinoleic acid). Oil content has to be determined using a reference method. In general, samples for analysis should be repesentative, homogeneous and free from impurities, especially metallic objects.

Pulsed low-resolution NMR spectrometer

Suitable for measurement of the oil and moisture contents of oilseeds with the following characteristics.

a) Proton resonance frequency: about 10 MHz,

b) Production of 90° and 180° RF pulses for the spin-echo method,

c) Magnetization time or pulse repetition time: 1s,

d) Excitation time (e) or pulse time: $e = 5\mu s$ approximately for the 90° pulse, $e = 10\mu s$ aproximately for the 180° pulse,

e) Dead time (t) = 20μs,

f) Total dead time (Tt), comprising the excitation time (*e*) and the dead time (*t*), where: T*t* = 25 μs for 90° pulses, T*t* = 30 μs for 180° pulses, *t* = 20μs.

g) NMR signal sampling: 1) 50μs after the 90° pulse on the FID, 2)7000 μs after the 90° pulse on the spin-echo.

h) Interval between the 90° and the 180° pulse: 3500 μs. Interval between the 180° pulse and the echo maximum: 3500 μs (the echo maximum corresponds to the oil signal only).

i) Number of NMR measurements per sample: 16 to 25 (i.e. the total time of measurement is between 16s and 25s).

k) Facility for storage of calibration curves (straight lines, least-squares method) of moisture and oil contents.

l) Homogeneity of magnetic field for a desirable sample volume of 35 ml.

m) Start system, which automatically starts the measurement after the introduction of the sample tube.

n) Temperature-regulated magnet, in order to maintain all the NMR parameters stable.

Other apparatus/ equipment

1. Sample tubes, made of glass, suitable for use with the NMR spectrometer.
2. Analytical balance, electronic, with an accuracy of ± 0.01g, preferably linked to the NMR spectrometer so that the sample mass is recorded directly by the NMR.
3. Drying oven, capable of being maintained at 103°C ± 2°C. Dishes, made of glass or metal, of diameter 7cm to 10cm, and provided with lids.
4. Desiccator, containing an efficient desiccant.

Procedure

1. Use the set-up parameters of the NMR spectrometer recommended by the manufacturer and optimize them by preliminary tests. For all calibration and measurement operations, follow the user's manual. Ensure that all operations during calibration and measurement are carried out under the same conditions and, in particular, at the same temperature (± 2 °C).

2. A minimum of three calibration samples is necessary, although six to eight are desirable. Enter the parameters for the measurements (pulse sequence, attenuation, etc) of the oil or moisture content into the NMR spectrometer and specify a number under which the calibration curve is to be stored. The total measurement time shall be a minimum of 16s. Set the apparatus to the calibration mode. Introduce a portion of the first calibration sample into a tared sample tube up to the optimum height and transfer the sample mass from the balance to the NMR apparatus.

3. Enter the value of either the moisture content (as a percentage by mass) or the oil content (as a percentage by mass) into the spectrometer. Introduce the sample tube containing the first calibration sample into the measuring head and determine the moisture or oil content. Repeat the steps for the two (or more) calibration samples. Calculate automatically or manually the calibration parameters of the calibration curve. The correlation coefficient shall normally be greater than 0.95. By following this procedure the calibration curves for moisture and oil contents can be prepared. These curves are used to obtain either moisture or oil content of unknown samples.

Note

1. *Calibration curves need to be prepared once every year or when the instrument is disturbed.*

2. *This method applies to seeds/samples with a moisture content of less than 10%.*

3. *For seeds with higher moisture contents, drying is necessary before the oil content can be determined by pulsed NMR.*

Reference

ISO 8292: 1991, 1992. Section 2.22 Determination of Solid Fat Content (Pulsed NMR Method). Methods of Analysis of Fats and Fatty Oils – Part 2: Other Methods, BSI Standards: London. pp 1-4.

Specific gravity/ density (of fats and oils)

Principle

Specific gravity of any substance is it's weight of a unit volume, in comparison to that of water. The weight of 1ml of water is taken as 1g. In general oils have densities lower than water (< 1).

Reagents and equipment

1. Specific gravity bottle or pycnometer.
2. Analytical balance.
3. Oil sample.
4. Thermometer.

Procedure

1. The oil or fat sample may be preserved under an atmosphere of nitrogen in the cold. The sample should be allowed to reach the room tempertature and mixed well before being subjected to the various analytical tests.
2. The specific gravity of a liquid is the weight of a given volume of liquid at the specified temperature, compared with the weight of an equal volume of water at the same temperature, all weights being taken in air.
3. The specific gravity bottle (or pycnometer) is weighed empty (W_B). The bottle is then filled completely with the oil/ fat sample and weighed (W_1).
4. The oil is removed and after cleaning, the bottle is filled completely with distilled water and weighed (Ww).
5. The temperature of the liquid is noted.

Calculation

$$\text{Specific gravity} = \frac{W1 - WB}{Ww - WB}$$

The temperature at which the specific gravity is determined should be specified. The density generally decreases with increase in temperature.

Colour of the oil

Principle

The method determines the colour of oils by comparison with Lovibond glasses of known colour characteristics. The colour is expressed as the sum total of the yellow and red slides used to match the colour of the oil in a cell of the specified size in the Lovibond Tintometer.

Apparatus

1. Lovibond Tintometer.
2. Glass Cells.

Procedure

1. Filter the oil through a filter paper (or centrifuge the oil) to remove any impurities.
2. Clean the glass cell with carbon tetrachloride and dry.
3. Fill it with the oil and place the cell in position in the tintometer.
4. Match the colour with sliding red, yellow and blue colours.

Report the colour of the oil in terms of Lovibond units

Colour reading = Y+5 R or Y+10 R.

Where, Y+5 R is the mode of expressing the colour of light coloured oils;

Y+10 R is for the dark-coloured oils.

Y= sum total of yellow slides used, R= sum total of red slides used.

Fresh oils are generally light coloured while preserved or oxidized oils are darker in colour.

Free fatty acids (FFA)/ acid value

Principle

Fats in grains are readily broken down by lipases into free fatty acids and glycerol during storage, particularly when the temperature and moisture contents are high. Fat hydrolysis gets accelerated due to mold growth in seeds and oil. The breakdown of polar lipids is more rapid. The acid value of a fat is the number of mg of KOH required to neutralise the free acid in 1g of the substance. Oils with more free fatty acid (1-2%) is unsuitable for edible purposes. Stored oils especially under high moisture conditions deteriorate quickly.

Reagents and equipment

1. 0.1 N KOH or N/10 aqueous NaOH solution accurately standardized with N/10 potassium hydrogen phthalate using phenolphalein indicator. Light pink colour is the end point. Dilute the alkali to N/40 solution.

2. Solvent: Mix 1litre ethanol and 1 litre diethyl ether in equal proportions. Neutralize shortly before use with N/10 NaOH solution using phenolphthalein as an indicator.
3. 1% phenolphthalein in alcohol.
4. Balance.
5. Burette.

Procedure:

1. About 10g of the oil or fat is weighed, accurately into a 250 ml conical flask to which is added 50 ml of a mixture of equal volumes of alcohol and ether previously neutralised, after the addition of 1 ml of phenolphthalien solution.
2. If necessary, the contents may be warmed in a water bath until the substance is completely dissolved. The solution is titrated with N/10 KOH with constant shaking until pink colour persists for 15 sec. The titre value in ml (V) is noted.

Calculation

Acid Value = $\frac{56.1VN}{W}$ Where,

V= Volume in ml of Standard Potassium hydroxide or Sodium Hydroxide used

N= Normality of Sodium hydroxide/ Potassium hydroxide solution; and

W= Weight in g of the oil sample.

Free fatty acid as Oleic acid % = $\frac{28.2\ VN}{W}$

Free fatty acids as Lauric acid % = $\frac{20.0\ VN}{W}$

Free fatty acids as Ricinoleic acid % = $\frac{29.8\ VN}{W}$

Free fatty acids as Palmitic acid % = $\frac{25.6VN}{W}$

Free fatty acids (colorimetric method)

Principle

Cupric acetate reacts with the free fatty acids dissolved in benzene solution to give the blue coloured copper salts of fatty acids. The colour is measured to quantify the free fatty acids.

Reagents

1. Cupric acetate-pyridine reagent: 5% (w/v) aqueous solution of cupric acetate is prepared. The pH of the solution is adjusted to 6.0-6.2 with pyridine before making the final volume.
2. Chloroform.
3. Benzene.

Apparatus

1. Screw-capped culture tubes.
2. Water bath.
3. Centrifuge.
4. Cyclomixer.
5. Colorimeter or spectrophotometer.

Procedure

1. 1-3 mg of lipid sample in chloroform is taken in the screw-capped culture tube (30ml capacity). The extracted sample is evaporated to dryness.
2. Five ml benzene is added and swirled to dissolve the lipid sample. One ml of cupric acetate-pyridine reagent is added to each tube. The contents are shaken in a cyclomixer for 2 minutes and then centrifuged for 5 min to get two layers.
3. The upper layer is separated with the help of a dropper and the absorbance is read at 715 nm.
4. A standard curve is prepared by taking palmitic acid in the range of 2-14 micromoles.
5. The concentration of fatty acids is read from the graph. Further calculations are made based the concentration read from the graph and the weight of lipid.

Refractive index

Principle

The refractive index (n) of a substance is the ratio of the velocity of light in vacuum to its velocity in the substance. It may also be defined as the ratio of the sine of angle of incidence to the sine of the angle of refraction. Refractive index generally varies with temperature and wave length of light. Refractive index of the oil sample is measured with the help of a suitable refractometer. It increases with the increase in unsaturation and also chain length of fatty acids in oils. It is useful in understanding the structure and purity of oils/fats.

Reagents and equipment

1. Refractometer: Abbe refractometer - Refractive indices vary between 1.4220 to 1.4859 and can be read from the instrument to the fourth decimal place accurately.
2. The temperature of the refractometer should be controlled to within $\pm 0.1^{0}C$ and for this purpose it should be provided with a thermostatically controlled water-bath and to circulate water through the instrument.
3. Alcohol or ether.
4. Fat or oil sample.

Calibration of the instrument

The instrument is calibrated with a glass prism of known refractive index (an optical contact with the prism being made by a drop of Bromonaphthalene) or using distilled water which has a refractive index of 1.3330 at $20.0^{0}C$.

Procedure

1. Clean the refractometer with alcohol and ether. A drop of oil or fat (in case of a solid fat the temperature should be suitably adjusted by circulating hot water) is placed on the prism. The prism is closed by the ground glass-half of the instrument. The dispersion screw is adjusted so that no colour line appears between the dark and illuminated halves. The dark line is adjusted exactly on the cross wires and the refractive index is read on the scale.
2. Usually commercial instruments are constructed for use with white light but are calibrated to give the rafractive index in terms of sodium light of wavelength, 589.3 nm at a temperature of $20^{o}C$ unless otherwise specified. The correction factor for temperature is 0.000365. This value needs to be

added for temperature above 25°C or deducted for temperatures below 25°C.

Butyro-refractometer

Readings of this butyro-refractometer have an arbitrary scale from 0-100 corresponding to refractive indices 1.422 to 1.489. This instrument is simple in construction and easy to operate with less cumbersome scale. The readings of solid fats need to be corrected by adding/ deducting 0.55 for each degree increase or decrease to the desired temperature. For oils this value is 0.58. But for corn oil this value is 0.62. The temperatures for fats (solid) is 40°C and for oils (liquid) it is 25°C.

Determination of viscosity

Principle

Flow is a property of a liquid and the flow rate is a measure of the viscosity of the liquid.

Reagents

1. Ostwald viscometer
2. Acetone
3. Water bath
4. Rubber tube
5. Stop watch.

Procedure

1. Wash the Ostwald viscometer with acetone and dry it by passing air through it thoroughly.
2. Clamp the clean and dry viscometer exactly vertical in water bath in such a way that upper "U" marks of the viscometer should be in the water bath.
3. Fill a definite volume of liquid sample into the viscometer through one end. Wait for 5-10min for equilibration with temperature of the water bath.
4. Keep a stop watch ready. Suck the liquid with the help of rubber tube till the liquid reaches slightly above the "U" mark in the viscometer.

5. Then release the liquid to flow down in the viscometer. Start the stop watch when the liquid passes the upper "U" mark. Note the time of flow of the liquid.
6. Repeat the procedure thrice and note down mean flow time of the liquid.

Calculation

Express the viscosity at room temperature as Seybolt viscosity in seconds

Viscosity of castor oil

Principle

Castor oil is useful as a lubricating oil. It is a highly viscous oil. The hydroxyl fatty acid, ricinoleic acid is responsible for the high viscosity. The higher the viscosity, the more it is useful as a lubricating oil.

Apparatus

1. Redwood viscometer.
2. Stop watch.
3. Water bath.
4. Thermometer.
5. Oil sample (liquid).

Procedure

1. Level the instrument and make it horizontal.
2. Keep the brass ball in position so as to seal the orifice.
3. Pour the oil under test into the cup upto the tip of the indicator.
4. Keep the 50ml flask in position below the jet. Measure the temperature.
5. Raise the ball valve and suspend it from the thermometer bracket. Start the stop watch.
6. When the level of oil dropping into the flask reaches the 50ml mark, stop the watch and note the time in seconds.
7. Replace the ball valve in position to seal the cup to prevent overflow of oil.

8. Refill the oil up to the indicator tip of the cup. Repeat the experiment to get nearly reproducible results. Report the viscosity as "t" seconds at T°C. This is the viscosity at room temperature T°C.
9. Express the viscosity as Seybolt viscosity in seconds.
10. If necessary determine the viscosities at elevated temperatures using the water bath.

Saponification value

Principle

The saponification value is the number of mg of potassium hydroxide required to saponify 1g of oil/fat. The saponification value is an index of mean molecular weight of the fatty acids of the glycerides. Lower saponification value indicates higher molecular weight of fatty acids and vice-versa. The oil sample is saponified by refluxing with a known excess of ethanolic KOH. The alkali required for saponification is determined by titration of the excess potassium hydroxide with standard hydrochloric acid.

$$C_3H_5(C_{17}H_{35}COOH)_3 + 3KOH \leftrightarrow C_3H_5(OH)_3 + 3C_{17}H_{35}COOK$$

Glycerol tristearate Glycerol Potassium stearate

Reagents and equipment

1. KOH solution (0.5 N): 28 g KOH pellets are dissolved in 20 ml of water to which is added sufficient alcohol (95%) to make upto 1litre. The solution is allowed to stand overnight and the clear supernatant is used for the estimation. Store in brown bottle.
2. 0.5 N HCl.
3. 1% Phenolphthalien solution in 95% alcohol.
4. 250ml capacity conical flask with ground glass joints.
5. Air condenser(65 cm to one meter length).
6. Reflux condenser, to fit the flask.
7. Hot water bath or thermostatic electric hot plate

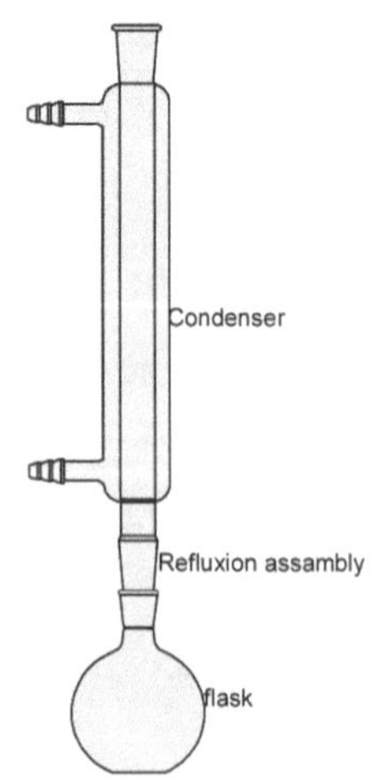

Procedure

1. About 2g of the oil/fat is weighed accurately in a 250ml round bottom flask.
2. Twenty five ml of the alcohoic KOH solution is added, a reflux air condenser is attached and the flask is kept on a boiling water bath for 1 hr. The contents of the flask are frequently mixed (a refluxion assembly also can be used).
3. When the saponification is complete, oily matter disappers and the solution becomes clear. Clarity may be achieved within one hour of boiling.
4. While the solution is still hot, 1 ml of phenolphthalein solution is added and the excess alkali titrated against 0.5 N HCl (B).
5. The experiment is repeated without the oil or fat to obtain the blank value (A).

Calculation

Since 1 ml of 0.5 N HCl is equivlent to 0.02805g of KOH, the following equation is used to calculate the saponification value.

$$\text{Saponification value (S.V.)} = \frac{56.1 \text{ x N x } (A\text{-}B)}{W}$$

Where N = Normality of HCl, **A** = Titre value with blank

B = Titre value with sample, W = Weight of oil

If the normality of HCl is 'N' then the above equation is multiplied by a factor of N/0.5.

Unsaponifiable matter

Principle

Unsaponifiable matter is the fraction of substances in oils and fats which is not saponified by caustic alkali, but is soluble in ordinary fat solvents. It includes sterols, higher aliphatic alcohols, pigments, vitamins and hydrocarbons.

Reagents and equipment

1. Alcoholic Potassium hydroxide solution: Dissolve 7 to 8 g of potassium hydroxide in an equal quantity of distilled water and add sufficient aldehyde free ethyl alcohol and make up to 100ml.

2. Ethyl alcohol (95%).
3. Phenolphthalein indicator solution: Dissolve one gram of phenolphathalein in 100 ml of ethylalcohol.
4. Petroleum ether (40-60^{0}C): Analytical reagent grade.
5. Aqueous alcohol: 10 per cent of ethyl alcohol in water.
6. Standard sodium hydroxide solution: Approximately 0.02N.
7. Acetone: Analytical reagent grade.
8. Anhydrous sodium sulphate.
9. Flat bottom flask or conical flask with a ground glass joint, 250 ml capacity.
10. Air condenser to fit the flask (or a refluxion assembly).
11. Separating funnel, 500ml capacity.

Procedure

1. Weigh accurately 2.0 to 2.5 g of the oil/fat sample into a 250ml conical flask. Add 25ml of alcoholic potassium hydroxide solution. Boil the contents under reflux air condenser (or assembly) for one hour or until the saponification is complete (complete saponification gives a homogeneous and transparent medium).
2. Care should be taken to avoid loss of ethyl alcohol during the saponification. Wash the condeser with about 10ml of ethyl alcohol.
3. Transfer the saponified mixture while still warm to a separting funnel, wash the flask first with some ethyl alcohol and then with cold water, using a total of 50ml of water to rinse the flask. Cool to 20 to 25^{0}C. Add to the flasks 50 ml of petroleum ether, shake vigorously, and allow the layers to separate.
4. Transfer the lower soap layer into another rseparating funnel and repeat the ether extraction three more times using 50 ml portions of petroleum ether.
5. Wash the combined ether extracts three times with 25ml portions of aqueous alcohol followed by washing with 25ml portions of distilled water to ensure that ether extract is free from alkali and dry it with anhydrous sodium sulphate.

6. Evaporate the ether extract on a water bath. Dry to constant weight, preferably in a vacuum oven at 70°C and not over 200mm of mercury. Cool in a desiccator and weigh.

7. After weighing dissolve the residue in 50ml of warm ethanol which has been neutralized to a phenolphthalein end point. Titrate with 0.02N NaOH (to calculate the free fatty acids).

Calculation

Unsaponifiable matter % $= \frac{100\,(A-B)}{W}$ where,

A = weight in g of the residue, B = weight in g of the free fatty acids in the extract, W = weight in g of the sample.

Iodine value (Wij's method)

Principle

The iodine value (I.V.) of a substance is the weight of iodine absorbed by 100 parts by weight of substance. This value indicates the degree of unsaturation of the oil. The stability or nutritional values of oil can be easily guessed from I.V. in the absence fatty acid profile information. Oils/fats having less than 90 I.V. are more saturated and highly stable while those with 90-120 have good storage and reasonable nutritional quality. Those with 120 and above I.V. are very good nutrionally, but do not have good keeping quality.

$C_{17}H_{33}COOH + I_2 \quad C_{17}H_{33}I_2COOH$

Oleic acid Diiodostearic acid

Reagents and equipment

1. Iodine monochloride solution: 5 ml iodine monochloride is dissolved in 500 ml glacial acetic acid (Wij's solution). If iodine monochloride is not readily available, the solution can be prepared by following any of the procdures given below:

   a) If ICl is not available the following procedure can be adopted. Iodine trichloride, 8g, is dissolved in about 200ml of glacial aceticacid with heating in a water bath, under moisture free condition (stoppered with a tube containing fused calcium chloride). Into another flask is taken 9g of iodine. Dissolve in 300 ml of carbon tetrachloride. The solutions are mixed and sufficient glacial acetic acid is added to make the volume 1litre.

b) Wij's solution: 13.0g iodine (resublimed) is dissolved in 1 l of acetic acid and pass dry Cl_2 to make the titre value with N/10 hypo double the original one.

2. Sodium thiosulphate (0.1N) :- Dissolve approximately 25 g.$Na_2 S_2O_3$ solution and accurately determine the normality using potassium dichromate solution.

3. Potassium iodide, 10% aqueous solution, add a pinch of $Na_2 CO_3$.

4. Starch (2%) indicator: Make a paste of 2g soluble starch with water and pour in 100 ml of boiling water. Keep on boiling for 2-3 min, cool and filter.

5. Burette and conical flasks etc.

Procedure

1. Weigh oil in a glass stoppered flask (about 2.5 g).
2. Add 15 ml carbon tetrachloride.
3. Add 20ml Wij's solution. Keep the mixture for 30 min to one hr in the dark.
4. Later, take out the mixture from dark.
5. Add 15 ml 10% KI and add 100 ml distilled water.
6. Titrate with N/10 $Na_2S_2O_3$. When light yellow colour develops, add starch indicator and again titrate till white colour appears at the bottom. Note down the titration value and calculate. Thorough shaking is essential during titration.

6. Conduct a blank titration without the oil.

Calculation

$$I.V. = \frac{12.69 \text{ x } N \text{ x } (B\text{-}S)}{W}$$

Where

N = Normality of thiosulphate, B = Blank titre value, S= Sample titre value, W = Weight of oil in g.

Iodine value (by Dam's reagent)

Principle

The oil sample is treated with bromine(pyridine dibromide) instead of iodine to estimate the unsaturation.

Reagents and equipment

1. Dam's reagent (0.05N pyridine dibromide solution): Add a solution of 2.0 g (2.06 ml) of pyridine in 5 ml of glacial acetic acid to a solution of 2.5 g (1.85ml) concentrated sulphuric acid in 5ml glacial acetic acid. Cool and add 2.0 g (0.63 ml) bromine. Dilute the solution to 500 ml with glacial acetic acid.
2. KI solution 10%: 10g KI in 100 ml distilled water.
3. Starch Indicator (1%): 1 g soluble starch in 100 ml of 13% KCl, boil and cool.
3. Burette and flasks etc.
4. Standard $Na_2S_2O_3$ (hypo) : 0.02N hypo.

Procedure

1. To a known (5.0 ml) aliquot of lipid solution in $CHCl_3$ containing a known amount (2-5 mg) of lipid, add 5.0 ml of the pyridine dibromide solution in a 50 ml glass stoppered Erlenmeyer flask.
2. Mix and leave at room temperature in the dark for 15 min.
3. Add 0.5 ml KI solution, 0.5ml H_2O and a few drops of starch indicator and titrate the liberated iodine with 0.02N hypo.
4. A blank consisting of 5 ml of $CHCl_3$ alone is run simultaneously.

Calculation

$$\text{I.V. (iodine value)} = \frac{A - B \; x \, 1.27}{WX5}$$

where, A= blank titration, B= sample titration, W= weight of lipid in mg.

Note

Using I. V. the number of double bonds can be calculated. Also I. V. can be calculated from the value of refractive index and the fatty acid composition, namely, the percentages of oleic, linoleic and linolenic acids.

$$\text{No. of double bonds per molecule} = \frac{I.V. X M.W}{127 X 2}$$

$N_D^{40} = 1.4515 + 0.0001171$ I.V.

$$I.V = \frac{\text{Refractive index} - 1.4515 \; x \; 10000}{1.171}$$

I.V. = 18:1 x 0.8601 + 18:2 x 1.7321

18:1 = % oleic acid and 18:2 = % linoleic acid

Thiocyanogen Number

Principle

The thiocyanogen number is useful in calculating 18:1, 18:2 and saturated fatty acid concentrations in an oil sample.

Reagents and equipment

1. Lead thiocyanate: To a solution of 250g KSCN in 500ml of water add slowly with stirring a solution of $Pb(CH_3COO)_2$ $3H_2O$ in 500ml water. Filter through a Buchner funnel and wash with water, ethanol and water. Dry by suction on funnel itself and then for 8 to 10 days over P_2O_5. The lead thiocyanate should have a greenish or yellowish white colour and should be used within two months.
2. Anhydrous acetic acid.
3. Thiocyanogen solution(0.2N): To 50g dry lead thiocyanate in 500ml anhydrous acetic acid in a glass stoppered bottle add slowly 5.1ml bromine in 500ml anhydrous acetic acid. Filter and store filtrate (thiocyanogen solution) in brown bottle at 15-20°C.
4. Burette and flasks etc.
5. Buchner funnel and flask.

Procedure

1. Weigh 0.1-0.3g of oil sample into a glass stoppered flask. Add 25ml thiocyanogen solution and dissolve by gentle shaking.
2. After standing for 24hrs in the dark (18-20°C), add 1g dry KI powder, dilute with 30ml water.

3. Titrate the liberated iodine with 0.1N hypo using starch as indicator.
4. Blank: Conduct three blank titrations simultaneously.

Calculation

$$\text{Thiocyanogen Number} = 0.01269(B - U)X \frac{100}{G}$$

B= blank titre value, U= sample titre value

G = weight of sample taken

Peroxide value (P.V.)

Principle

The peroxide value is the quantity of those substances in the sample, expressed in terms of milliequivalents of active oxygen per kg, which oxidize potassium iodide under the operating conditions described.

Reagents and equipment

1. Chloroform, analytical reagent quality, free from oxygen by bubbling a current of pure, dry inert gas through it.
2. Glacial acetic acid, analytical reagent quality, free from oxygen.
3. Potassium iodide, saturated aqueous solution, recently prepared, free from iodine and iodates.
4. Sodium thiosulphate, 0.01 or 0.002 N accurately standardized aqueous solution, standardized just before use.
5. Starch solution, 10g/litre aqueous dispersion, recently prepared from natural soluble starch.
6. 3-ml glass scoop.
7. Flasks, with ground necks and stoppers, of about 250 ml capacity, dried before hand and filled with a pure, dry inert gas (nitrogen or, preferably, carbon dioxide).
8. 25-or 50-ml burette, graduated in 0.1 ml.

Sample

Take care that the sample is taken and stored away from the light, kept cold and contained in completely filled glass containers, hermetically sealed with ground-glass or cork stoppers.

Procedure

1. The test has to be carried out in diffuse daylight or in artificial light. Weigh in a glass scoop or in a flask to the nearest 0.001 g, a mass of the sample in accordance with the following table, according to the expected peroxide value:

Expected peroxide value (m.e.q.)	Mass of test portion (g)
0 to 12	6.0 to 2.0
12 to 20	2.0 to 1.2
20 to 30	1.2 to 0.8
30 to 50	0.8 to 0.5
50 to 90	0.5 to 0.3

2. Unstopper a flask and introduce the glass scoop containing the test portion. Add 10 ml of chloroform. Dissolve the test portion rapidly by stirring. Add 15 ml of acetic acid , then 1ml of potassium iodide solution, insert the stopper quickly, shake for 1 minute, and leave for exactly 5 minutes away from the light at a temperature from 15 to 25°C.

3. Add about 75 ml of distilled water. Titrate the liberated iodine with the sodium thiosulphate solution (0.002 N solution for expected values less than 12, and 0.01 N solution for expected values above 12) shaking vigorously, using starch solution as indicator. Have a duplicate analysis and a blank determination.

Calculation

The peroxide value (P.V.), expressed in milliequivalents of active oxygen per kg, is given by the formula:

$$P.V. = \frac{V \times T \times 1000}{m}$$

where, V is the number of ml of the standardized sodium thiosulphate solution used for the test, corrected to take into account blank test.

T is the exact normality of the sodium thiosulphate solution.

m is the mass, in g, of the sample.

Take the mean of two determinations.

References

Cox, H. E. and Pearson, D. (1962) *The Chemical Analysis of Foods* Chemical Publishing Co Inc New York p 421.

Paquot,1979. Standard Methods for the Analysis of Oils, Fats and Derivatives. Pergamon Press, Oxford. U.K. pp.1-170.

Hydroxyl value

Principle

Hydroxyl value is a measure of hydroxyl content of oils and fats. It is measured by reacting with acetic anhydride and titrating the excess acetic anhydride as acetic acid with standard alkali.

Reagents and equipment

1. Reflux assembly
2. Water bath
3. Burette, flasks etc.
4. Pyridine: acetic anhydride (3:1 v/v).
5. Neutralized butanol.
6. 0.5N alcoholic KOH solution.
7. Phenolphthalein indicator.

Procedure

1. The amount of sample to be weighed and the amount of acetylating agent to be added depend on the hydroxyl value of the sample.

Expected OH^- value	Weight of sample to be taken
0-20	10
20-50	5
50-100	3
100-200	2

2. Weigh a suitable quantity of the sample and add 5ml of pyridine: aetic anhydride (3:1 v/v) into a refluxion flask. Heat over a water bath for 10 min.
3. The condenser and sides of the flask are washed with 25ml neutralized butanol.
4. Titrate against 0.5N alcoholic KOH using phenolphthalein as indicator.
5. Perform a blank titration (without sample).

Calculation

$$\text{Hydroyl value} = \left(B + \frac{WA}{C} - S\right) X\ 56.1\ X\ \frac{N}{W}$$

Where W = Weight of the sample taken,

B = ml of alkali required for the blank,

S = ml of alkali required for the sample,

A = ml of alkali equivalent to acidity,

C = weight of the sample used for the acdity titration. and

N = normality of the alkali.

Note

Hydroxyl value plays an important role in castor oil based industries as castor oil is rich in ricinoleic acid (a mono hydroxyl oleic acid). The hydroxyl group imparts higher viscosity. Castor oil and its products are value added lubricants, especially in the aviation field.

Determination of polar material

Principle

Lipids are fractionated on a silica gel column using using different solvents. The fraction eluted with diethyl ether are the polar material in the total lipids. This information is useful in evaluating the quality with respect to its edibility.

Reagents and equipment

1. Reagent A: 87% Petroleum ether+ 13% diethyl ether.
2. Reagent B: Diethyl ether.

3. Silica gel.
4. Glass column.

Procedure

1. The fat or oil sample (1-2 g) to be analysed is placed on the the silica gel column (without any pre-treatment).
2. The column is eluted with reagent A . the non-polar lipids get eluted.
3. The column is next eluted with reagent B. the polar lipids get eluted.
4. The fractions are collected separately. The solvents evaporated, and the residued weighed after drying.

Calculation

The calculations are made based on the sample weight and the fraction weight. The values are expressed as percent of the sample.

Fatty Acid Methyl Esters (FAME) (BF_3 method)

Principle

Fatty acids in biological material can easily be identified and quantified by gas-liquid chromatography of their methyl esters prepared from total lipids, or other lipid fractions. The methyl esters can be extracted into hexane, and quantitative measurements of each fatty acid can be made with a gas liquid chromatograph (G.L.C.) equipped with appropriate columns.

Scope

The method is applicable to common fats, oils and fatty acids. Unsaponifiables are not removed and, if present in large amounts, they may interfere with subsequent analyses. The procedure will result in partial or complete destruction of epoxy, hydroperoxy, cyclopropenyl, cyclopropyl and possibly hydroxyl groups, and is not suitable for the preparation of methyl esters of fatty acids containing these groups.

Apparatus

1. 50 ml and 125 ml Flat bottom boiling flasks or Erlenmeyer flasks with 19/38 or 24/40 ground glass outer necks.
2. Water cooled condensers (Liebig) 20 or 30 cm Jacket, with 19/38 or 24/ 40 ground glass inner joint.

3. 250 ml separatory funnels.
4. 200 ml boiling flask for solvent removal.
5. Boiling chips – free of fat.

Reagents

1. BF_3:Methanol reagent (125g/l of methanol), commercial or may be prepared using the gas and methanol.
2. Sodium hydroxide, 0.5 N in methanol.
3. Sodium chloride, saturated solution in water.
4. Petroleum ether, redistilled (30-60°C).
5. Heptane, gas chromatographically pure.
6. Sodium sulphate, anhydrous, A.R.
7. Methyl red indicator, 0.1% in 60% ethanol.
8. Nitrogen gas, high purity.

Procedure

Accurate weighing is not required. Sample size need be known only to determine the size of flask and amounts of reagents as per the following table:

Sample (mg)	Flask (ml)	NaOH (0.5N-ml)	BF_3:Methanol (ml)
100-250	50	4	5
250-500	50	6	7
500-750	125	8	9
750-1000	125	10	12

1. **For fatty acids:** Introduce the fatty acids into the 50 or 125 ml. reaction flask. Add the specified amount of BF_3-methanol reagent, attach a condenser and boil for two minutes. Add 2 to 5 ml of heptane through the condenser and boil for one minute longer. Remove from heat, remove condenser and add enough saturated salt solution to float the heptane solution of the methyl esters into the neck of the flask. Transfer about 1 ml of the heptane solution into a test tube and add a small amount of anhydrous sodium sulphate. The dry heptane solution may then be injected directly into a gas chromatograph.

2. **For fats and oils:** Introduce the fat into the 50 or 125 ml reaction flask. Add the specified amount of 0.5 N methanolic sodium hydroxide and add a boiling chip. Attach a condenser and heat the mixture on a steam bath until the fat globules go into solution. This step takes 5-10 minutes. Add the specified amount of BF_3:methanol reagent through the condenser and boil for two minutes. Add 2 to 5 ml of heptane through the condenser and boil for one minute longer. Remove from heat, remove condenser and add enough saturated salt solution to float the heptane solution of the methyl esters into the neck of the flask. Transfer about 1 ml of the heptane solution into a test tube and add a small amount of anhydrous sodium sulphate. The dry heptane solution may then be injected directly into a gas chromatograph.

Note

1. *There is danger of losing some of the more volatile esters if the solvent romoval step is prolonged or if too vigorous a stream of nitrogen gas is used. The method may be extended to fatty acids with 8-carbon atoms if the solvent is not completely removed.*

2. *The methyl esters should be analyzed as soon as possible. They may be kept in an atmosphere of nitrogen in a screw cap vial at 2^0C for 24 hours. For longer storage they should be sealed in a glass ampoule under vacuum and placed in a freezer.*

Other Methods of FAME Preparation

FAME can be prepared using many different procedures. The following is a list of some of them. The analyst can select any one depending on the ease and convenience or the facilities available in the laboratory.

1. Simple method: Into a test tube, take one drop of oil, add two drops of CCl_4 and one ml of 1% sodium methoxide (0.5 g Na dissolved in 100 ml methanol). Stopper the test tube and heat on a water bath for 3min or shake the contents vigorously for 5min by hand or in a wrist action shaker. Allow to separate the layers and use the upper layer for analysis of fatty acids in a GLC.

2. Reflux (50 mg fat) with 5 ml of concentrated H_2SO_4:toluene:methanol (1:10:20) for one hr. Add 5ml H_2O and 5ml hexane. Separate upper hexane layer, dry over 3 to 5g anhydrous Na_2SO_4. Resulting dry solution is ready for injection to a gas chromatograph.

3. FFA alone: Diazomethane is added in slight excess to the fatty acid solution in methanol (5ml) and allowed to react for no more than 5 min. Reagent excess is destroyed by adding a few drops of 30% formic acid in methanol. The resulting solution can be directly used for GLC.

4. The oil (20 mg) in 2 ml hexane is taken in a stoppered test tube. Add 0.5M Sodium methoxide in methanol (0.1ml) and shake for 5min at room temperature, then add 5µl acetic acid and 1g powdered anhydrous $CaCl_2$. Allow to stand the mixture for 1hr. Centrifuge at 2000-3000 rpm for 2-3 min. An aliquot of the supernatant can be used for GLC.

5. KOH Method: 100 mg oil + 10ml hexane + 0.5ml methanolic KOH (2N). Mix and invert. An aliquot of the supernatant can be used for GLC (Paquot,1979).

6. Castor methyl esters: Oil (50mg) in methanol (5ml), add 0.5M Sodium methoxide in methanol (1ml). Shake for 5min while heating, cool and add glacial acetic acid (0.1 ml). Later add 0.5g Na_2SO_4 (anhydrous). Use the clear methanolic solution for injection into G.L.C. This is suitable for samples containing hydroxyl fatty acids.

7. Take 1 g finely powdered oleogenous sample and extract the lipids with 20ml of $CHCl_3$: ispropanol (2:1 v/v). Separate the extract and evaporate the solvent completely and dissolve the lipids in 5 ml 5% HCl in methanol (super dry) and 5 ml dry benzene. Reflux at 80-100°C for 2 hr. Cool and add 2 volumes of distilled water and neutralize with Na_2CO_3. Extract with 3 ml portions each of petroleum ether (40-60°C) thrice. Pool the extracts and dry with anhydrous Na_2SO_4. If necessary, neutralize with solid Na_2CO_3. Finally concentrate the supernatant to 1ml. Use this solution for injection to the gas chromatograph.

8. Transfer 0.5 to 1 mg of the purified total lipids to glass-stoppered vials, and evaporate the solvent under a stream of nitrogen at 30 to 37°C. Add 4 ml of 5% (wt/vol) sulfuric acid in methanol to the dried lipid, and heat for 2hr at 70°C. After cooling, add 4 ml of n-hexane and shake the tube for 10min. Remove and save the hexane phase. Re-extract the methanol phase with 4 ml of hexane. The pooled hexane phases are used for gas chromatographic analysis.

Reference

Paquot,1979. Standard Methods for the Analysis of Oils, Fats and Derivatives. Pergamon Press, Oxford. U.K. pp.1-170.

Fatty acid composition (GLC Method)

Principle

The separation of compounds in a gas chromatograph depends on the difference in partition coefficients between the liquid (stationary) and gaseous (mobile) phases of the constituents in a mixture. The constituents with a high affinity for the stationary phase will tend to remain in it for a long time. These will travel slowly through the column. Those with little or no affinity for the stationary phase will travel faster and get eluted quickly. Methyl esters of fatty acids are separated based on their affinities and relative mobilities between the stationary and mobile phases and determined quantitatively (Metcalfe, *et al.,* 1966, Young, *et al.,* 1984).

Scope

The method is applicable to methyl esters of fatty acids having 8 to 24 carbon atoms. The method permits quantitative separation of mixtures containing saturated and unsaturated methyl esters.

Apparatus

1. The gas chromatographic instrument should have as a minimum the following characteristics.

   a. Column oven, capable of constant temperature operation to ± 0.0°C., at a temperature between 170 and 210°C.

   b. Sample inlet port, independently heated to about 50°C higher temperature than the column oven.

   c. Columns, 5 to 10 feet long 1/8 inch to ¼ inch in outside diameter, madeof stainless steel, glass, copper, or aluminum tubing. Columns of 1/8 inch O.D. are packed with 8 to 12% polyester liquid phase coated on 80 to 100 mesh acid washed chromosorb W or the equivalent. Diethylene glycol succinate (DEGS) is the recommended liquid phase although for specific separations other liquid phases are employed. It is recommended that the analysis be made on two columns, each with different separating properties (polyesters of differing polarity) to check for the presence of coincident peaks

   d. Detectors, thermal conductivity (T.C.) or flame ionization (F.I.D), if separately thermostated, should be maintained above 50°C of (hotter) column temperature.

e. Recorder, 0 to 1.0, 2.5 or 5.0 mv range 1 second full scale deflection with a chart speed of ½ to 1 inch per minute. Attenuator switch to change recorder range as required. The 0-1.0 mv recorder is used when a thermal conductivity detector is employed; any one other recorder may be used with the flame ionization detector. (computer aided equipments are now-a-days used to carry out the job of recorders).

f. Gases: For T.C.: Helium, minimum purity 99.95 mol % For F.I.D.: Carrier gas: Helium, nitrogen, or argon, purity 99.95 mol %, Hydrogen (for flame), of pruity 99.95 mol % and Air (dry) - (G.C grade).

g. **Syringe** for injecting sample 1 to 5 μl capacity with good reproducibility

Procedure

1. With carrier gas flowing through the apparatus, adjust to operating temperature and record a base line to check for stability of the instrument. With F.I.D., it is recommended that the apparatus be operated at medium sensitivity rather than near its maximum sensitivity. A new column should be conditioned by holding it at about 10°C above operating temperature with helium flowing for at least 24 hours. The column should be disconnected from the detector during the conditioning period.

2. Proper gas flow rate and temperature will permit elution of methyl linolenate, and shorter chain length esters in less than 30 minutes. When esters of fatty acids of greater chain length are present, the gas flow and/ or temperature should be increased so that the retention time of the last component is reduced. In addition, the largest sample size consistent with attenuation requiremens should be used in order to better detect and quantitate these slow moving esters. A constant gas flow should be maintained throughout the analysis. When using a T.C. detector, the gas flow should be measured at the exit with a soap bubble flow meter or other suitable device. It is more difficult to measure the flow in an instrument equipped with a flame ionization detector. It is better to follow the manufacturer's instructions.

3. Measure the methyl ester sample in a syringe. When using a T.C. detector, the usual sample size is between 0.5 and 4 μl. For F.I.D., about 0.1 to 0.01 μl of actual esters are measured but this is usually in a suitable solvent so that 0.5 to 4 μl of solution is injected. Pierce the septum of the

sample inlet port and quickly discharge the sample. Withdraw the needle and note on the recorder chart the peak due to air when using T.C. or solvent when using F.I.D., which marks the sample introduction reference point. Sample size must be adjusted so that the major peak is not attenuated more than eight times.

4. Watch the recorder pen to see that the peaks do not go off scale. Change the setting of the attenuator as necessary to keep the peaks on the chart paper. Mark attenuator setting on the chart.

5. After all the peaks have been traced and the pen has returned to base line, remove the chart for calculation

Calculations

1. Identify the peaks by relative position on the chart. The esters appear on the chromatogram in order of increasing number of carbon atoms and of increasing unsaturation for the same number of carbon atoms. The fatty acid methyl esters elute in the order – palmitic (16:0), stearic (18:0), oleic (18:1). linoleic (18:2), arachidic (20:0). eicosenoic (20:1), behenic (22:0), and lignoceric (24:0), in a polar column. The C20 saturated (arachidic) ester usually appears before C18:3 (linolenic) ester but may be coincident or appear after C18:3 on some columns (Coincident peaks are usually revealed by repeating the analysis at a different column temperatures or employing a second column with a different polyester).

   With constant operating conditions, the retention times (or chart distances) from the air (solvent) peak to various sample component peaks can be used for identification of the peaks. However, relative retentions are more reproducible. Relative retentions are determined by dividing the observed retention time for each peak by the retention time observed for the peak of methyl palmitate (or any other selected standard peak). Compare the observed retention times or relative retention times with those calculated from known mixtures run periodically on the same column under the same conditions.

2. Determine the area of each peak. If the instrument is equipped with an electro-mechanical or electronic integrator, the area is best measured by following the manufacturer's instructions. When an integrator is employed, the base line must remain constant or the integrator must be equipped with a base line corrector. Otherwise the area is obtained by drawing lines tangent to the sides of the peak and intersecting the base line.

Determine the area of the resulting triangle by multiplying the height by one-half the base. For an attenuated peak the outer sides of the peak must be full chart span and the tangents drawn intersecting the base line should utilize the upper two-thirds to obtain the peak width. Determine the area by multiplying the height by one-half the base. The height must be corrected for attenuation including the correction required if the base line is not the recorder zero. Divide the area of each component by its calibration factor. The percentage of each component is calculated from the ratio of each area to the sum of the areas under all of the component peaks and reported as percent by weight.

Area = 1.064 X h X $W_{1/2}$ where 'h' is the peak height and $W_{1/2}$ the width of the peak at half the maximum height (Littlewood, 1970).

3. Calibration factors are determined relative to methyl palmitate to correct for nonlinearity of instrument response and for molecular weight differences. Such factors are determined by analyzing known mixtures preferably having composition similar to that of the unknown sample. Divide the area of each peak by the true weight percentage of that component, then by dividing each value by the value for methyl palmitate, the calibration factors are obtained.

4. Instrument and column performance are monitored by noting the separation of the oleate and stearate ester peaks which is expressed as peak resolution.

$$\text{Peak Resolution} = 2Y/ S+O$$

Where: Y = the distance between the peak maxima for stearate and oleate esters and S = the base width of the stearate peak, O is the base width of the oleate peak

These values should be determined on a sample containing approximately equal quantities of oleate and stearate esters using a sample size such that peak heights are 24-50% of the chart width. If the peak resolution is equal to or greater than 1.0, then the column and instrument are in satisfactory condition. All columns when used will show a gradual loss in peak resolution. When the value becomes less than 1.0, a new column should be installed.

Precision

1. Two single determinations of major components (>5%) performed in one laboratory shall not differ by more than 1.0 percentage unit.

2. Two single determinations performed in different laboratories shall not differ by more than 3.0 percentage units.

References

Metcalfe, L.D., Schmitz, A. and Pelka, J.R. 1966. Rapid Preparation of Fatty Acid Esters from Lipids for Gas Chromatographic Analysis. Analtycal Chemistry. 38:514-415.

Littlewood, A.B. 1970. Gas Chromatography-Principles, Technique and Applications. Academic Press. pp. 289.

Young, C.T., Mazingo R.W., Sanders, T.H. and Taber, R.A. 1984. Quality Methods. American Peanut Research and Education Society, 376 Ag. Hall, Oklahoma State University, Stillwater, OK 74078, U.S.A

Polymers and heated products

Principle

The heated products and polymers in an oil sample are estimated using the GLC elution and fractionation.

Reagents equipment

1. Triheptadecanoin.
2. Hexane.
3. 0.5N alcoholic NaOH.
4. Micro refluxion assembly.
5. BF_3- methanol.
6. Separating funnel.
7. Petroleum ether (40-60^{0}C).
8. Anhydrous sodium sulphate.
9. G.L.C. with FID, DEGS column and integrator.

Procedure

1. Weigh 40mg oil and 10mg triheptadecanoin(internal standard) into 50ml Erlenmeyer flasks.
2. Dissolve the lipid in hexane and dilute to 50ml.
3. Transfer 10ml to conical flask and evaporate under a stream of N_2.
4. Add 4ml 0.5N alcoholic NaOH solution and a boiling chip.

5. Reflux for 10min.
6. Add 5ml BF_3- methanol and boil for another 2min.
7. Add 2-5ml petroleum ether (40-60^0 C) and boil for 1min.
8. Cool and transfer to a separating funnel. Shake and extract with 10ml petroleum ether. Drain the lower methanolic layer and re-extract it with petroleum ether.
9. Combine the petroleum ether layers and wash with water. Dry over anhydrous sodium sulphate.
10. Evaporate the solvent and redissolve in 1ml hexane.
11. Analyse the the sample in a G.L.C. with F.I.D. and an integrator or computer.

Calculation

$$\% \text{ Non-elution material} = \frac{Pa' \, x \left(W' + \dfrac{W}{W'} \right) - PA \, x \, 100}{Pa' \, x \, (W' + W / W') - 1}$$

Where Pa' = peak area of internal standard,

W' and W = weghts of internal standard and sample respectively and

P_A = total area of chromatogram

Poly unsaturated fatty acids (P.U.F.) (Lipoxygenase method)

Principle

Lipoxidase catalyzes the reaction between O_2 and cis-poly unsaturated fatty acids to give the conjugated diene- hydroperoxides of fatty acids. Their absorption at 234nm is a measure of fatty acids containing double bonds.

Reagents

1. Potassium borate buffer: 1M: p^H 9.0: Dissolve 61.9g boric acid and 25g KOH in about 800ml water by warming on a steam bath with stirring. Cool and adjust to P^H 9.0 with 0.1N HCl or 1NKOH and then make up to 1000 ml.
2. Potassium borate buffer: p^H 9.0; 0.2M. Dilute 200ml of the above solution (1) to one litre.

3. Lipoxydase (stock solution - 1mg protein/ml): Dissolve 10.0mg lipoxydase in ice cold potassium borate buffer: p^H 9.0; 0.2M (2 above).
4. Lipoxydase dilute: 0.2mg protein/ml, dilute 2ml stock solution(3 above) with 8.0 ml ice cold buffer (Potassium borate buffer: p^H 9.0; 0.2M).
5. Lipoxydase, boiled dilute solution 0.2mg protein/ml: pipet 5ml of dilute lipoxydase solution into a test tube. Place in a boiling water bath for 5min. The enzyme gets deactivated.
6. HCl 0.5N: Dilute 50ml concentrated HCl to1000ml with distilled water.
7. KOH 10N: Dissolve 56g KOH in about 80ml water. Cool and make up to 100ml.
8. Alcoholic KOH 0.5N: Mix 0.5ml 10N KOH with 9.5ml ethanol. This solution needs to be prepared everyday.
9. Alcoholic KOH 1.5N: Mix 1.5ml10N KOH with 8.5ml ethanol. Prepare afresh when needed.
10. Spectrophotometer with quartz cuvettes.

Procedure

1. Weigh 1-5g seed sample. Saponify in a 200ml volumetric flask with 4ml 1.5N alcoholic KOH by heating gently on a steam bath for 2hrs. Allow to stand in the dark at room temperature for 24hrs. Add 40ml 1.0 M buffer and 6ml 0.5NHCl. Dilute to 200ml with water and filter through a dry filter paper. Pipet 3.0ml of sample into two test tubes. Use one as control and the other as experimental sample tube.
2. Control: Add 0.1ml boiled enzyme and mix.
3. To the experimental (sample tube) add 0.1ml dilute enzyme and mix. Allow to stand at room temperature and read the O.D.(E) at 234nm against control

Calculation

$$\% \text{ P.U.F.} = \frac{\text{E x 3964}}{\text{W}}$$

W = μg sample in 3.0ml solution, the factor 3964 is obtained as follows:

$$3964 = \frac{3.1 \times 1000 \times 100 \times 3}{3.0 \times 78.2}$$

3.1/3.0 = dilution factor (3.0ml sample + 0.1 ml enzyme) 78.2 = Extinction coefficient of 1g P.U.F. per litre with 1cm light path at 234nm, 1000 = conversion factor from g/litre to μg/ ml, 100 = conversion to percentage, 3 = volume of sample

Lipid profile studies

Principle

The mixture of lipids in a sample are separated on a TLC plate. They are visualized by spraying with sulphuric acid followed by charring (300°C). The saturated and unsaturated lipids are separated on $AgNO_3$ coated TLC plate. Quantification of the lipid components is carried out using a densitometer.

Reagents and equipment

1. Chloroform,
2. Isopropanol,
3. Acetone,
4. Heptane,
5. H_2SO_4,
6. Cyclohexane,
7. $AgNO_3$.
8. Balance,
9. Thin layer applicator set,
10. Hot air oven,
11. Hot air blower
12. TLC solvent tank,
13. Chromatography sprayer,
14. Densitometer.

Procedure

Extraction of lipids

1. Take 1.0g seed in a mortar, powder finely and add 20 ml of chloroform: isopropanol (2:1). Mix thoroughly in a homogenizer and filter/centrifuge and decant.

2. Re-extract twice with 10 ml of solvent each time and pool all the filtrates/ supernatants. Concentrate the supernatant/ filtrate under vacuum to 1ml. (A).

TLC

1. Spot 10 µl on a silica gel G TLC plate and develop using acetone: heptane (2:88) solvent system. Spray with 20% aqueous H_2SO_4.
2. Char it in an oven at high temperature (300^0 C) and measure the concentration of different spots densitometrically.

Argentation TLC (for mono, di, & tri unsaturated lipids)

1. Evaporate the solvent from extract (A) and re-dissolve it in 5 ml of cyclohexane.
2. Spot 5 µl of the solution on a silica gel G $AgNO_3$ plate (5% $AgNO_3$ to silica G – see below) along with marker solution of a standard lipid.
3. Develop the plate in chloroform: methanol (98:2) in dark.
4. Spray with 20% aqueous H_2SO_4 and char in an oven at 300^0C.
5. Measure the concentration of different spots densitometrically.

Argentation TLC (plate preparation)

1. Take 2g $AgNO_3$ and dissolve in 81 ml distilled water and add to 40 g silica gel G. Shake and prepare a slurry.
2. Prepare the 0.25 mm TLC plate using a spreader.
3. Use immediately and do not expose to light.

Glycolipids

Principle

Glycoproteins are estimated through the hexoses present in them. The hexose content is multiplied by 4.45 to obtain the concentration of glycolipids.

Reagents

1. 2% phenol.
2. 60% phenol.
3. Galactose standard: 1mg/ml solution.

4. Conc. Sulphuric acid.
5. Spectrophotometer with quartz cuvettes.

Procedure

1. Standard curve: 20-200μg of galactose standards are taken in separate test tubes and 1ml of 2% phenol is added followed by 4ml conc. Sulphuric acid. Measure the absorbance after 15min at 480nm against an aqueous blank. Plot a graph of concentration vs absorbance.
2. Hydrolyse the glycolipid fraction with 2ml 2N sulphuric acid for 2hr.
3. 4ml chloroform is added and the mixture centrifuged.
4. Pipet 1ml aqueous layer and add 50 μl 80% phenol and 4ml conc. Sulphuric acid.
5. Measure the absorbance at 480nm after 15min.
6. Read the concentration from the standard graph and make calculations based on the sample extract and dilutions if any, etc.

Phospho lipids

Principle

These compounds contain a P atom and are constituents of membranes. The quantity of phosphorus serves as an indirect measure of the phospholipid content.

Reagents and equipments

1. 2.5% Ammonium molybdate: Dissolve 25 g of reagent grade ammonium molybdate in 200 ml H_2O, Transfer to 1 lit volumetric flask containing 300 ml of 10N H_2SO_4 and make up.
2. Fiske- Subbarao reagent: To 195 ml of 15% sodium bisulphite solution add 0.5g of 1, 2, 4 amino naphthosulfonic acid. Add 5 ml of 20% sodium sulfite and mix well (filter and store in an amber bottle, stable for 4 weeks).
3. Standard P solution : Dissolve 0.351 g monopotassium phosphate in 10ml of 10N H_2SO_4 and make upto 1litre in volumetric flask with D.W. (0.4mg P/5ml).
4. Laboratory oven.

5. Spectrophotometer with quartz cuvettes.
6. Water bath.

Procedure

1. Evaporate the lipid sample to dryness, add 0.5 ml of 10 N H_2SO_4. Samples are placed in an oven at 150 –160°C for 3 hr.
2. Two drops of fuming HNO_3 are added. Keep in an oven for 90 min to complete combustion.
3. Remove and cool. Ammonium molybdate solution, 4.6 ml, + 0.2ml of Fiske- Subbarao reagent are added.
4. Tubes are covered with marbles, heated on a water bath for 7 min.
5. The O.D. at 660 nm recorded.

Calculation

Phospholipid = O.D. value X 25

Note

The color is proportional to the concentration of P up to 1.5μ moles in the reaction mixture.

Hydrocarbon content

Principle

The total lipids are initially extracted and fractionated on a silicic acid column. The hydrocarbons from the eluate are precipitated from the acetone extract at -15°C. The quantity is weighed after drying to calculate the hydrocarbon content.

Reagents and equipment

1. Petroleum ether 40-60°C.
2. Acetone.
3. Soxhlet extractor.
4. Rotary evaporator.
5. Glass column for chromatography.
6. Freezing mixture.

Procedure

1. Extract finely powdered plant sample in a soxhlet extractor with petroleum ether(40^0 -60^0C).
2. Filter the extract and evaporate the solvent in a rotary evaporator.
3. Dry the residue in an oven at 100^0C. Record the weight of the residue.
4. Dissolve the residue in a small quantity of petroleum ether.
5. Column chromatography: Methanol washed and dried silicic acid (100 mesh) is mixed with a minimum quantity of petroleum ether and a column is prepared. Apply the petroleum ether solution to the top of the column(1g extractives to 100g silicic acid). Elute with 200ml petroleum ether.
6. The petroleum ether extractives are dissolved in a minimum quantity of acetone. This solution is cooled at -15°C in a freezing mixture.
7. The solid hydrocarbon is filtered and dried. It is weighed.
8. The calculations are made based on the quantity of sample and dilutions made, if any.

G.L.C. of Hydrocarbons

Principle

The total lipids are initially extracted and the hydrocarbons are fractionated on a slicic acid column. The eluted total hydrocarbons are separated using GLC and identified based on their RT values.

Reagents and equipments

1. Petroleum ether 40-60°C
2. Acetone.
3. Soxhlet apparatus.
4. Rotary flash evaporator.
5. Glass column for chromatography.
6. G.C. Equipment with dexil GC column and FID.
7. Decalin solvent.
8. A micro syringe.
9. Appropriate hydrocarbon standards.

Procedure

1. Extract finely powdered plant sample in a soxhlet extractor with petroleum ether(40° -60° C).
2. Filter the extract and evaporate the solvent in a rotary evaporator.
3. Dry the residue in an oven at 100°C. Record the weight of the residue.
4. Dissolve the residue in a small quantity of petroleum ether.
5. Column chromatography: Methanol washed and dried silicic acid (100mesh) is mixed with a minimum quantity of petroleum ether and a column is prepared. Apply the petroleum ether solution (from 4 above) to the top of the column(1g extractives to 100g silicic acid). Elute with 200ml petroleum ether. Evaporate the solvent.
6. The petroleum ether extractives are dissolved in a minimum quantity of decalin.
7. Inject 1µl of the solution to the GLC with dexil 300GC (60-80 mesh) and FID. Nitrogen at 40ml/min is used as the carrier gas. The column is maintained at 330°C.
8. Identification is based on known RT values of standards.
9. Quantification is based on the peak areas.

Total sterols

Principle

The sterols in the sample are extracted from the acidified lipid fraction into hexane. They are precipitated as digitonides and estimated gravimetrically.

Reagents

1. Petroleum ether 40-60°C.
2. Conc H_2SO_4.
3. Ethanol.
4. Hexane.
5. Methanol.
6. Digitonin – 2% solution in ethanol.

Procedure

1. Weigh about 3g of the plant sample and extract with petroleum ether in a soxhlet extractor.
2. Dry the extract and redissolve in 50ml ethanol containing 0.3ml H_2SO_4 and reflux for 12hr.
3. The mixture is neutralized with NaOH and extracted with 75ml portions of hexane four times. Add water to have two clear layers.
4. Pool the hexane fractions. Evaporate to dryness. Re-dissolve the residue in 20ml hot methanol. Take into a centrifuge tube. Add 5ml hot 2% digitonin in ethanol. Boil and add 5ml water. Cool to room temperature.
5. Centrifuge at15000g for 30min. Wash thrice with 80% ethanol and thrice with 30ml diethyl ether .
6. Dry the centrifuge at 70^{0}C for 12hr and weigh.

Calculation

The sterol weight is 25.3% of the weight of the digitonides

Sterols- chromatography

Principle

The sterols in the sample are extracted from the acidified lipid fraction into hexane. They are precipitated as digitonides and estimated

Reagents

1. Methanol.
2. Benzene.
3. T.L.C chamber, plates, spreader etc.
4. Silica gel G plates.
5. Rhodamine 6G (1% alcoholic solution).
6. U.V. chamber.
7. Rotary flash evaporator or hot water bath.
8. Standard sterols (ppm solutions).
9. Petroleum ether 40-60^{0}C.

10. Conc. H_2SO_4.
11. Ethanol.
12. n Hexane.
13. Chloroform.
14. Suitable standards like β-sitosterol, stigmasterol, cycloartenol, etc. (10μg/ml solutions).

Procedure

1. Weigh about 3g of the plant sample and extract with petroleum ether in a Soxhlet extractor.
2. Dry the extract and redissolve in 50ml ethanol containing 0.3ml H_2SO_4 and reflux for 12hr.
3. The mixture is neutralized with NaOH and extracted with 75ml portions of hexane four times. Add water to have two clear layers.
4. Pool the hexane fractions. Evaporate to dryness.
5. The sterol is dissolved in benzene and spotted on a silicagel G plate impregnated with rhodamine 6G. β-sitosterol, stigmasterol cycloartenol etc. are spotted on the plate.
6. The plate is developed with chloroform for one hr.
7. Examine the plate under U.V. light.
8. Identify the spots in comparison to the standards.
9. Quantification based on intensity and area of the spots can be attempted.

Resins and waxes

Principle

The alcoholic extract of the plant sample is washed with water and re-extracted with alcohol to measure the resins.

Reagents and equipment

1. Acid washed sand.
2. 95% ethanol.
3. Anhydrous $CaCl_2$ (Desiccator).

4. Water bath.
5. Soxhlet extractor.

Procedure

1. Weigh 3g finely powdered dry plant sample. Mix with 30g acid washed sand.
2. Transfer to a thimble and extract with 95% ethanol for 24hr in a soxhlet extractor.
3. Filter the alcoholic extract into a 250ml beaker. Wash the flask and filter system with boiling alcohol.
4. Evaporate the alcoholic filtrate to dryness on a steam bath.
5. To the resinous extract add 30 – 40ml of water at 45^{0}C. Mix and filter.
6. Place the filter paper of the above filtration in a 250ml beaker containing the resinous material and add 75ml 95% ethanol.
7. Cover the beaker with a watch glass and heat on steam bath till the resinous material is dissolved.
8. Wash the filter paper(after taking out) with 95% ethanol and add the washings to the beaker.
9. Evaporate the pooled alcoholic filtrate in a weighed 50ml flask on a steam bath to dryness.
10. Dry the 50ml Erlenmeyer flask in a desiccator over anhydrous $CaCl_2$ and weigh.

Calculation

$$\% \text{ resins and waxes} = \frac{\text{wt of resins and waxes x 100}}{\text{wt of dry sample}}$$

Total volatile bases

Principle

The sample is steam distilled and the volatile bases collected into an excess of acid. The acid consumed by the bases furnishes the quantity of bases in the sample.

Reagents and equipment

1. Standard Na_3PO_4 solution(1N).
2. 0.1N HCl.
3. 0.1N NaOH.
4. Methyl red -methylene blue indicator: 0.2% methyl red + 0.1% methylene blue in alcohol (1:1).
5. Kjeldahl flask and steam distillation set-up.

Procedure

1. Weigh 5g finely powdered plant sample into a 800ml kjeldahl flask.
2. Add 75 ml standard Na_3PO_4 solution and connect the flask to an apparatus arranged for distillation in a current of steam.
3. Collect the distillate in a 1000ml Erlenmeyer flask containing an excess of 0.1N HCl (25ml).
4. Collect the distillate into a flask. Arrange for boiling in kjeldahl flask in such a way that 800ml distillate is collected in 45min. See that volume in the kjeldahl flask remains constant throughout the distillation.
5. Titrate the contents in the Erlenmeyer flask with 0.1N NaOH using 7-8 drops of methyl red -methylene blue indicator. End point is red violet to green colour.

Calculation

$$\text{\% Volatile bases as ammonia} = \frac{\text{ml 0.1N HCl required x 0.17032}}{\text{Wt of dry sample}}$$

Volatile acids

Principle

The volatile acids in the sample are steam distilled and collected into an excess NaOH. The NaOH consumed is a measure of the volatile acids.

Reagents and equipment

1. Tartaric acid
2. 0.1N NaOH solution
3. 0.1N H_2SO_4 solution
4. Kjeldahl flask and steam distillation set-up
5. Phenolphthalein indicator

Procedure

1. Weigh 5 g finely powdered plant sample into a 800 ml kjeldahl flask.
2. Add 100 ml water and 2g tartaric acid and connect the flask to an apparatus arranged for distillation in a current of steam.
3. Collect the distillate in a 1000 ml Erlenmeyer flask containing an excess of 0.1N NaOH (25 ml).
4. Collect the distillate into a flask. Arrange for boiling in kjeldahl flask in such a way that 700ml distillate is collected in 45min. See that volume in the kjeldahl flask remains constant throughout the distillation.
5. Titrate the flask (with excess NaOH) with 0.1N H_2SO_4 using phenolphalein indicator.

Calculation

$$\text{Total volatile acids (as acetic acid)} = \frac{\text{ml of 0.1N NaOH required x 0.006 x 100}}{\text{Wt of dry sample}}$$

4

AMINO ACIDS AND PROTEINS

The word protein was first coined in 1838 to emphasize the importance of this class of molecules. The word is derived from the Greek word proteios which means "of the first - rank". Proteins are polymers/macro molecules made up of several amino acids. When there are more than 50 amino acids in a chain, the polypeptide is normally called a protein. Each protein in the cells of living beings has a unique sequence of amino acids that determines its biological function. The proteins are folded into specific defined structures, which are maintained by large number of relatively weak bonds. The three dimensional structure is well defined and suited to it's specific function.

Very small changes in the structure can modify the function. Hence, it is important to know the structure of proteins to understand their behavior and function. The biological activity of proteins depends on maintenance of folded conformation. Proteins fold into well defined three dimensional shapes and they are able to recognize their corresponding substrates or antigen molecules and bind them tightly. The protein structure has been classified into four different levels based on the folded conformation of the protein.

Amino acids are the building blocks of proteins. They occur in all living plant cells as constituents of proteins and as free acids or amides. More than 200 naturally occurring amino acids have been identified in higher plants. Around 20 different amino acids have been observed to be incorporated into proteins. They have two functional groups, an amino group ($-NH_2$) and a carboxylic group($-COOH$). Both these groups are attached to the α carbon atom only. Amino acids with this structure are called α amino acids or 2-aminoacids.

The unique characteristics of the 20 amino acids are due to the side chain(R), which can be an alkyl, hydroxyl, thiol, amino, sulphide, aromatic, or heterocyclic group. The amino acids, namely, arginine, histidine, isoleucine, leucine, lysine, methionine, phenylalanine, threonine, tryptophan and valine are called essential amino acids. They cannot be synthesized by the human body. Hence they have to be supplemented through the diet.

Essential and non essential aminoacids

Essential amino acids	Non-essential amono acids
Arginine	Alanine
Histidine (for children only)	Aspargine
Isoleucine	Aspartic acid
Leucine	Cysteine
Lysine	Glutamic acid
Methionine	Glutamine
Phenyl alanine	Glycine
Threonine	Proline
Tryptophan	Serine
Valine	Tyrosine

Separation of amino acids

Amino acids can be separated based on their isoelectric points using a chromatographic technique called **electrophorsis**. A buffered amino acid mixture is applied to the gel on a thin plate or piece of filter paper, which is connected to two electrodes. A voltage applied to the electrodes causes the positively charged amino acids to move towards cathode while the negatively charged amino acids move towards the anode. Amino acids at the isoelectric point with zero charge do not move. Sufficient time is allowed for the amino acids to move under the electric field and the gel or paper is removed. Ninhydrin is sprayed to visualize the amino acids. They are identified based on their direction and movement (R_f values – movement relative to the solvent front). They can be recovered from the gel or paper by cutting out the spots and extracting them individually into suitable solvents. The identities can be further confirmed using spectroscopic techniques.

For example, we can separate a mixture of amino acids, namely, lysine, valine and aspartic acid in a buffer at p^H 6.0. Aspartic acid moves towards the anode, lysine moves towards the cathode and valine, being a zwitter ion at p^H 6.0, does not move and remains at the place of spotting. Electrophoresis is a widely used technique for separation of charged compounds.

Estimation of amino acids

Amino acids can be estimated by some of the following procedures.

1. Nitrous acid reacts with the amino group liberating N_2 and the corresponding hydroxyl acid.

   $RCHNH_2COOH + OHNO \rightarrow RCHOHCOOH + N_2 + H_2O$

   This reaction is the basis for the Van Slyke method for determining the number of free amino groups in a protein.

2. Formaldehyde reacts with amino group to form a methylene compound

   $NH_2RCHCOOH + HCHO \rightarrow CH_2{=}NRCHCOOH + H_2O$

   The resulting acid can be titrated to estimate the amino acids. This is called "Formal titration". The amino acids react with other aldehydes to form Schiffs bases.

3. *Ninhydrin test*: Ninhydrin is an oxidizing agent which oxidatively de-aminates the alpha-amino groups of amino acids. It is very important for the detection and the quantitative analysis of amino acids. Ninhydrin also reacts with primary amines. However the formation of carbon dioxide is quite diagnostic for amino acids. Alpha amino acids yield a purple substance (Ruhemann's purple) that absorbs maximally at 570 nm. Imino acids (proline, hydroxyproline) yield a yellow product.

   When a solution of amino acid is heated with ninhydrin, the amino acid is oxidatively de-aminated to produce ammonia and a keto acid. The keto acid is decarboxylated to produce an aldehyde with one carbon atom less than the parent amino acid. The net reaction is that, ninhydrin oxidatively de-aminates and decarboxylates α -amino acids to CO_2, NH3 and an aldehyde. The reduced ninhydrin then reacts with the liberated ammonia and another molecule of intact ninhydrin to produce a purple colored compound known as Ruhemann's purple. This ninhydrin reaction is employed in the quantitative determination of amino acids. Proteins and peptides that have free amino groups (in the side chain) will also react and give color with ninhydrin.

4. Amino acids react with dinitro fluorobenzene, phenyl iso thiocyanate and dansyl chloride, which are also useful in their quantitative estimation.

5. Tyrosine, tryptophan and phenyl alanine, the aromatic amino acids absorb at 280 nm. Most proteins contain tyrosine and this property is used for estimation of proteins by U.V. spectrophotometry.

Protein quality evaluation methods

Proteins are vital to the living processes and carry out a wide range of functions essential for the sustenance of life. In judging the adequacy of dietary proteins to meet the human needs, not only the quantity, but also the nutritional quality of the dietary protein is important. Proteins present in different foods vary in their nutritional quality because of differences in their amino acid composition. The human body derives the essential amino acids from the dietary protein. The quality of the dietary protein depends on its essential amino acid composition and its digestibility. The best quality protein is the one which provides essential amino acid pattern very close to the pattern of reference proteins such as egg protein or milk protein. The quality of a protein is evaluated by the following methods.

Biological methods: Proteins of raw grains (particularly legumes) are less digestible than that of animal foods. Generally, the low digestibility of plant proteins is due to the presence of trypsin inhibitors which are destroyed on cooking. The overall quality of a protein can be determined by biological methods with laboratory animals like rats as follows;

a) *Protein efficiency ratio (PER):* The gain in the weight of young animals per unit weight of protein consumed is measured and the value obtained is designated as PER. Protein efficiency ratio is defined as gain in body weight by protein in take.

$$\text{PER} = \frac{\text{gain in body weight } (\text{g})}{\text{protein in take} (\text{g})}$$

b) *Digestibility coefficient (DC):* Dietary proteins are hydrolyzed to amino acids on digestion by various proteolytic enzymes. Proteins differ in the digestibility depending on their source. Since only amino acids are absorbed into the blood stream, undigested portion of the protein is excreted in the fecal matter. The term digestibility coefficient of protein refers to the percentage of the ingested protein absorbed into the blood stream after the process of digestion is complete. When an animal is fed on nitrogen free diet, certain amount of nitrogen is excreted in the fecal matter. This is derived mainly from the digestive juices. This is called endogenous fecal nitrogen. When a protein food is given, the nitrogen found in feces consists of both endogenous nitrogen and food nitrogen lost in digestion. To find out nitrogen lost in digestion, endogenous faecal nitrogen should also be determined. For the determination of the digestibility coefficient, the data required are 1. Food nitrogen intake–*In,* 2. Total fecal nitrogen excreted–*Fn* and 3. Endogenous fecal nitrogen – *Fe*

$$DC = \frac{In - (Fn - Fe)}{In} x100$$

Where Fn – Fe is the food nitrogen lost in digestion.

The digestibility coefficient of proteins is influenced by several factors, such as indigestible carbohydrates and proteolytic enzyme inhibitors etc.

c) *Biological value (BV):* The amount of nitrogen (N) in the diet and in excreta of adult animals is measured and the percentage of N retained by animals from out of N absorbed from the diet is calculated. The value thus obtained is the biological value of the protein. It measures the quantity of dietary protein utilized by the animal for meeting its protein needs for sustenance and growth.

1. Food nitrogen intake – *In,* 2. Total fecal nitrogen excreted – *Fn,* 3. Endogenous fecal nitrogen – *Fe,* 4. Total urinary nitrogen excreted – *Un,* 5. Endogenous urinary nitrogen - *Ue*

$$B\,V = \frac{In - (Fn - Fe) - (Un - Ue)}{In - (Fn - Fe)} X100$$

Calculation of protein content

The protein content of samples is determined on the basis of total nitrogen content. Kjeldahl (or similar) method has been almost universally applied to determine nitrogen content. Nitrogen content is then multiplied by a factor to arrive at protein content. This approach is based on two assumptions: that carbohydrates and fats do not contain nitrogen, and that nearly all of the nitrogen in the sample is present as amino acids in proteins. On the basis of early determinations, the average nitrogen (N) content of proteins was found to be about 16 percent, which led to use of the calculation N x 6.25 (1/0.16 = 6.25) to convert nitrogen content into protein content.

The nitrogen is present not only in proteins, but also in nucleotides, creatine and choline. Also the nitrogen content in amino acids vary depending on the number of amino groups present. The nitrogen content of proteins actually varies from about 13 to 19 percent. This would equate to nitrogen conversion factors ranging from 5.26 (1/0.19) to 7.69 (1/0.13). Hence the conversion factors for determining protein content vary with respect to the type of protein in the sample. The different factors are presented in the Table.

Table. Conversion factors for proteins

Organic item	Factor	Organic item	Factor
Eggs	6.25	Meat	6.25
Milk	6.38	Barley	5.83
Corn(maize)	5.83	Millets	5.83
Oats	5.83	Rice	5.95
Rye	5.83	Sorghum	6.25
Wheat(whole)	5.83	Wheat bran	6.31
Wheat endosperm	5.70	Castor bean	5.30
Jack, lima, mung	6.25	Soybean	5.70
Peanut	5.46	Velvet bean	6.25

Proteins through amino acid composition

Proteins are made up of chains of amino acids joined by peptide bonds. They can be hydrolysed to their component amino acids, which can then be measured by ion-exchange, gas-liquid or high-performance liquid chromatography. The sum of the amino acids then represents the protein content (by weight) of the food. This is sometimes referred to as a "true protein". The advantage of this approach is that it requires no assumptions about, or knowledge of, either the non- protein nitrogen(NPN) content of the food or the relative proportions of specific amino acids.

Sequencing of amino acids

Protein/amino acid sequencing technique is followed to determine the primary structure of proteins. The different steps involved are 1. The number of polypeptide chains in the protein. 2. Cleavage of the disulphide bonds. 3. Amino aid composition of polypeptides. 4. Specific cleavage reactions are used to fragment big units and 5. The structure is put together based on the amino acid arrangement.

There are several effective methods by which the polypeptide end groups of protein may be identified. The most effective method in identification of N-terminal residue is Edman degradation method named after its inventor Edman. Phenylisothiocynate reacts with the N- terminal amino groups of proteins under mildly alkaline conditions to form their phenylthiocarbamyl derivative. This product on reaction with an anhydrous strong acid such as trifluoroacetic acid (TFA, F_3CCOOH), forms thiazolinone derivative with the cleavage of a peptide bond involving the carboxyl group of the N terminal residue. The treatment with TFA does not affect the other peptide bonds leaving a peptide chain with n-1 amino acid residues. The Edman degradation therefore releases the N-terminal amino acid residue but leaves intact, the rest of the polypeptide chain.

The treatment of the thiazolinone derivative with aqueous acid leads to the formation of phenylthiohydantoin (PTH) derivative of N-terminal amino acid which can be identified by chromatography. The residual peptide chain can be now submitted to a new coupling reaction and the next amino acid can be identified.

$$C_6H_5-N=C=S \quad + \quad H_2N-\underset{R1}{\underset{|}{CH}}-\underset{O}{\underset{||}{C}}-NH-\underset{R2}{\underset{|}{CH}}-\underset{NH-}{\underset{|}{C}}=O$$

Phenyl isothio-
cyanate

Peptide

$$\longrightarrow C_6H_5-NH-\underset{S}{\underset{||}{C}}-NH-\underset{R1}{\underset{|}{CH}}-\underset{O}{\underset{||}{C}}-NH-\underset{R2}{\underset{|}{CH}}-\underset{O}{\underset{||}{C}}-NH-$$

Phenyl isothio carbamyl peptide

Other procedures followed include C-terminal analysis and disulphide cleavage. C-terminal analysis is carried out using carboxy peptidases which cleave the C-terminal residue of the polypeptide chain. Disulphide bonds are formed between two cysteine residues. Detergents are useful in cleavage of the disulphide bonds.

Purification techniques

The aim of protein purification is to isolate a specific protein from all the others in a mixture of proteins. A combination of fractionation techniques is used for this purpose. The properties based on which the proteins are separated are the size, charge on the molecule and the specific binding ability. The following techniques are used for purification of proteins.

a) **Salting in and salting out:** Increase in the solubility of a protein by addition of small quantities of sodium chloride is called as salting in and this is due to an increase in the ionic strength of the solution. On the other hand, when excess of salt is added to the solution, there is decrease in the solubility and this is called as salting out. Salting out occurs due to hydrophobic effect. Both the hydrophobic and hydrophilic amino acids are present in protein molecules.

When the salt concentration is increased, a competition develops for the water between the protein and the salt. Some of the water molecules are attracted by the salt ions, which decreases the number of water molecules available to interact with the charged part of the protein. As the salt concentration increases, the water on protein is removed thus exposing the hydrophobic area of protein molecule. These hydrophobic areas on the protein molecule get attracted to each other by hydrophobic effect. This results in increase in weight of the molecule and its aggregation. Larger the surface hydrophobic area on a protein molecule, quicker will be the precipitation of the protein at a lower concentration of the salt. Salting out is also used to concentrate the proteins from a dilute solution. The precipitated protein is re-dissolved in a small quantity of solvent.

b) **Dialysis:** Proteins can be separated by dialysis through a semi permeable membrane such as cellulose membrane which has pores in it. Bigger molecules are retained in the bag and smaller molecules pass through the membranes. As the size of protein molecules are above 10kDa, this technique is not so effective. Further, at equilibrium the quantity of small molecules inside and outside the bag are equal. Several charges are needed to reduce the quantity of small molecules.

c) **Gel filtration or Size Exclusion Chromatography:** This chromatographic technique is based upon the use of a porous gel in the form of insoluble beads placed into a column. When a solution of proteins is passed through the column, a protein of small size is likely to enter the pores. Small proteins can penetrate into the pores of the beads and, therefore, are retarded in their rate of travel through the column. The larger proteins will move through the gaps present in the column and are likely to move fast and are collected first when compared to the smaller ones. Different beads with different pore sizes can be used depending upon the desired protein size separation profile. Polymer beads made up of dextrose, agarose or polyacrylamide are generally used and the commercial names for few polymer beads are Sephadex, Sepharose and Biogel respectively.

d) **Ion Exchange Chromatography**: Each individual protein exhibits a distinct overall net charge at a given pH. Some proteins will be negatively charged and some will be positively charged at the same pH. This property of proteins is the basis for ion exchange chromatography. The basic principle is, like charges repel and unlike charges attract each other. Fine cellulose resins are used that are either negatively (cation exchanger) or positively (anion exchanger) charged. Proteins of opposite charge to the resin are retained as a solution of proteins is passed through the column. The bound proteins are then eluted by passing a solution of ions bearing a charge opposite to that of the column. By utilizing a gradient of increasing ionic strength, proteins with increasing affinity for the resin are progressively eluted. CM (carboxy methyl)-Cellulose and DEAE (diethyl aminoethyl)-Cellulose are examples of negatively charged and positively charged resins.

DEAE- Cellulose is positively charged and is called an anion exchanger. CM - cellulose is negatively charged and acts as a cation exchanger. Proteins can be bound to the resin and can be displaced by increasing the strength of buffer and or changing the p^H. Proteins are eluted from a column in order of least strongly bound to the most strongly bound.

Crude protein content

Principle

The biological material (plant products or tissues) is boiled (digested) with concentrated sulfuric acid. The nitrogenous compounds are converted to ammonium sulphate. The ammonium sulphate is then hydrolysed with strong alkali to ammonia which is distilled and trapped into boric acid. It is then titrated against standard acid (Mckenzie and Wallace, 1954).

$$(NH_4)_2SO_4 + 2NaOH \rightarrow Na_2SO_4 + 2H_2O + 2NH_3$$

$$3NH_3 + H_3BO_3 \rightarrow (NH_4)_3BO_3$$

$$(NH_4)_3BO_3 + 3\,HCl \rightarrow 3NH_4Cl + H_3BO_3$$

Reagents and equipment

1. Concentrated sulphuric acid
2. Copper sulphate (A.R.)
3. Sodium Hydroxide (40%): Dissolve 400 g of NaOH in distilled water and make up to 1 litre.

4. Potassium sulphate (A.R.)
5. Mixed indicator: Dissolve 0.5 g of bromocrsol green and 0.1 g of methyl red in 100 ml of 95% alcohol and adjust the pH to 4.5 with dilute NaOH or HCl.
6. Boric acid (4%): Dissolve 40 g of boric acid (A.R.) in distilled water and make up the volume to 1 litre.
7. 0.01 N HCl: Standardise against 0.01 N NaOH which has been standaridised against 0.01 N oxalic acid.
8. Kjeldahl flasks.
9. Heaters/Burners.
10. Micro Kjeldahl distillation apparatus.

Procedure

1. Weigh accurately 0.5 to 1.5 g of the sample and transfer to a 300 ml Kjeldahl flask. Add 25 ml concentrated H_2SO_4, 10g. potassium sulphate and a few crystals of copper sulphate (0.5 to 1.0 g) as catalyst. Potassium sulphate is used to raise boiling temperature of H_2SO_4.
2. Digest till the sample is clear and no carbonisation remains.
3. Cool, add a few ml of distilled water and transfer the sample to a 250 ml. volumetric flask. Rinse many times and transfer the washings to the volumetric flask. Make up to volume when contents have cooled to room temperature.
4. Take 1-5 ml aliquots into micro-kjeldajahl distillation apparatus and carry out the steam distillation.
5. Dip the tip of the condenser in 5ml boric acid solution with a drop of mixed indicator, contained in a 50 ml graduated conical flask. Before dipping the tip of the condenser in boric acid, have a blank distillation to clear the apparatus of any contamination with ammonia.
6. Place 2-5 ml aliquots of digested sample in the distillation chamber giving a washing with 1-2 ml distilled water. Before lowering the sample aliquot in the distillation chamber, dip the tip of the condenser in boric acid. Add sufficient concentrated NaOH to the digest in the chamber to more than neutralize the amount of acid present (generally 3-4ml 40% NaOH is sufficient).

7. Start steam distillation and steam distill the sample until 20 ml distillate has been received in the receiver. Lower the receiver and wash the tip of the condenser with distilled water.

8. Before adding another aliquot of the digest into the distillation chamber, disconnect the flow of the steam into the chamber by pressing rubber pipe leading from boiling water flask to the distillation chamber. The residue of the sample will be removed automatically by siphoning in a few moments. A few rinsings should be given by the addition of distilled water into the chamber, starting distillation and siphoning off the chamber contents.

9. Titrate the contents of the receiver flask with 0.01 N HCI to light pink end point. Also test the accuracy by using standard ammonium sulphate solution.

Calculations

1ml.0.01 N HCI = 0.140 mg N.

$$\% N = \frac{0.14 X T X 100}{1000 X W X 100-M}$$

Where,

T= Titre value, W= weight of air dry sample, and M= moisture percent in the sample.

Reference

McKenzie, H.A. and Wallace, H.S. 1954. The Kjeldahl's Determination of Nitrogen. Australian Journal of Chemistry. 17:55.

Total Nitrogen by microprocedure

Principle

Nitrogen content in the samle is estimated by the blue color developed by the phenol –nitroprusside reagent.

Reagents and equipment

1. $(NH_4)_2 SO_4$ (0.05 mM): 33.04 mg $(NH_4)_2SO_4$ made upto 500ml in distilled water (1.0 ml = 14 mg N).

2. $HClO_4$ –72% .
3. Phenol (0.6M) Sodium Nitroprusside (1mM): 28.2g phenol and 130 mg of $Na_2[Fe(NO)(CN)_5]$ made up to 500ml in distilled water.
4. Trisodium Phosphate (0.6M): 107.5 g Na_2HPO_4: 12 H_2O and 12 g NaOH made up to 500 ml in distilled water (in a polythene bottle).
5. NaOH (0.75N) NaOCl (0.03M): 15.0g NaOH and 1.1 g NaOCl made up to 500 ml in distilled (in a polythene bottle)
6. Vortex mixer.
7. Spectrophotometer with quartz cuvettes.
8. Polytetra fluoro ethylene - anti-bumping rod

Procedure

1. Place the sample (or standard) containing 1-14 μg N in a rimless glass test tube (13 x 130 mm, 1.5 mm wall thickness) and evaporate the solvent to dryness under N_2.
2. Add 0.05ml $HClO_4$ and one piece (3 x 3 mm) of polytetra fluoro ethylene anti-bumping rod. Heat the tube on a digestion rack for 30-45 min so that the acid boils gently and refluxes about 5 cm up the tube.
3. Cool and add on a vortex mixer 0.25 ml phenol-nitroprusside solution, 2.5ml trisodium phosphate solution and 0.3 ml of NaOH-NaOCl solution. Stand at room temperature for 30 min. Read absorbence (blue colour at 635 mμ. Calibrate a graph with aliquots of standard $(NH_4)_2SO_4$ –2.8-14.0 μg N.

Calculation

$$\mu g\ N = \frac{\text{absorbence of sample}}{\text{absorbence standard}} \times \mu g\ N \text{ in standard}$$

Reference

Sloane and Stanley. (1967) Biochem J. 104: 293

Protein N and total N (micro-Kjeldahl method)

Principle

The nitrogen in the sample is converted to ammonia. In the presence of acid, it becomes an ammonium salt, for example ammonium sulphate with H_2SO_4.

The ammonia is steam distilled and measured by titration with an acid to estimate the nitrogen content.

Reagents and equipments

1. Salicylic – sulphuric acid mixture: 5 g of pure salicylic acid is dissolved in 100 ml of concentrated H_2SO_4.
2. Catalyst: 1g of $CuSO_4$ $5H_2O$, 8g of K_2SO_4 and 1g of SeO_2 are ground to a fine powder separately in a mortar and then mixed together .
3. Indicator (Conway): A mixture of 6 ml methyl red (0.16% in 95% ethanol) 12 ml bromo-cresol green (0.04% in water) and 6 ml 95% ethanol (turns faint pink at end point pH 4.9).
4. HCl (N/28): 31.8 ml of concentrated HCl is diluted to 10 litre with water and standardized with standard Na_2CO_3 solution .
5. Boric acid (2%) : 20 g boric acid (A.R) is dissolved in water and made up to 1 liter .
6. Standard $Na_2 CO_3$ solution.
7. Kjeldahl flasks.
8. Heaters/Burners.
9. Micro Kheldahl distillation apparatus.

Procedure

Determination of protein – N

1. 0.2 to 0.5g of plant sample is taken in a 100 ml Pyrex beaker. About 20 ml of 0.5% acetic acid solution is added. The beaker is heated to boil on a hot plate to extract non-colloidal soluble nitrogenous components and to precipitate proteins.
2. The solution is filtered through Whatman no. 44 filter paper. The residue is washed several times with hot 0.5% acetic acid solution until the filtrate is colourless.
3. Filter with the residue is dried in an oven and then folded and placed in a 50 ml micro Kjeldahl flask and proceed for digestion.

Digestion

1. 6 ml of salicylic acid-H_2SO_4 mixture is added to the filter with residue in 50 ml Kjeldahl flask. Approximately 0.5 g of sodium thiosulphate (powder) is added after allowing the salicylic-sulfuric acid mixture to soak the plant sample for half an hour (30min).
2. The flask is heated in a fume hood until fumes appear. The flame is removed and the flask is allowed to cool for 5 minutes. A small amount of catalyst is added and the digestion continued.
3. Blank using only reagents without sample powder is also run simultaneously with every set of ten determinations.
4. After cooling the digest, 2 to 3 ml of distilled water is streamed along the sides of the flask and the contents are mixed.
5. The flask is closed with a rubber stopper and allowed to cool on a wooden stand before distillation.
6. The total contents can be taken for distillation in a Kjeldahl distillation apparatus or made up to a suitable volume and aliquots taken for distillation in a micro Kjeldahl distillation assembly.

Total N

1. Weigh about 0.4 g of the seed / plant sample into a Kjeldahl flask. Add 100 ml concentrated H_2SO_4. Add 1ml 10% Cu SO_4 and 1g K_2 SO_4. The flask is heated in a fume hood until coplete digestion takes place. The flame is removed and the flask is allowed to cool for 5 minutes.
2. The flask is closed with a rubber stopper and allowed to cool on a wooden stand before distillation. The total contents can be taken for distillation in a Kjeldahl distillation apparatus or made up to a suitable volume and aliquots taken for distillation in a micro Kjeldahl distillation assembly.

Distillation

1. 10 ml of above digested sample is put in a distillation chamber. Add 15-20 ml of 30 % NaOH. In a 100 ml of conical flask take 5ml of 2% boric acid and add a few drops of Conway indicator.
2. Start distillation and collect 20-25 ml of distillate into boric acid in the conical flask.
3. Titrate the collected distillate with N/28 HCl. 2 to 3 ml of distilled water is streamed along the sides of the flask and the contents are mixed prior to titration.

Calculation

1ml of N/28 HCl = 0.5 mg N

$$\% \text{ Nitrogen in the sample} = \frac{5 \times T \times 100}{(A\,D\,M) \times (100-M)}$$

Where T= Titre value (of N/28 HCl) – blank.

A D M = Air dry material of plant sample taken.

M = % moisture in the plant sample

Soluble protein

Principle

The sample is extracted with dilute sidium hydroxide (0.2%) and the protein estimated by Kjeldahl procedure.

Reagents and equipment

1. 0.2% Na OH.
2. Centrifuge .
3. Kjeldahl set up (digestion and distillation).
4. Flask shaker.

Procedure

1. Take 12.5g finely powdered sample
2. Add 100ml 0.2% Na OH.
3. Shake for one hr.
4. Pipet 10ml of clear liquid.
5. Add 1ml conc. H_2SO_4.
6. The flask is heated in a fume hood until fumes appear. The flame is removed and the flask is allowed to cool for 5 minutes. A small amount of catalyst is added and the digestion continued.
7. Blank using only reagents without sample powder is also run simultaneously with every set of ten determinations.
8. After cooling the digest, 2 to 3 ml of distilled water is streamed along the sides of the flask and the contents are mixed.

9. The flask is closed with a rubber stopper and allowed to cool on a wooden stand before distillation.

10. The total contents can be taken for distillation in a Kjeldahl distillation apparatus or made up to a suitable volume and aliquots taken for distillation in a micro Kjeldahl distillation assembly.

11. 10 ml of above digested sample is put in a distillation chamber. Add 15-20 ml of 30 % NaOH. In a 100 ml of conical flask take 5ml of 2% boric acid and add a few drops of Conway indicator.

12. Start distillation and collect 20-25 ml of distillate into boric acid in the conical flask.

13. Titrate the collected distillate with N/28 HCl. 2 to 3 ml of distilled water is streamed along the sides of the flask and the contents are mixed prior to titration.

Calculation

1ml of N/28 HCl = 0.5 mg N

$$\% \text{ Nitrogen in the sample} = \frac{5 \times T \times 100}{(A\,D\,M) \times (100-M)}$$

Where T = Titre value (of N/28 HCl) – blank.

A D M = Air dry material of plant sample taken.

M = % moisture in the plant sample.

Protein content (Lowry *et al.*, 1951)

Principle

The most widely used colorimetric method for estimation of proteins is that of Lowry *et al*, (1951), using the Folin reaction. The blue color is due to the reaction of protein with copper ion in alkaline solution (the biuret reaction) and the reduction of the phosphomolybdate- phosphotungstic acid in the Folin reagent by the aromatic amino acids in the treated protein.

Reagent and equipment

1. Reagent A: 2% Na_2CO_3 (sodium carbonate) in 0.1 N NaOH.
2. Reagent B: 0.5% $CuSO_4$ $5H_2O$ in 1% sodium citrate.

3. Reagent C: 1 ml of reagent B mixed with 50 ml of reagent A.
4. Reagent D: Folin Ciocalteu reagent: Into a 1500 ml Florence flask introduce 100g of sodium tungstate ($Na_2 WO_4 2H_2O$), 25 g of sodium molybdate($Na_2 MoO_4 2H_2O$), 700 ml of water, 50 ml of 85 percent phosphoric acid, and 100 ml of concentrated hydrochloric acid, and reflux gently for 10 hr. Add 1 of lithium sulphate, 50 ml of water and a few drops of bromine. Boil the mixture for 15 min without condenser, to remove excess bromide. Cool, dilute to 1 litre, and filter. The reagent should not have greenish tinge. Protect from dust. Dilute a portion with an equal volume of water before use.
5. Standard solution: Bovine serum albumin is used for preparing a standard curve. Weigh 100 mg bovine serum albumin, dissolve in 1N NaOH and make upto 100 ml in a volumetric flask. Take 10 ml and dilute to 100 ml in 1N NaOH.
6. Centrifuge.
7. Spectrophotometer with quartz cuvettes.

Procedure

a. Weigh 12.5 g of finely powdered plant/seed / cake(defatted) sample. Add 250 ml of 0.2% NaOH. Shake for one hr. Centrifuge 50 ml of the suspension.
b. Take 5 ml and dilute to 100 ml with distilled water.
c. To 0.5 ml of the diluted protein solution, 5 ml of reagent C is added.
d. The solution is allowed to stand for 10 min at room temperature and 0.5 ml of reagent D is pipetted rapidly into it with thorough mixing.
e. The absorbance is read at 520 nm after 30 min.
f. The protein content of the homogenate is then calculated from the absorbance using the standard curve.

Standard curve

g. Take 0.1, 0.2, 0.4, 0.6, 0.8, 1.0, 1.5 and 2.0 ml of diluted bovine serum albumin solution (10-200 ppm). Add 5 ml of reagent C.
h. The solution is allowed to stand for 10 min at room temperature. 0.5 ml of reagent D is pipetted rapidly into the mixture with thorough mixing.
i. The absorbance is read at 520 nm after 30 min.

j. The standard curve is prepared by plottting the concentration on X axis and absorbance on the Y axis.

Calculation

$$\% \text{ Protein} = \frac{\text{ppm X } 2.5}{\text{W } (100\text{-M})}$$

Where

ppm = value read from standard graph, W = weight of sample, M = moisture %

Note

1. *When proteins are insoluble in dilute alkali, it is necessary to heat the suspension at 90°C in 1 N NaOH for 10 min to obtain complete solubilization. The standard should be treated in the same way because this treatment reduces the intensity of the color. When this procedure is applied to dissolve or extract proteins, a reagent containing 20 g of* Na_2CO_3 *in 1 liter of water (no alkali) is substituted for reagent A. The protein analysis procedure is otherwise identical.*
2. *The same quantities of various proteins give different readings with this reagent because the colour depends in part on the aromatic residues in the protein molecule. The standared curve should be prepared with samples of known concentration of the protein of interest for reliable results.*

Reference

Lowry. O.H., Resebrough, N.J., Farr, A.L. and Randall, R.J. 1951. Protein Measurement with Folin-Phenol Reagent. Journal of Biological Chemistry. 193:265.

Protein content (biuret method)

Principle

Peptide bonds of proteins form blue-colored chelates with copper ions in alkaline solutions. This reaction is much less sensitive than the Lowry procedure, but has the advantage that it is less time-consuming and can be used in the presence of amonium salts and other materials that interfere with the Lowry procedure. There is also less variability between proteins in the amount of color formed per unit weight of protein.

The procedure involves the use of tartrate to form a copper complex that is soluble in NaOH. The protein displaces the copper ion from the complex to form a copper protein complex with a different colour and absorption intensity.

Reagents and equipment

1. Biuret reagent: Dissove 1.5 g of cupric sulfate ($CuSO_4 5H_2O$) and 6 g of sodium potassium tartrate, $4H_2O$ in 500 ml of water. Add with stirring, 300 ml of a 10% (wt/vol) solution of sodium hydroxide in water. Adjust the volume of the mixture to 1 litter with distilled water. Discard if a reddish or black precipitate appears.
2. 0.2% NaOH solution.
3. Ultraviolet spectrophotometer with cuvettes of quartz or silica.
4. Incubator.
5. Standard solution: Bovine serum albumin is used for preparing a standard curve. Weigh 100 mg bovine serum albumin, dissolve in 1N NaOH and make upto 100 ml in a volumetric flask. Take 10 ml and dilute to 100 ml in 1N NaOH.

Procedure

1. Weigh 12.5 g of finely powdered plant/seed / cake(defatted) sample. Add 250 ml of 0.2% NaOH. Shake for one hr.
2. Centrifuge 50 ml of the suspension.
3. Take 5 ml and dilute to 100 ml with distilled water. This dilute protein extract can be used for further analysis.
4. Add 4.0 ml of the above biuret reagent to a neutralized sample (1.0ml) containing 1 to 10 mg of protein. Mix thoroughly, and incubate at room timperature for 30 min. Time and temperature for all samples, standards, and blanks should be the same.
5. Measure the absorbance at 500 nm.
6. A blank is prepared using 1.0 ml of water or aqueous solution used to dissolve proteins and 4.0 ml of biuret reagnt.
7. A standard curve can be prepared using any completely soluble protein such as crystallinc bovinc scrum albumin. The response is not linear with protein concentration because of competition between the tartrate and protein for copper ions. Take 10-200 ppm of the standard solution and add 4ml of biuret solution. Mix thoroughly and mesure the aborbance at 500 nm. Plot a of graph of concentration of protein vs absorbance.

8. From the concentration read from the standard curve, calculate the % preteins in the sample from the concentration read from the graph. Take into dilutions etc. made.

Protein content (by ultraviolet absorption)

Principle

The absorption maximum of proteins is 280nm and is due to the presence of the aromatic amino acids, tyrosine and tryptophan which are present in all proteins. The absorption of ultraviolet light is a suitable means of estimating proteins in solution if the protein solution does not contain more than 20% by weight of other ultraviolet light absorbance compounds, such as nucleic acids or phenols. The solution also should not be turbid.

Reagent and equipment

1. Reagent A: 0. 2% NaOH.
2. Ultraviolet spectrophotometer. Calibrate with an absorbance standard.
3. Cuvettes: Quartz or silica, which do not prevent transmitance of ultraviolet light.
4. Centrifuge.

Procedure

1. Weigh 12.5 g of finely powdered plant/oilseed / cake (defatted) sample. Add 250 ml of 0.2% NaOH. Shake for one hr. Centrifuge 50 ml of the suspension.
2. Take 5 ml and dilute to 100 ml with distilled water. This dilute protein extract can be used for further analysis.
3. Absorbance measurements are made at 280 nm and 260 nm with the protein dissolved in a suitable buffer.
4. A solvent blank containing the buffer is used to zero the instrument at each wavelength.
5. If the ratio of absorbance at 280 nm to absorbance at 260 nm is not greater than 1.70, except with solutions known to contain pure protein, the following equation can be used to calculate the concentration of protein.

Calculation

Protein concentration (mg/ml) = 1.45 A_{280} – 0.74A_{260}

Protein content (Coomassie blue method)

Principle

Coomassie dye(blue) turns blue when the dye binds to amino groups of proteins. This procedure is more sensitive and can be performed more rapidly.

Reagents and equipment

1. Coomassie blue reagent: Coomassie blue G250 is dissolved in 50 ml of 95% ethanol, add 100ml of 85% phosphoric acid, make up to one litre.
2. 0.2% NaOH solution.
3. Centrifuge.
4. Spectrophotometer.
5. Standard solution: Bovine serum albumin is used for preparing a standard curve. Weigh 100 mg bovine serum albumin, dissolve in 1N NaOH and make upto 100 ml in a volumetric flask. Take 10 ml and dilute to 100 ml in 1N NaOH.

Procedure

1. Weigh 12.5 g of finely powdered plant/oilseed /cake (defatted) sample. Add 250 ml of 0.2% NaOH. Shake for one hr. Centrifuge 50 ml of the suspension.
2. Take 5 ml and dilute to 100 ml with distilled water. This dilute protein extract can be used for further analysis.
3. Take a solution of protein (10-100µg) in 0.1ml in a test tube. Add 5ml of the Coomassie blue reagent and thoroughly mix the contents.
4. Prepare a blank without protein, but 0.2% NaOH diluted as above and 5ml Coomassie blue reagent.
5. Measure absorbance at 595nm after 2min (preferably within one hr) against the blank prepared above.
6. The protein content of the sample is measured from a standard graph prepared using bovine serum albumin solution.

7. Calculations are made based on the concentration read from the graph and weight of sample and dilutions made.

Protein assay (by SDS - PAGE)

Principle

The SDS-polyacrylamide gel electrophoresis (SDS-PAGE) is a versatile technique for the characterization of proteins both quantitatively and qualitatively. It is an easy and quick method to quantify a particular protein at microgram level from a mixture. This is done by scanning the gel (strips) and by densitometry of stained bands on it. Similarly, radioactivity-labeled polypeptides electrophoresed on SDS-PAGE can also be quantitated by densitometric scanning of the fluorographic plate.

The percent absorption of incident light is directly proportional to the color intensity of the protein-dye complex on the gel and is directly related to the protein concentration. Similarly, the intensity of darkening of the X-ray plate is directly proportional to the radioactivity in the protein in fluorographic plates.

Reagents and equipment

1. A spectrophotometer with suitable scanning facility and chart recorder (or integerator facility).
2. Protein Stain (Quantitative): 0.2% Proceion Navy MXRB dye in Methanol : Acetic Acid : Water (5:1:4). Dissolve the dye first in the methanol and then proceed. Prepare fresh everytime.
3. Destaining Solution: Methanol : Acetic Acid : Water (1:1:8).
4. For Fluorograph Scan: Fluorograph Plate.
5. Electrophoresis apparatus.
6. Bovine serum albumin standard: 100ppm solution in 1N NaOH.

Procedure

Protein species regardless of their size has an equivalent charge density and is driven through gel with the same force. However, because the poly acrylamide is highly cross linked, larger proteins are held up to a greater degree than the smaller proteins.

1. After the electrophoresis, immerse the gel in Proceion Navy dye solution and shake until the proteins are completely stained (for a period of 2h).

2. Destain the gel until the background is colorless.
3. Scan the gel at 580nm to measure the degree of dye bound by each band of protein. Depending upon the type of equipment available for scanning, the whole gel is used or each lane is cut out and scanned individually. The total absorption by the dye in each band is proportional to the area of the peak in the scan profile.
4. Each peak in the scan profile is traced using a planimeter to determine the area under it. Otherwise, each peak in the chart may be cut out and weighed. When an integrator is interposed, the area under each peak is automatically calculated.
5. A curve is obtained by plotting A_{580} *vs* amount of protein used as standard. Bovine serum albumin at different known concentrations co-electrophoresed in different lanes in the same gel is also used to construct the standard curve. It should however, be noted that the protein both in the standard and under examination to have equal dye-binding property.
6. Scanning Fluorographic Plate: Scan the individual lane strip or the whole fluorographic plate at 620nm as described above. The standard curve is obtaining using a radioactively labeled standard protein whose concentration and radioactivity are known.

References

Carlier, A R, Manickam, A and Peumans, W. J. (1980) Planta 149. 227.

Smith, B J, Toogood, C and Johns, E. W. (1980) J Chromatogr. 200. 200.

Casein content of milk

Principle

Casein is insoluble at its isoelectric point. The protein can be precipitated by adjusting the p^H and the casein estimated gravimetrically.

Reagents and equipment

1. Milk.
2. 10% acetic acid.
3. 1M sodium acetate.
4. 95% ethanol.

5. Ether.
6. Centrifuge.
7. Incubator.

Procedure

1. Take 30ml milk and 30ml water in a 250ml beaker and warm to 40°C.
2. Add dropwise while stirring 2ml of 10% acetic acid.
3. Allow the mixture to stand for 5min and then add 1M sodium acetate buffer.
4. Cool and allow to stand for 5min.
5. Centrifuge and wash the preicipitate with a small amount of water. Re-cetrifuge if needed.
6. Suspend the precipitate in 10ml 95% alcohol and centrifuge.
7. Wash the precipitate with 10ml ether.
8. Remove ether either by gentle heating or by pessing between sheets of filter paper.
9. Weigh the dry casein and calculate the percentage.

Note

Assume density of milk as 1.0 g/cc. The casein content of milk is roughly around 3.5%

NH_4SO_4 fractionation of proteins

The solubility of proteins is markedly affected by the ionic strength of the medium. As the ionic strength is increased, protein solubility at first increases. This is referred to as 'salting in'. However, beyond a certain point the solubility begins to decrease and this is known as 'salting out'.

At low ionic strengths the activity coefficients of the ionizable groups of the proteins are decreased so that their effective concentration is decreased. This is because the ionizable groups become surrounded by counter ions which prevent interaction between the ionizable groups. Thus protein-protein interactions are decreased and the solubility is increased.

At high ionic strengths much water becomes bound by the added ions that are not enough remains to properly hydrate the proteins. As a result, protein-protein interactions exceed protein-water interactions and the solubility decreases. Because of differences in structure and amino acid sequence, proteins differ in their salting in and salting out behavior. This forms the basis for the fractional precipitation of proteins by means of salt.

Ammonium sulphate is a particularly useful salt for the fractional precipitation of proteins. It is available in highly purified form, has great solubility allowing for significant changes in the ionic strength and is inexpensive. Changes in the ammonium sulphate concentration of a solution can be brought about either by adding solid substance or by adding a solution of known saturation, generally, a fully saturated (100%) solution.

Free amino acid content (With ninhydrin reagent)

Principle

Protein free amino acids are made to react with ninhydrin and the colour complex is estimated colorimetrically (Yemm and Cooking, 1954).

Reagents and equipment

1. Leucine standard: Dissolve 100 mg of leucine in distilled water and make upto 100 ml in a volumetric flask. Dilute this stock solution 1:10 to get 100 ppm solution.
2. Citrate buffer (pH5.0): Weigh 21g citric acid and dissolve in 1N NaOH and make upto 500 ml with distilled water. This solution could be stored in the cold with a thymol crystal added to prevent bacterial growth.
3. Ninhydrin reagent: Dissolve 4.0g of ninhydrin in 100 ml of 2-methoxy ethanol.
4. $SnCl_2$: Dissolve 800mg of hydrated stannous chloride in 500 ml of citrate buffer.
5. Diluent : Isopropanol : Distilled water (1:1).
6. Spectrophotometer with quartz cuvettes.

Procedure

Standard curve

1. Take 9 test tubes and add 0.0, 0.1, 0.2, 0.3, 0.4, 0.5, 0.6, 0.8, and 1.0ml of 100 ppm leucine solution. Add 1.0, 0.9, 0.8, 0.7, 0.6, 0.5, 0.4, 0.2, and 0.0 ml of citrate buffer to the above tubes.

2. Add 0.5 ml of ninhydrin reagent (equal proportions of reagents 3 and 4 above mixed just before use). Shake well, heat on a boiling water bath for 20 min. Make up to 5ml with diluent.
3. Cool and measure the absorbence at 570 nm in a spectrophotometer. Plot a graph of the readings against concentrations in ppm.

Extraction of amino acids

1. Weigh a suitable quantity of the plant sample (0.1g kernel/ seed or defatted meal or 0.5 g of leaf powder). Macerate the sample in 10ml of 75-80% aqueous ethanol.
2. Filter and re-extract the residue twice more with 5ml portions of aqueous ethanol and pool all the filtrates. Make up to a suitable volume (20 or 25ml).

Amino acid estimation

1. Take 0.5 ml of the extract into a test tube. Add 1 ml citrate buffer and 0.5 ml of ninhydrin reagent (equal portions of reagents 3 and 4 above, mixed just before use).
2. Shake well and heat to boiling on a water bath for 20 min. Make up to 5ml with diluent.
3. Cool and measure the absorbence at 570 nm in a spectrophotometer. Read the concentration from the standard curve prepared above.

Calculation

Make the calculations based on the quantity of sample taken and dilution of the extracts using the concentration value obtained from the standard curve.

Reference

Yemm, E.W. and Cooking, E.C. 1954. The Determination of Amino Acids with Ninhydrin. Analyst. 80: 209-13.

Available lysine

Principle

Lysine is the limiting amino acid in the cereal grains. Assessment of lysine in cereal grains for nutritional quality is hence one of the procedures adopted for screening varieties. FDNB (1fluoro 2,4-dinitrobenzene) reagent reacts with lysine producing a colour. It is measured at 435nm.

Reagents

1. FDNB Reagent: 2.5%(v/v) solution in ethanol. Prepare afresh.
2. Ethanol.
3. Ether .
4. 8.1N HCl.
5. 1N HCl.
6. 8% $NaHCO_3$.

Procedure

1. Sample containing 30-50 mg protein is shaken gently with 8ml 8% $NaHCO_3$ for 10min.
2. FDNB reagent 0.3 ml in 12 ml ethanol is added. Shake for 2hrs.
3. Evaporate ethanol on water bath.
4. Add 24 ml 8.1N HCl to the residue and reflux for 16hr on a sand bath.
5. Filter and wash with water. Dilute an aliquot such that 2 ml has 50μg available lysine.
6. Take 2 ml in a graduated stoppered tube, add 5 ml ether, suck off ether.
7. Remove ether on water bath, make up to10 ml with 1N HCl.
8. Meaure absorbency at 435 nm.
9. Blank: 1ml dilute solution, extract with ether, neutralize with NaOH, add 2ml carbonate buffer at p^H 8.5 , also add 0.05 ml methyl chloroformate and 0.75 ml conc HCl. Extract twice with ether, remove ether over water bath and make up to 10ml. measure at 435nm.
10. Also measure FDNB lysine of known concentration at 435nm.

Calculation

Available lysine/ 16g N =

$$\frac{\text{absorbancy of sample x 39.85 x 125}}{\text{absorbancy of standard x 30 x 0.5 x protein in sample}}$$

Estimation of lysine

Principle

Though an ideal simple method for direct estimation of lysine is yet to be found, the given procedure is sufficient for routine screening. The protein in the grain sample is hydrolyzed with a proteolytic enzyme, papain. The alpha-amino group of the derived amino acids are made to form a complex with copper. The e-amino group of lysine which does not couple with copper is made to form e-dinitropyridyl derivative of lysine with 2-chloro-3, 5-dinitropyridine. The excess pyridine is removed with ethyl acetate and the color of e-dinitropyridyl derivative is read at 390 nm.

Reagents

1. Solution A: 2.89 $CuCl_2.2H_2O$ in 100 ml water.
2. Solution B: 13.6g $Na_3PO_4.12H_2O$ in 200 ml water.
3. Sodium Borate Buffer (0.05M pH 9.0).
4. Copper Phosphate Reagent: Pour solution A (100 ml) to B (200 ml) with swirling, centrifuge and discard the supernatant. Re-suspend the pellet three times in 15 ml borate buffer and centrifuge after each suspension. After third washing resuspend the pellet in 80 ml of borate buffer. The reagent must be prepared fresh every seven days.
5. 3% solution of 2-Chloro-3, 5-Dinitropyridine in methanol. Prepare afresh just prior to use.
6. 0.05M Sodium Carbonate Buffer (pH 9.0).
7. Amino Acid Mixture: Grind in a motor 30 mg alanine, 50 mg glutamic acid, 60 mg aspartic acid, 20 mg cystein, 300 mg glutamic acid, 40 mg glycine, 30 mg histidine, 30 mg isoleucine, 80mg leucine, 30 mg methionine, 40 mg phenylalanine, 80 mg proline, 50 mg serine, 30 mg threonine, 30 mg tyrosine and 40 mg valine. Dissolve 100 mg of this mixture in 10 ml of sodium carbonate buffer (0.05M, pH 9.0).
8. Dissolve 400mg technical grade papain in 100ml 0.1M sodium acetate buffer (pH 7.0).
9. 1.2N HCl.
10. Ethyl acetate.

Procedure

1. To 100mg of defatted grain sample add 5ml of papain solution and incubate overnight at 65°C. Cool to room temperature, centrifuge and decant the clear digest.
2. To one ml digest taken in a centrifuge tube add 0.5ml carbonate buffer and 0.5ml copper phosphate suspension.
3. Shake the mixture for 5min in a vortex mix and centrifuge.
4. To one ml supernatant add 0.1ml pyridine reagent, mix well and shake for 2 hours.
5. Add 5ml of 1.2N HCl and mix.
6. Extract three times with 5ml ethyl acetate and discard the ethyl acetate (top) layer.
7. Read the absorbance of aqueous layer at 390nm.
8. Prepare a blank with 5ml papain alone repeating steps 1 to 7.
9. Dissolve 62.5mg lysine monohydrochloride in 50ml carbonate buffer (1mg lysine/ml). Pipette out 0.2, 0.4, 0.6, 0.8 and 1ml and make up to 1ml with carbonate buffer. Add 4ml papain to each tube and mix, pipette out one ml from each and add 0.5ml of amino acid mixture and 0.5ml of copper phosphate suspension. Carry out steps 3 to 7. The standard curve represents absorbance values for 40, 80, 120, 160 and 200mg lysine.

Calculation

Prepare a standard curve from the readings of the standard lysine. Subtract the absorbance of the blank from that of the sample and calculate the lysine content in the aliquot from the graph.

Lysine content of the sample (g per 16g N) =

$$\frac{\text{Lysine value from graph in mg x 0.16}}{\text{\% N in the sample}}$$

Reference

Mertz, E T, Jambunathan, R and Misra, P S (1975) In: Protein quality, Agric. Experiment Stn. Bull No. 70 Purdue Univ. USA p 11.

Theymoli, Balasubramanian and Sadasivam, S (1987) Plant Foods Hum Nutr 37 41.

Cystine content

Principle

The sulphur in cystine is reduced by hydrazine hydrate to hydrogen sulphate. The H_2S formed is converted to bismuth sulphide which is measured colorimetrically.

Reagents and equipment

1. Hydrazine hydrate 99-100% purity.
2. Acetate buffer 0.2N: Dissolve 19.5g sodium acetate in water and add 30.5ml acetic acid. Make up to 100ml.
3. Bismuth nitrate solution: Dissolve 2.2g Bismuth nitrate($5H_2O$) in 250ml of 3.2% solution of mannitol in water. Add 80ml glycerol and 360ml of 2.5% solution of gum arabic in water. Dilute to one litre with 0.2N acetate buffer (p^H 4.4) and filter.
4. Sulphuric acid 6N: Dilute 167ml conc H_2SO_4 in water and make up to one litre.
5. Balance.
6. Nitrogen gas (cylinder).
7. Hot air oven.
8. Spectrophotometer.
9. Glass manifold.

Procedure

1. Weigh 40-50mg defatted plant/ seed sample and transfer to a screw cap tube.
2. Add 1ml distilled water and 2ml hydrazine hydrate.
3. Make the final concentration of sample such that it is 50-100% (w/v).
4. Run cystine standard 0-50µg.
5. Hydrolyze the tubes at 120^{0}C for 18hr in an oven.
6. Cool and then add 1ml water and connect to glass manifold.
7. Arrange a flow of nitrogen gas.
8. Add 5ml H_2SO_4 that generates H_2S gas.

9. Trap the H_2S gas into bismuth nitrate solution for 15-20min.
10. Measure the colour of bismuth sulphide at 400nm.
11. Calculate cystine concentration from the standard graph prepared.
12. Express the values as g/16gN.

Calculation

$$\mu g \text{ cystine } g/16g\ N = \frac{\text{Absorbanc}(\text{graph}) \text{ x } 100}{\text{sample wt x } 1000}$$

Methionine content (Nitroprusside method)

Principle

The method is based on nitroprusside-methionine reaction. The colour developed is measured at 520 nm. The amino acid is extracted into tris buffer and the estimation is carried out.

Reagents and equipment

1. Glycine 1% solution (1g in 100ml water).
2. Sodium nitroprusside 1% solution (1g in 100ml water).
3. Orthophosphoric acid 88%, standard grade.
4. 1.0M Tris buffer p^H 7.2: Dissolve 121.4 g trishydroxymethyl aminomethane in water and add 83 ml HCl and make up to one litre.
5. Papain.
6. 0.1N HCl: Mix 8.3 ml conc. HCl and make up to one litre.
7. 5N NaOH: Dissolve 200 g NaOH in water and make up to one litre.
8. Balance.
9. Water bath.
10. Vortex mixer.
11. Spectrophotometer.
12. Incubator.

Procedure

1. Weigh 750 mg of defatted plant sample into a test tube.
2. Add 10 ml Tris buffer.
3. Add 100 mg papain.
4. Incubate at 55°C for 1hr.
5. Remove and add 4-5 drops of orthophosphoric acid to stop enzymatic hydrolysis.
6. Transfer to a 25 ml volumetric flask and make up to volume.
7. Shake thoroughly and filter through No.41 filter paper into a test tube.
8. Pipet out 5 ml into a test tube.
9. Add 1ml of 5N NaOH followed by 1ml 1% glycine and 1ml 1% sodium nitroprusside.
10. Run every set of samples with a reagent blank and a standard check sample.
11. After shaking, keep tubes in a water bath at 40°C for 10min.
12. Cool the contents and add 6ml 88% orthophosphoric acid.
13. Shake thoroughly to get red colour.
14. Cool to room temperature and read at 520nm in a spectrophotometer.
15. Prepare methionine standards 0.5, 1.0, 1.5, 2.0, 2.5 mg/ml and develop colour as asbove. Prepare a curve. Read sample concentration.
16. Express the values g/100 g protein.

Proline content (Ninhydrin procedure)

Principle

Proline, a basic amino acid has an important role in plant stress. Proline is extracted by sulphosalicylic acid and made to react with ninhydrin to estimate its content.

Reagents and equipment

1. Ninhydrin reagent: Warm1.25g ninhydrin in 30ml glacial acetic acid and 20ml 6M phosphoric acid to dissolve the reagent. Store at 4°C and use it on the same day.

2. 3% aqueous salicylic acid.
3. Glacial acetic acid.
4. Proline.
5. Toluene.
6. Water bath.
7. Ice bath.
8. Spectrophotometer with cuvettes.

Procedure

1. Extract 0.5g of the plant sample by grinding in 10ml 3% aqueous salicylic acid.
2. Filter through No.2 filter paper.
3. Take 2ml filtrate in a test tube and add 2ml glacial acetic acid and 2ml ninhydrin reagent.
4. Heat in a water bath for 1hr at 100^{0}C.
5. Terminate reaction by putting the test tube in an ice bath.
6. Add 4ml toluene and stir well for 20-30 sec.
7. Separate the toluene layer and warm the contents to room temperature.
8. Measure the red colour at 520nm.
9. Prepare a standard graph using proline solutions of appropriate concentration.
10. Read the sample concentration from the graph based on the absorbance.

Calculation

$$\mu \text{ moles } \frac{\text{proline}}{\text{g}} \text{ sample} =$$

$$\mu\text{g proline x ml toluene} \frac{\text{x5}}{\text{115.5x wt of sample}}$$

where 115.5 is the molecular weight of proline

Reference

Bates, L S, Waldeen, R P and Teare, I. D. 1973: Plant Soil 39: 205

Tryptophan content

Principle

The indole ring of tryptophan gives an orange red colour with ferric chloride under acidic conditions. It is measured at 545nm. Determination of this essential amino acids is useful in identifying superior grain types.

Reagents and equipment

1. Papain solution: Dissolve 400mg papain in 100ml 0.1N sodium acetate buffer at p^H 7.0. Prepare solution afresh.
2. Reagent A: Dissolve 135mg $FeCl_3$ $6H_2O$ in 0.25ml water and dilute to 500ml with glacial acetic acid containing 2% acetic anhydride.
3. Reagent B: 30NH_2SO_4 for N determination.
4. Reagent C: Mix equal volumes of 2 and 3 (reagents A & B), one hr prior to use.
5. Standard tryptophan: Dissolve 5mg tryptophan in 100 ml water (50µg/ml).
6. Incubator.
7. Vortex mixer.
8. Spectrophotometer with cuvettes.

Procedure

1. Weigh 100 mg defatted plant sample into a small test tube.
2. Add 5 ml papain solution and shake well.
3. Incubate at 65^{0}C overnight.
4. Cool the contents to room temperature, centrifuge and collect the clear supernatant.
5. Take one ml supernatant and add 4ml reagent C.
6. Mix in a vortex mixer and incubate at 65^{0}C for 15 min.
7. Cool and measure the colour at 545 nm.
8. Run a blank side by side with papain alone.
9. Standard tryptophan: Pipet 0, 0.2, 0.4, 0.6, 0.8, and 1.0 ml standard tryptophan. Make upto one ml. Develop colour by following steps 5-7.

Calculation

Tryptophan content in the sample g/16 g N =

$$\frac{\text{tryptophan conc. as read from graph in } \mu g \times 0.096}{\% \text{ N in the sample}}$$

Reference

Balasubramanian, T. and Sadasivam, S.1987. Plant Foods Hum Nutr. 37: 41

Protein digestibility (*In Vitro*)

Principle

Pepsin enzyme is added to protein sample to simulate the digestion. After a specific digestion period, the enzyme activity is arrested and the decrease in protein content is used to measure the digestibility.

Reagents and equipment

1. Pepsin: Dissolve 40mg pepsin in 0.1N HCl and make upto 100 ml.
2. 10%Trichloroacetic acid (TCA): Dissolve10g trichloroacetic acid in 100 ml water.
3. Trichloroacetic acid (5%) : Dissolve 5g trichloroacetic acid in 100ml water.
4. Borate buffer: a)Boric acid (0.2M) : Dissolve 12.4 g boric acid in water and make upto 1 litre. b) Borax solution (0.05M): Dissolve 19.05g borax in water and make upto 1 litre. Borate buffer is prepared by adding 140ml boric acid (0.2M), 50 ml D.W. and a few drops of borax (0.05M) to adjust the pH to 6.8.
5. Borate buffer (0.1M, pH 6.8) containing 0.025 M calcium chloride is prepared by dissolving 3.6755 g calcium chloride in 1 litre of the buffer.
6. Pancreatin: Dissolve 50mg pancreatin in 10 ml of 0.1 M borate buffer (pH6.8) containing calcium chloride.
7. Centrifuge.
8. Water bath.
9. Incubator.

Procedure

1. Two hundred mg of dried plant/seed sample is taken and incubated with 5ml of pepsin at 37°C for 16 hrs in a water bath shaker. A few drops of toluene is added to check the growth of microbes.
2. After neutralisation with 0.2N sodium hydroxide and addition of 2ml of panceratin, the digestion mixture is incubated for an additional 24 hrs at 37°C in a waterbath shaker.
3. Enzyme blanks are prepared by incubation under the described conditions with the protein sample omitted.
4. At the end of the incubation period the enzymes are inactivated by addition of 7ml of 10% TCA.
5. The contents are centrifuged at 12,000 rpm for 20 min.
6. The supernatants are pooled and volume made upto 25ml with 5% TCA.
7. Five ml aliquot is taken and dried at 80-90°C in a hot air oven and the nitrogen content is determined (Kjeldahl method or any other procedure).
8. The protein content is obtained by multiplying the nitrogen content with 6.25.

Calculations

$$\text{Protein digestibility (\%)} = \frac{\text{Digested protein}}{\text{Total protein}} \times 100$$

Amino acid analysis (Automatic Analyser)

Principle

Amino acids actually account for the real quality of proteins in any sample. The protein is converted to amino acids by hydrolysis with 6N HCl. The amino acids are passed through an ion exchange column in an automatic amino acid analyser. The colour developed with the ninhydrin reagent is measured to estimate the quantity of the separated individual amino acids (in successive eluted fractions).

Reagents and equipment

1. Balance.
2. Flash evaporator.

3. Refluxing unit.
4. P^H Meter.
5. Automatic amino acid analyser.
6. P^H 3.25 - 0.20 N sodium citrate (1% iso propanol added).
7. P^H 3.95 - 0.40 N sodium citrate (1% iso propanol added).
8. P^H 6.40 - 1.00 N sodium citrate (1% iso propanol added).

Preparation of sodium citrate buffers for amino acid analysis

P^H	Buffer conc. (N)	Sodium Citrate $2H_2O$(g)	NaCl (g)	Conc. HCl (ml)	Thiodiglycol (ml)	Octanoic Acid (ml)	Final volume (ml)
2.20	0.2	19.6	-	16.5	5.0	0.1	1000
3.25	0.2	78.4	-	50.3	10.0	0.4	4000
3.95	0.4	78.4	46.8	35.2	10.0	0.4	4000
6.40	1.0	78.4	187.2	1.0	-	0.4	4000
4.25	0.2	78.4	-	33.5	10.0	0.4	4000
5.26	0.158	61.98	-	11.8	-	0.4	4000

9. 0.2N NaOH: Dissolve 8g NaOH pellets in deionized water and make upto 1litre.
10. Sodium acetate buffer: a) add 2720g of sodium acetate, $3H_2O$ in 2.5 litre deionized water in a beaker. Stir with water to dissolve the salt. b) transfer to 5 litre volumetric flask and add 500ml glacial acetic acid. c) cool to room temperature and make up to 5 litres. d) filter the buffer through a 0.4 to 0.6μm filter and store at room temperature.
11. Ninhydrin reagent: Dissolve 76g ninhydrin in 3.8 litre solvent (950ml sodium acetate buffer + 2.85 litre methyl cellosolve). Add one vial (7.5ml) titanous chloride. Keep the bottle under nitrogen atmosphere.
12. Standard amino acids: Dissolve 46.8mg cysteic acid, 41.4mg methionine sulfoxide, 45.3mg methionine sulfone and 32.795 mg norleucine together in 0.2N sodium citrate buffer P^H 2.2 and make up to 100ml with the buffer. This gives a conc. of 2.5μmoles/ml of each amino acid. Add 5ml of commercial standard to 5ml of the above solution and make upto 25ml with 0.2N sodium citrate buffer P^H 2.2.

Procedure

Sample preparation

1. Weigh 50mg defatted sample into a 150ml flask and add 50ml 6N HCl.
2. Reflux for 24hr, cool and flash evaporate the acid. Repeat thrice with washings and dry completely.
3. Dissolve the contents in 0.2N sodium citrate buffer, P^H 2.2 and add 5μmoles/ml norleucine(internal standard). Cereals: 4.5ml buffer+ 0.5ml norleucine and for legumes: 9.0ml buffer + 1.0ml norleucine.
4. Filter through Whatman No.41 filter paper and use 0.1ml for analysis in the amino acid analyzer.

Calculation

Run the standards along with samples (0.5 μmoles/ml). In the case of protein hydrolyzate, the amino acids are eluted in the following order – cysteic acid, methionine sulfoxide, aspartic acid, methionine sulfone, threonine, serine, glutamic acid, proline, glycine, alanine, half cystine, valine, methionine, isoleucine, leucine, norleucine, tyrosine, phenylalanine, histidine, lysine, ammonia, arginine.

Each amino acid concentration is calculated separately

For example the calculation of aspartic acid is furnished below:

Wt of sample = 50mg

Sample volume = 10ml

Protein % in sample = 25

Peak area of sample = 1.8651

Peak area of standard = 0.92

Concentration of standard = 50 nano moles

Concentration of sample (aspartic acid) =

$$1.8651 \times \frac{50}{0.92} = 101.36 \textit{ nano moles}$$

Molecular weight of aspartic acid = 133.1

Concentration of aspartic acid in μg = 13. 492 μg

g/100g sample of aspartic acid =

$$\frac{13.492\ x\ 10\ x\ 100}{0.05\ x\ 0.1\ x\ 1000\ x\ 1000} = 2.698$$

g/100g protein of aspartic acid = $2.698 \text{ x } \frac{100}{25} = 10.79$

In a similar way the concentration of other amino acids can be calculated.

5

ENZYMES

Enzymes are organic catalysts produced by living organisms. They make possible biochemical reactions. Hence they are called biocatalysts. They are mostly a specialized class of proteins. However, there are some non-protein enzymes like ribozymes, which are RNAses. Enzymes speed up reactions at ambient temperatures. This is in contrast to chemical reactions like saponification of lipids which take place when lipids are boiled for a few hours with concentrated alkali. Enzymes, namely, lipases, hydrolyze lipids at body temperatures in minutes. They are superior to chemical catalysts by a factor of 10^7 to 10^{14}. An enzyme may be a simple protein or a complex protein. They are produced by cells and are mostly globular proteins. Each cell contains an estimated 3000 different enzymes.

The names of enzymes describe the compound or the reaction that is catalyzed. The actual names of enzymes are derived by replacing the name of the reaction or the reacting compound with the suffix *ase*. For example, an oxidase catalyzes an oxidation reaction, and a dehydrogenase removes hydrogen atoms. Sucrase hydrolyses sucrose to glucose and fructose while lipases hydrolyse lipids to fatty acids and glycerol. In olden nomenclature the names of enzymes end with the suffix *in* for example papain, pepsin, rennin and trypsin.

Enzyme units: The amounts of enzymes are expressed as molar amounts, or measured in terms of activity, in enzyme units.

Enzyme activity: Enzyme activity is a measure of the quantity of active enzyme present and is thus dependent on conditions, which should be specified. Enzyme activity = moles of substrate converted per unit time = rate × reaction volume. The SI unit is the katal, 1 katal = 1 mol s^{-1}, but this is an excessively large unit. A more practical and commonly used value is enzyme unit (U) = 1 μmol min^{-1}. 1 U corresponds to 16.67 nanokatals.

Specific activity: The specific activity of an enzyme is another common unit. This is the activity of an enzyme per milligram of total protein (expressed in μmol min^{-1} mg^{-1}). Specific activity gives a measurement of enzyme purity in the mixture. Specific activity is equal to the rate of reaction multiplied by the volume of reaction divided by the mass of total protein. The SI unit is katal kg^{-1}, but a more practical unit is μmol mg^{-1} min^{-1}.

Factors influencing enzyme assays

- **Salt Concentration**: Most enzymes cannot tolerate extremely high salt concentrations. The ions interfere with the weak ionic bonds of proteins. Typical enzymes are active in salt concentrations of 1-500 mM. As usual there are exceptions such as the halophilic algae and bacteria.

- **Effects of Temperature**: All enzymes work within a range of temperature specific to the organism. Increases in temperature generally lead to increases in reaction rates. The "optimum" temperature for human enzymes is usually between 35 and 40°C. The average temperature for humans is 37°C. Human enzymes start to denature quickly at temperatures above 40°C.

- **Effects of pH**: Most enzymes are sensitive to pH and have specific ranges of activity. All have an optimum pH. The pH can stop enzyme activity by denaturating the three-dimensional shape of the enzyme by breaking ionic, and hydrogen bonds. Most enzymes function between a pH of 6 and 8; however, pepsin in the stomach works best at a pH of 2 and trypsin at a pH of 8.

- **Substrate Saturation**: Increasing the substrate concentration increases the rate of reaction (enzyme activity). However, enzyme saturation limits reaction rates. An enzyme is saturated when the active sites of all the molecules are occupied most of the time. At the saturation point, the reaction will not speed up, no matter how much additional substrate is added. The graph of the reaction rate will be a plateau.

- **Level of crowding:** Large amounts of macromolecules in a solution will alter the rates and equilibrium constants of enzyme reactions, through an effect called macromolecular crowding.

Preparation of the crude extract for enzyme studies

Tissue (without seed coat) is hand homogenised in chilled glass mortar, with the following extraction medium at pH 7.6, using quartz as an abrasive. The extraction medium: 0.1 M tris – HCl buffer: 0.4M sucrose: 0.01M KCl; 10 mM Mg Cl_2 : 0.01M EDTA (Ethylene diamine tetra-acetic acid); 0.01 M mercaptoethanol and 2.5 % PVP (Polyvinyl pyrrolidone). The homogenate is filtered through four layers of muslin cloth and centrifuged at 10,000 x g for 20 min at 0-4^{0}C. The supernatant is stored in cold for further studies.

Amylase

Principle

Amylase activity is measured based on the amount of starch it can hydrolyze in a specified time. The amount of starch is estimated from the color it produces with a dilute iodine solution.

Reagents and equipment

1. Standard starch solution: 0.8g starch and 0.1g benzoic acid are dissolved in boiling water, cooled and made up to 100ml.
2. 0.1M phosphate buffer (p^H 7.4): 19ml of 0.2M NaH_2PO_4 and 81ml of 0.2 M Na_2HPO_4 are mixed and diluted to 200ml with distilled water.
3. Iodine solution: 1.0g iodine and 2.0g KI are dissolved in distilled water and made up to 100ml.
4. 5% H_2SO_4 .
5. Spectrophotometer with cuvettes.

Procedure

1. Take 0.5ml of starch solution in a large test tube and add 0.5ml phosphate buffer.
2. Add 0.2ml of sample (containing amylase enzyme) and incubate at 37^{0}C for 30min.
3. Add 2ml 5% H_2SO_4 and 1ml iodine solution and make up to 50ml.
4. A control tube without starch solution is prepared.
5. The color (OD) is measured in a colorimeter/spectrophotometer at 640nm against the reagent blank.

Calculation

$$\text{Amylase content} = \frac{\text{OD of control} - \text{OD of sample X 4 X 10 units}}{\text{OD control X 0.2}}$$

1unit = amount of starch in mg hydrolyzed by 10ml sample in 20 min.

Reference

Am. J. Clin. Path. 27, 714 (1957).

Salivary amylase activity (Iodimetric procedure)

Principle

Salivary amylase catalyses the hydrolysis of starch to maltose and dextrins. The complete disappearance of starch is measured by the iodimetric method and the amylase activity is estimated.

Reagents and equipment

1. 1% starch solution.
2. 1% sodium chloride solution.
3. Dilute iodine solution.
4. Phosphate buffer p^H 6.6.
5. Water bath (38°C).
6. Stop clock.

Procedure

1. Pipet exactly 1ml saliva into a 100ml measuring jar. Dilute to 100ml mark and mix well.
2. Pipet 5ml 1% soluble starch into a test tube, add 2ml of 1% NaCl + 2ml buffer at p^H 6.6. Mix well and place in a water bath at 38±1°C.
3. Take 10 test tubes each with 2ml of pale yellow iodine solution.
4. Add 1ml dilute saliva to the starch mixture and put the tubes in a water bath at 38±1°C. Note the time of addition of saliva.
5. At the end of each min, take out two drops of reaction mixture and add to one of the tubes of iodine (10 times). Record the time when no change in colour appears in iodine solution (achromic point).

6. Repeat the experiment using 2ml water instead of sodium chloride. Observe the colour.

Calculation

One amylase unit is the amount of amylase which catalyses the hydrolysis of 5ml 1% starch solution in 10 min.

$$\text{Amylase units} = 100 \times \frac{10}{T} \text{ min}$$

T = reaction time for complete disappearance of starch.

100 = dilution factor.

Cellulase activity

Principle

The production of reducing sugar (glucose) due to the cellulolytic activity is measured by the dinitrosalicylic acid method.

Reagents

1. Sodium citrate buffer 0.1M p^H 5.0.
2. Carboxy methyl cellulose(CMC) 1% solution: 1g in 100ml citrate buffer.
3. Dinitrosalicylic acid (DNS) reagent: Dissolve 1g DNS, 200mg crystalline phenol and 50mg sodium sulfite in 100ml 1% NaOH. Store at 4^0C. Preferably prepare afresh.
4. Rochelle salt(potassium sodium tartrate) - 40%.
5. Water bath.
6. Spectrophotometer with quartz cuvettes.

Procedure

1. Pipet 0.45ml 1% CMC solution at 55^0C and 0.05ml enzyme extract.
2. Incubate at 55^0C for 15min.
3. Immediately add 0.5ml DNS reagent.
4. Heat the mixture on boiling water bath for 5min.
5. Add 1.0 ml Rochelle salt (40%) solution while still hot.

6. Cool to room temperature.
7. Add water to make up to 5 ml.
8. Measure absorbance at 540 nm. Measure concentration of sample from the standard graph.
9. Prepare a standard graph with glucose (50-1000 μg/ml).

Calculation

Express enzyme activity as mg glucose released/min/mg protein.

Reference

Denison, D.A. and Koehn, R.D. (1977) Mycologia. LXIX, 692.

Catalase (Manometric method)

Principle

Catalase releases oxygen during a reaction with H_2O_2. The oxygen is measured manometrically.

Reagents and equipment

1. 0.1M phosphate buffer p^H 7.0.
2. Sample (enzyme preparation): 10 mg of the fresh tissue is crushed using a teflon pestle (or a grinder) with 4 ml 0.1M phosphate buffer p^H 7.0 for 2 min at 5°C. The homogenate is then centrifuged for 10min at 1000 rpm in a centrifuge and the supernatant is used for the assay.
3. 0.2M H_2O_2: 1.0 ml of H_2O_2 solution 20 volumes is diluted to 8.0ml with distilled water. The solution needs to be freshly prepared every time.
4. Warburg apparatus.

Procedure

1 The Warburg apparatus is used with air as gas phase. 0.2 ml of 0.2M H_2O_2 and 0.1M phosphate buffer p^H 7.0 are added to the main vessel of the Warburg apparatus. The enzyme preparation is added slowly from the side arm after equilibration of system for 5min at 24°C.

2 Manometric readings are taken at one minute intervals, 4 times which would be linear with time.

Calculation

Catalase activity is expressed as µl O_2 liberated/mg of sample N/hr at 24^{0}C.

Catalase (Colorimetric method)

Principle

The U.V. light absorption of H_2O_2 solution can be measured at 240 nm. The absorption decreases with time. This decrease is utilized to record the enzyme activity.

Reagents and equipment

1. Phosphate buffer (0.067M p^H 7.0): Dissolve 3.522g KH_2PO_4 and 7.268g Na_2HPO_4 $2H_2O$ in water and make up to one litre. When diluted 10 times the molarity becomes 1/150M.
2. Hydrogen peroxide-phosphate buffer: Dilute 0.16 ml H_2O_2 (10% w/v) to 100 ml with phosphate buffer. Prepare freshly. The absorbance of the solution need to be around 0.5 at 240nm with 1cm path length.
3. Spectrophotometer with quartz cuvettes.
4. Centrifuge.

Procedure

1. Enzyme extract: Homogenize plant tissue in a blender with I/150M phosphate buffer at 1-4^{0}C and centrifuge. Stir the sediment with cold phosphate buffer. Allow to stand in the cold and shake regularly. Repeat extraction twice more. The combined supernatants can be used for the assay.
2. Read absorbance against a control (enzyme extract + H_2O_2 free phosphate buffer M/15) at room temperature and wavelength 240nm. Pipet into the experimental cuvette 3ml hydrogen peroxide-phosphate buffer. Mix in 0.01-0.04ml sample and stir. Note the time needed for decrease in absorbance from 0.45 to 0.40. If the time is more than 60sec, the measurements need to be repeated.

Calculation

1g sample homogenized to 20ml, then 1ml diluted to 10 ml with water. 0.01 ml taken for enzyme assay. For Ex. the absorbance at 240 nm decreased from 0.45 to 0.40 in 30 sec.

17/30 = 0.57 units in the assay mixture (or)

0.57 x10/0.01 = 570 units/ml extract

1.14 x 10^4 units/g tissue

The concentration of Hydrogen peroxide can be calculated using the extinction coefficient 0.036/ µmoles/ml.

Esterase

Principle

The esterase activity is measured from the amount of acid it splits during the reaction.

Reagents

1. 1% Phenolphthalein in alcohol.
2. N/20 NaOH.
3. M/3 Potassium phosphate buffer p^H 7.0. mix 10ml M/3K_2HPO_4 and 3ml M/3 KH_2PO_4.
4. Ethyl butyrate.
5. Incubator.

Procedure

1. To 1ml of enzyme (extract) solution in a conical flask add 3ml distilled water. Also add 1ml ethyl butyrate and 0.5ml M/3 Potassium phosphate buffer. The contents are mixed by shaking and kept in an incubator at 37^0C for 24 hr. After the incubation, add 5 drops of 1% phenolphthalein in alcohol. Titrate with N/20 NaOH till a permanent pink colour appears.
2. A control tube is similarly used without the addition of the enzyme extract.

Calculation

$$\text{Units/ml} = \frac{\text{ml N/20 NaOH for the test sample - N/20 NaOH for control}}{2}$$

1 unit = acid released to neutralize 1 ml of N/10 NaOH

Lipase activity (P^H measurement)

Principle

Lipase catalyses the hydrolysis of glycerides to fatty acids and glycerol. Change in p^H due to formation of fatty acids over time is measured to assay lipase activity.

Reagents and equipment

1. Broad beans or any other oleaginous material (sample).
2. Olive oil.
3. 0.1M HCl.
4. 0.1M NaOH.
5. Oleic acid.
6. P^H meter.
7. Magnetic stirrer.

Procedure

1. Grind about 20 g of the sample. Add 100 ml water in 5 ml aliquots. Grind thoroughly to make smooth paste.
2. Centrifuge for 20 min. Three layers are formed.
3. Pipet off the upper creamy layer and place in a stoppered conical flask.
4. Discard the middle aqueous layer.
5. Re- extract the residue with a further 100 ml water.
6. Again centrifuge for 20 min.
7. Pipet the upper layer and combine with the first portion (crude enzyme extract).
8. Take 20 ml olive oil, 18 ml water and 2 ml 0.1M NaOH and 0.5 ml oleic acid to a separating funnel and shake vigorously.
9. Using a P^H meter adjust the P^H of emulsion to 7.0.
10. Pipet 10 ml emulsion to a 50 ml beaker. Stir with a magnetic stirrer.
11. Add 2ml enzyme suspension to the substrate and record the time and P^H.
12. Tabulate changes of P^H for a total time of 60 min.

13. Repeat the experiment with 2ml boiled enzyme suspension.
14. Plot a graph of lipase activity (change of P^H = sample- blank) against time.
15. From the change in P^H the concentration of the enzyme is calculated.

Lipase (Titrimetric method)

Principle

Lipase hydrolyses triglycerides to release free fatty acids (FFA) and glycerol. Lipases play an important role during germination to furnish energy to growing seedling. The free fatty acid content is used to quantify the lipase activity. The FFA content can be measured by titrating against an alkali.

Reagents

1. Substrate: 2ml of neutral vegetable oil. Make it into an emulsion with water and a suitable emulsifying agent (100 mg sodium taurocholate and 2 g gum Arabic).
2. 0.01N NaOH.
3. Enzyme source: Homogenize the plant sample with twice its volume of ice cold acetone. Filter and wash the powder successively with acetone: ether (1:1) and ether. Air dry the powder. This powder can be stored in a fridge.
4. Hot plate cum magnetic stirrer.
5. P^H Meter.

Procedure

1. Extract 1g acetone powder with 20ml ice-cold water or a buffer. Centrifuge and the supernatant serves as a source of enzyme.
2. Take 20 ml of substrate in a 500 ml beaker. Add 5ml phosphate buffer p^H 7.0.
3. Keep the beaker on a hot plate cum magnetic stirrer. Stir the contents slowly. Maintain temperature at 35^0 C. Adjust to p^H 7.0.
4. Add enzyme extract 0.5ml, immediately record p^H .
5. At frequent intervals of about 10min, add 0.01N NaOH to bring p^H to initial value. Continue titration up to 60 min.

6. Note the volume of alkali consumed.

Calculation

The enzyme acitivity is the amount of enzyme which releases one meq of FFA/ min/g sample.

Specific activity = meq/min/mg protein

$$\text{activity}(\text{meq})\text{per min per mg protein} = \frac{\text{volume of alkali}}{\text{wt of sample x time in min}}$$

Reference

Jaya raman, J (1981): in Laboratory Manual in Biochemistry. Wiley Eastern Limited, New Delhi, pp 133

Determination of lipase activity (Olive oil emulsion method)

Principle

The method consists of incubating the enzyme solution with an olive oil emulsion and titrating the free fatty acid produced with sodium hydroxide.

Reagents and equipment

1. Gum acacia (5 %): 5 g gum acacia is suspended in 100 ml of warm water containing 0.2 g sodium benzoate. It is mixed and kept over-night.
2. Olive oil emulsion: A mixture of equal parts olive oil and 5 % of gum acacia, (reagent 1) homogenised and kept in refrigerator.
3. Buffer solution: Anhydrous sodium phosphate (Na_2 HPO_4) - 4.7 g is dissolved in 25 ml distilled water and the final volume is made upto 100 ml with water.
4. 1 per cent phenolphthalein in 95 per cent alcohol.
5. 0.05 N Sodium hydroxide in distilled water.
6. Centrifuge.
7. Homogenizer.
8. Hypodermic syringe.
9. Incubator.

Procedure

Preparation of the homogenate

1. The plant samples are powdered and then homogenised in 0.1 M potassium phosphate buffer, pH 7.5. The buffer is thoroughly pre-chilled to avoid enzyme destruction and the homogenization chamber during homogenization is kept immersed in crushed ice.
2. The homogenization is carried out for 6 minutes, using three regimes of 2 min homogenization and 2 min rest to prevent rise in temperature.
3. The slurry is filtered through four layers of muslin cloth and is centrifuged in a refrigerated centrifuge at 20,000g for 30 minutes at 4^0C.
4. The supernatant is removed with a hypdermic syringe for the determination of the enzyme activity.

Assay of lipase activity

1. Disstilled water (3ml) and 1 ml of enzyme solution are added to two test tubes. One of these test tubes is placed in boiling water for 5 min and is then cooled to obtain the blank.
2. Buffer solution (0.5 ml) and 2 ml of olive oil emulsion are added to both the test tubes. The contents are shaken well and incubated at 37^0C for 24 hours.
3. 3 ml of 95% alcohol and 2 drops of phenolphthalein solution are then added and the contents are mixed.
4. Each tube is then titrated with 0.05N NaOH using phenolphthalein as an indicator until the appearance of a permanent pink colour (pH 10).

Calculation (Expression of the results)

The enzyme activity is expressed on the basis of ml of 0.05 N NaOH used to titrate the FFA per g of seed/sample material.

Lipase (Colorimetric method)

Principle

The free fatty acids are liberated by the lipolytic activity of the tissue homogenates when incubated with triolein or other naturally occurring oils. The fatty acids can be complexed with copper reagent and the content can be estimated by measuring the developed color.

Reagents and equipment

1. Reaction mixture: 20mg of the oil or synthetic glycerides, 5ml of 1.0 M triethanolamine (p^H 8.5), 2.5ml of 2.5ml 20% BSA and 2.5ml of 0.1M NaCl are mixed to obtain the reaction mixture. It is emulsified prior to enzyme reaction in a hand grinder.
2. Tissue preparation: Tissues are homogenized in a medium containing 0.25 M sucrose and 0.01M tris HCl at p^H 7.4. The homogenates are diluted to have about 10mg protein/ml. This preparation serves as the source of lipase.
3. Cupric acetate-pyridine reagent: 5% (w/v) aqueous solution of cupric acetate is prepared. The pH of the solution is adjusted to 6.0-6.2 with pyridine before making up the final volume.
4. Chloroform.
5. Benzene.
6. Screw-capped culture tubes.
7. Water bath.
8. Centrifuge.
9. Cyclomixer.
10. Colorimeter or spectrophotometer.

Procedure

1. The lipolytic activity is determined by adding 0.25 ml of diluted homogenate to 0.75ml of pre-incubated reaction mixture at 37^{0}C for 15 min.
2. The incubation is carried out for about 20 min with shaking at 37^{0}C.
3. The reaction is stopped by adding 8ml of 2:1 $CHCl_3$-CH_3OH mixture.
4. The activity is determined colorimetrically using copper reagent as detailed below.
5. 1-3 mg of lipid sample in chloroform is taken in the screw-capped culture tube (30 ml capacity). The lipid sample is evaporated to dryness.
6. Five ml benzene is added and swirled to dissolve the lipid sample. One ml of cupric acetate-pyridine reagent is added to each tube. The contents are shaken in a cyclomixer for 2 minutes and then centrifuged for 5 min to get two layers.

7. The upper layer is separated with the help of a dropper and the absorbance is read at 715 nm.
8. A standard curve is prepared by taking palmitic acid in the range of 2-14 micromoles.
9. The concentration of fatty acids is read from the graph and further calculations are made based the value.
10. Protein in the homogenate is determined by Lowry method (mentioned under proteins).

Calculation

Specific activity = μ moles of FFA liberated /mg protein/hr

Peroxidase

Peroxidase (POD) includes in its widest sense a group of specific enzymes such as NAD-Peroxidase, NADP-Peroxidase, fatty acid peroxidase etc. as well as a group of very non-specific enzymes from different sources which are simply known as POD. It catalyses the dehydrogenation of a large number of organic compounds such as phenols, aromatic amines, hydroquinones etc. POD occurs in animals, higher plants and other organisms. The best studied is horse radish POD.

Principle

Guaiacol is used as substrate for the assay of peroxidase.

Guaiacol + H_2O_2 + POD $\rightarrow$ Oxidized guiacol + $2H_2O$.

The resulting oxidized (dehydrogenated) guaiacol is probably more than one compound and depends on the reaction conditions. The rate of formation of guaiacol dehydrogenation product is a measure of the POD activity and can be assayed spectrophotometrically at 436 nm.

Reagents and equipment

1. Phosphate Buffer 0.1M (pH 7.0).
2. Guaiacol Solution 20 mM: Dissolve 240 mg guaiacol in water and make up to 100ml. It can be stored frozen for many months.
3. Hydrogen peroxide solution (0.042% = 12.3mM). Dilute 0.14ml of 30% H_2O_2 to 100 ml with water. The extinction of this solution should be 0.485 at 240 nm. Prepare freshly.

4. Enzyme Extract: Extract 1g of fresh plant tissue in 3 ml of 0.1M phosphate buffer pH 7 by grinding with a pre-cooled mortar and pestle. Centrifuge the homogenate at 18,000 g at 5°C for 15 min. Use the supernatant as enzyme source within 2-4h. Store on ice till the assay is carried out.
5. Spectrophotometer with quarz cuvettes.

Procedure

1. Pipette out 3 ml buffer solution, 0.05 ml guaiacol solution, 0.1 ml enzyme extract and 0.03 ml hydrogen peroxide solution in a cuvette. (Bring the buffer solution to 25°C before assay).
2. Mix well. Place the cuvette in the spectrophotometer (436nm).
3. Wait until the absorbance has increased by 0.05. Start a stop-watch and note time required in minutes (Δt) to increase the absorbance by 0.1.

Calculation

Since the extinction coefficient of guaiacol dehydrogenation product at 436nm under the conditions specified is 6.39 per micromole, the enzyme activity per liter of extract is calculated as below:

$$\text{Enzyme activity units/liter} = \frac{3.18 \times 0.1 \times 1000}{6.39 \times 1 \times \Delta t \times 0.1} = \frac{500}{\Delta t}$$

References

Putter, J (1974) In: Methods of Enzymatic Analysis 2 (Ed Bergmeyer) Academic Press New York p 685.

Malik, C P and Singh, M B (1980) In: Plant Enzymology and Histoenzymology. Kalyani Publishers New Delhi p 53.

Urease activity

Principle

Urease is a plant enzyme. It catalyzes the hydrolysis of urea to ammonia, CO_2 and water. The ammonia produced is determined by absorbing the liberated ammonia into saturated boric acid. The amount of ammonium borate formed is a measure of the urease activity.

Reagents and equipment

1. 0.1M urea solution.
2. Powdered plant sample.
3. Saturated boric acid.
4. 0.01M HCl solution.
5. 40% NaOH solution.
6. 10% trichloroacetic acid.
7. n- octanol.
8. Bromocresol green and methyl red indicator solutions.
9. Buffers pH 6.5.
10. Constant temperature water bath.
11. Incubator.
12. Titration assembly.

Procedure

1. The apparatus consists of two boiling tubes connected by glass tubing as in the figure. Boiling tube one contains the reaction mixture and tube two contains saturated boric acid.
2. Incubate for 30 min at 40^{0}C the reaction mixture containing 1. Substrate: 0.5 ml 0.1M urea solution, 2. Buffer: 5ml buffer at p^H6.5 and 3. Sample containing enzyme (5 ml).
3. Remove stopper from tube one and stop reaction with 5ml 10% trichloroacetic acid. Add 2-3 drops of octanol to prevent frothing during aeration.
4. Add 5ml 40% NaOH.
5. Add indicator (1:3 methyl red : bromocresol green) to tube 2.
6. Aerate the liberated ammonia into boric acid for 20 min.
7. Titrate ammonium borate against 0.01M HCl
8. Run a sample blank separately and deduct the value from that of sample.

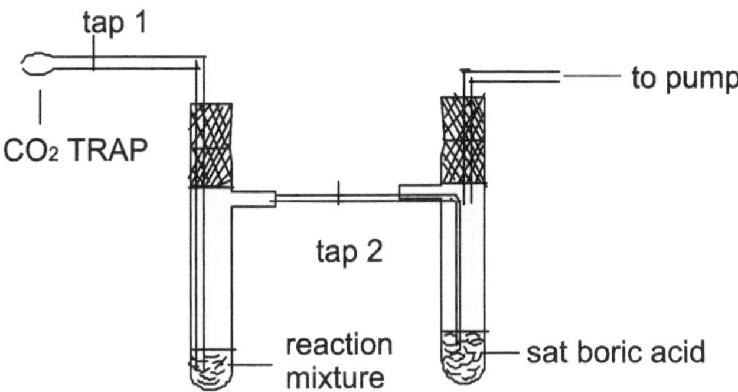

Fig. Apparatus for determination of urease activity

Calculation

Enzyme activity = µmoles NH_3 formed or the titre value

Nitrate reductase

Principle

Nitrate reductase (NR) utilizes the reduced form of pyridine nucleotides. NADH – dependent nitrate reductase is most prevalent in plants. NR activity can be measured by following the oxidation of NAD (P)H at 340 nm. NR activity is commonly measured by colorimetric determination of nitrite produced.

Reagents and equipment

1. Potassium phosphate buffer 0.1M p^H 7.5.
2. Potassium nitrate 0.1M: Dissolve 1.01g KNO_3 in 100 ml water.
3. NADH 2m M: Dissolve 14mg NADH disodium salt in 10 ml water.
4. Sulphanilamide1% w/v: Dissolve 1g sulphanilamide in 100 ml 2.4N HCl.
5. N- (1-naphthyl) ethylene diamine dihydrochloride 0.02%: Dissolve 20 mg in 100 ml water.
6. Potassium nitite standard solution 0.01M: Dissolve 851mg KNO_2 in 100 ml water in a volumetric flask. Dilute 10 ml to 100ml to have a working solution of KNO_2.
7. Enzyme extract: Homogenize a small quantity of the plant material in 6 volumes of a medium (1m M EDTA, 1-25m M cysteine and 25m M potassium phosphate adjusted to a final p^H 8.8 with KOH). Filter and

centrifuge for 15min at 30,000g. decant supernatant through glass wool and use for enzyme assay. The extraction should be carried out under ice cold conditions.

8. Incubator.
9. Spectrophotometer with quarz cuvettes.

Procedure

1. Pipet 0.5 ml phosphate buffer p^H 7.5 into a test tube.
2. Add 0.2 ml KNO_2 solution, 0.4 ml NADH solution and 0.7 ml water.
3. Initiate the raction by adding 0.2 ml enzyme extract. Have a control with water instead of enzyme extract.
4. Incubate at 30^{0}C for 15min.
5. Terminate the reaction by the rapid addition of 1ml suphanilamide followed by 1 ml N- (1-naphthyl) ethylene diamine dihydrochloride reagent.
6. Wait for 30 min.
7. Measure absorbance at 540 nm.
8. Prepare a standard graph with KNO_2 . Pipet 0.2 -1.0 ml aliquots of KNO_2 solutions into different tes tubes. Make up volumes to 2 ml with water. Follow steps from 6 to 7 . Prepare graph of concentration Vs absorbance.
9. Make calculations from the sample concentration read from the graph from the absorbance value obtained at 7 above.

Calculation

Activity is expressed as µmoles nitrite produced/mg protein or per g tissue.

Reference

Hageman, R.H. and Reed, A.J. (1980) Methods in Enzymology, Academic press. New York

Nitrite reductase

Principle

Nitrite is reduced to ammonia by nitrite reductase. The disappearance of nittite is measured in the reaction reduced methyl violegen is used as electron donor.

$$NO_2^- \xrightarrow[+\ 6e]{\text{Nitrite reductase}} NH_4^+$$

Reagents and equipment

1. Tris HCl buffer 0.5M p^H 7.5.
2. Sodium nitite solution: 43.2mg $NaNO_2$ in 20ml water.
3. Methyl viologen: 60.1 mg methyl viologen in 20ml water.
4. Sodium dithionite- bicarbonate solution: Dissolve 250mg $Na_2S_2O_4$ and 250mg $NaHCO_3$ in 10ml water.
5. Enzyme extract: Homogenize the plant tissue (10g/100ml) with Tris-HCl buffer (p^H 7.5) in a waring blender for 3min. Centrifuge and use the clear supernatant as enzyme extract.
6. N– (1-naphthyl) ethylene diamine dihydrochloride 6.02%: Dissolute 20g in 100 ml water.
7. Incubator.
8. Spectrophotometer with quarz cuvettes.

Procedure

1. Mix 6.25 ml Tris-HCl buffer, 2 ml $NaNO_2$ solution, 2 ml viologen solution and 14.75 ml water (reaction mixture).
2. Pipet 1.5 ml reaction mixture and 0.3ml enzyme extract into a test tube.
3. Run a blank alongside without enzyme.
4. Start the reaction by adding fresh sodium dithionite- bicarbonate solution.
5. Incubate for 15 min at 30°C.
6. Stop reaction by vigorous shaking till blue colour disappears.
7. Use 20 μl aliquot for nitrite determination.
8. Terminate the reaction by the rapid addition of 1ml suphanilamide followed by 1ml N-(1-naphthyl) ethylene diamine dihydrochloride reagent.
9. Wait for 30 min.
10. Measure absorbance at 540nm.
11. Estimate the quantity of nitrite that has disappeared using blank as a reference.

12. Prepare a standard graph with KNO_2 . Pipet 0.2 -1.0ml aliquots of KNO_2 solutions into different test tubes. Make up volumes to 2ml with water. Add 1 ml N– (1-naphthyl) ethylene diamine dihydrochloride reagent. Wait for 30 min. Measure absorbance at 540 nm.

13. Prepare a graph of concentration Vs absorbance.

Calculation

The enzyme activity is expressed as the amount of nitrite (μM) reduced/min/ mg protein

Nitrogenase

Principle

Acetylene is reduced to ethylene by nitrogenase. The ethylene produced is quantified in a gas chromatograph. The nitrogenase activity is expressed as 'n' moles ethylene produced in unit time per g dry nodules.

Reagents and equipment

1. Gas chromatograph with F.I.D.
2. Syringes (air tight).
3. Acetylene gas.
4. Ethylene gas.
5. Conical flasks, 100 ml with small mouth to fit serum caps.

Procedure

1. Take plants with nodule.
2. Excise the roots with nodules.
3. Place the root system into the 100ml conical flask.
4. Seal the flask with rubber septum.
5. Remove 10ml of air from the flask with an air tight syringe.
6. Inject 10ml acetylene into the flask.
7. Incubate for 30-60min at ambient temperatures.
8. Remove 0.5 to 1ml gas mixture from the flask with air tight syringe.

9. Inject the gas mixture to a GC porapak column at 60°C.
10. Locate the acetylene and ethylene peaks (1.3 and 1.8 min retention time respectively). Measure the peak heights and widths. Calculate peak areas.

Calculation

Peak area of sample is computed to concentration based on the area of a standard ethylene peak.

Activity of nitrogenase = nmole or µmole/ g sample/unit time.

Reference

Turner, G.L. and Gibson, A. H. (1980) In Methods for Evaluating Biological Nitrogen Fixation (Ed. by Bergersen, F.J.) John Wiley and Sons New York p 111.

Polyphenol oxidase

Principle

The intensely yellow 2-nitro-5- thiobenzoic acid (TNB) has an absorbance at 412 nm. It reacts with quinines generated by the enzyme on reaction with 4-methyl catechol. The decrease in yellow colour due to enzyme activity is measured.

Reagents and equipment

1. Citrate phosphate buffer (02M p^H 6.0): Take 6.008 g of citric acid , 3.893g potassium dihydrogen phosphate, 1.79 g boric acid and 5.226 g of pure diethyl barbituric acid and dissolve in water and make up to one litre. Take 100 ml of this solution and add 38.9 ml of 0.2M sodium hydroxide solution (carbonate free) to obtain p^H 6.0 citrate buffer.
2. 2-nitro -5- thiobenzoic acid anion (TNB): Add 30mg sodium borohydride to a suspension of Ellman's reagent (5,5-dithiobis -2-nitrobenzoic acid - 19 mg in 10 ml water) within 1hr, the disulfide is quantitatively reduced to the intensely yellow, water soluble thiol. This solution is stable for one week at 4°C.
3. Quinine solution: Dissolve 4-methyl-1,2- benzoquinone in double distilled water in 50 ml volumetric flask. Para benzoquinone solution also is prepared similarly. Both are stable for 30 min.

4. Substrate solution: 4-methyl catechol (2m M) for catechol oxidase assay. Quinol (1:4 dihydroxy benzene 2mM) for laccase assay.

5. Spectrophotometer with cuvettes.

Procedure

1. Enzyme extract: Homogenize 25 g frozen tissue in two portions of 100ml each cold acetone. Centrifuge or filter under vaccum. The homogenate is air dried to remove acetone. The resulting dry powder is weighed and stored in a freezer. Mix 100 mg acetone powder with 6.5 ml 0.2 M citrate phosphate buffer, 1ml 1% Triton X-100, 6.5 ml water and 500 mg polyamide. Shake for 1hr and filter. The filtrate can be used as the enzyme source.

2. Pipet into a cuvette 1.4ml citrate phosphate buffer, 0.5ml TNB and 1ml substrate solution.

3. Add 0.1ml enzyme solution and immediately note down absorbance at 412nm.

4. Follow the decrease in absorbance at 30sec intervals and note down.

Calculation

Read the change in absorbance per min from a graph.

Calculate the enzyme units using the equation:

$$\text{Units in the test} = K\frac{\Delta A}{\min}$$

Where , K is 0.272 for catechol oxidase and 0.242 for laccase (one unit of either catechol oxidase or laccase is defined as the enzyme which transforms 1μ mol of dihydric phenol to 1μ mol of quinine per min under the assay conditions. One unit is equivalent to the consumption of 1μ mol of TNB). ΔA is the change in absorbance per min.

Reference

Esterbaner, H., Swarzl, E and Hayn, M (1977) Anal. Biochem. 77: 486.

Acid phosphatase

Principle

Phosphatases liberate inorganic phosphate from organic phosphate esters. The enzyme phosphatase hydrolyzes p-nitrophenol phosphate. The released p-nitrophenol is yellow in colour and can be measured at 405nm.

Reagents and equipment

1. Sodium hydroxide 0.085N: 0.85 g NaOH IN 250ml water.
2. Substrate solution: Dissolve 1.49 g EDTA, 0.84 g citric acid and 0.03 g p-nitrophenyl phosphate in 100 ml water. Adjust p^H to 5.3.
3. Standard: Dissolve 69.75 g p-nitrophenol in 5ml distilled water (100 m M).
4. Spectrophotometer with cuvettes.
5. Incubator.

Procedure

1. Enzyme extract: Homogenize 1g fresh tissue in 10ml ice cold 50m M citrate buffer (p^H 5.3) in a chill mortar and pestle. Centrifuge at 10,000g for 10min. The supernatant serves as the enzyme source.
2. Incubate 3ml substrate at 37^{0}C for 5min.
3. Add 0.5ml enzyme extract and mix well.
4. Remove immediately 0.05ml and mix with 9.5ml NaOH 0.085N. This is zero time assay (blank).
5. Incubate the remaining solution of substrate and enzyme for 15min at 37^{0}C.
6. Draw 0.5ml sample and mix it with 9.5ml NaOH 0.085N solution.
7. Measure the absorbance of blank and incubated tubes at 405nm.
8. Take 0.2 to 1.0ml (4-20m M) of the standard, dilute to 10ml with NaOH solution. Read the colour and draw the standard curve.

Calculation

Specific activity is expressed as m moles p-nitrophenol released per mg protein per min.

Reference

Sadasivam, S. and Manickam, A. 1996. Biochemical Methods. New Age International (P) Limited, New Delhi. pp121-122.

Papain

Principle

Papain, a plant protease, is present in papaya latex. An artificial substrate benzoyl, L- arginine p-nitoanilide (BAPNA) is hydrolysed by papain and colour intensity of the product p-nitroaniline is measured at 410nm.

Reagents and equipment

1. Tris-HCl buffer 50 m M (p^H 7.5): dissolve 605 mg of tris in 50 ml distilled water. Adjust to p^H 7.5 with0.05N HCl and make up to 100 ml.
2. To the above buffer,100 ml add 87.8 mg cysteine hydrochloride (0.005M) and 74.4 mg of EDTA(0.002M) and dissolve thoroughly.
3. Buffered substrate: Dissolve 43.5mg BAPNA in 1ml demethyl sulfoxide and make up to 100 ml with Tris – HCl buffer containing 5m M cysteine and 2 m M EDTA.
4. Acetic acid 30%.
5. Spectrophotometer with cuvettes.
6. Incubator.

Procedure

1. Enzyme source: Crude latex 100 mg/ml.
2. Pipet 0.5 ml of enzyme solution into a test tube. (Can pipet varying levels of enzyme solution for more replications).
3. Make up to 1.0 ml with Tris – HCl buffer.
4. Add 5 ml substrate solution.
5. Incubate for 25 min at 25^0C.
6. Terminate the enzyme reaction by adding 1ml of 30% acetic acid.
7. Measure the absorbance of released p-nitroaniline at 410 nm against a control (without the enzyme solution).

Calculation

Molar extinction coeff. of p-nitroaniline = 8,800/ 1 mole/cm, 1μ mole/ml = 8.8

If the absorbance is 8.8 the amount of p-nitroaniline = 1μ mole/ml

If X is the absorbance of the sample = $\frac{1\mu\ mole / ml}{8.8} x\ X$

$= X \frac{\mu\ Mole / ml}{8.8}$ This is for one ml of sample for 25min,

for 7ml (volume of assay mixture) = $\frac{X\ \mu\ mole\ x7.0}{8.8\ x\ 25(\text{min})}$ the assay mixture contains 0.5ml of the enzyme

$= \frac{X\mu\ mole\ x\ 7.0}{8.8\ x\ 0.5\ x\ 25}$ μmole of p-nitroaniline released per min

Activity of enzyme/g can also be calculated.

If protein content of the sample is known specific activity can be calculated.

Reference

In Methods in Enzymology; 19 (ed: Perlman Lorand, L.) Academic Press, New York,pp 228

Sucrose synthase

Principle

Sucrose synthase catalyses the following reaction in plants.

$$\text{UDP - Glucose + Fructose} \xrightleftharpoons{\text{Sucose synthase}} \text{Sucose + UDP}$$

The cleavage activity of sucrose is easily assayed by coupling the formation of UDP – glucose to the reduction of NAD^+ in the presence of excess UDP- glucose dchydrogcnasc. The change in absorbance at 340nm is measured to estimate sucrose synthase activity.

Reagents and equipment

1. HEPES-KOH buffer 0.1M p^H 7.5.
2. Sucrose 0.5M (17.11g / 100ml).
3. UDP 0.01M (9.95mg/ml).
4. UDP- glucose dehydrogenase (0.25mg/ml).
5. Spectrophotometer with cuvettes.
6. Centrifuge.

Procedure

1. Enzyme extract: Homogenize 10g plant tissue in 20ml 10mM potassium phosphate buffer p^H 7.2 containing 1m M EDTA and 5m M 2-mercaptoethanol. Centrifuge at 30,000g for 15min at 4^0C. the supernatant serves as the source of enzyme.
2. Pipet reagents as mentioned below:

Reagent	Assay	Blank(ml)
HEPES-KOH	0.2	0.3
Sucrose	0.2	0.2
UDP	0.2	0.2
Enzyme extract	0.2	0.2
UDP-glucose DH	0.1	0.1
NAD^+	0.1	-

3. Set the spectrophotometer to zero absorbance at 340 nm without adding NAD in the test against blank.
4. Add NAD quickly to the test solution, mix well and record the initial absorbance and set a timer.
5. Record the decrease in A_{340} every min till further reaction does not occur.

Calculation

The change in A_{340} is 12.0 for each µmole of UDPG/ ml. Express enzyme activity as µmole of UDPG formed/ mg protein.

Reference

Morrel and Copeland (1985): Pant Physiology. 78: 149.

Ascorbic Acid Oxidase

Principle

Ascorbic acid oxidase is widespread in plant tissues. The role of this enzyme is to regulate the levels of oxidized and reduced glutathione and NADPH. The activity of this enzyme is increased during infection. The method used for assay of ascorbic acid oxidase depends on the fact that within a certain range of enzyme concentrations, the rate of oxygen consumption during ascorbic acid oxidation is proportional to the amount of enzyme present. Ascorbic acid has an absorption maximum at 265nm. Hence, the decrease in the absorption peak of the acid due to oxidation by ascorbic acid oxidase is followed spectrophotometrically.

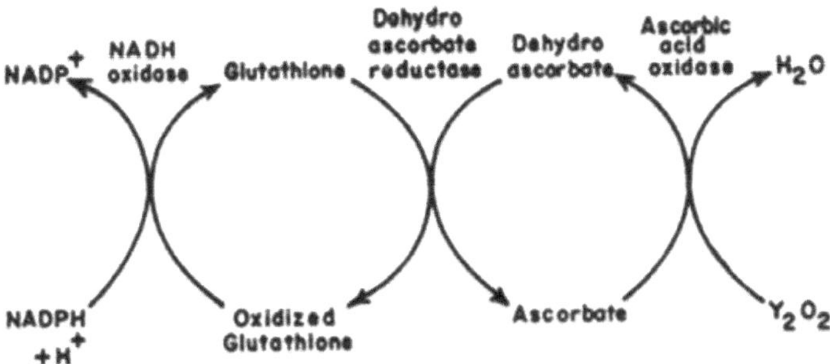

Reagents and equipment

1. Phosphate buffer: 0.1M (pH 5.6 and 6.5 separately).

2 7.8 g of mono basic sodium phosphate is dissolved in 1000 ml water to get solution A. 53.65 g of Na_2HPO_4, $7H_2O$ is dissolved in 1000 ml water to get solution B. 93.5 ml of solution A and 6.5 ml of solution B are mixed to get pH 5.6 phosphate buffer. 68.5 ml of solution A and 31.5 ml solution B are mixed to get pH *6.5* phosphate buffer.

2. *Substrate Solution:* Dissolve 8.8mg ascorbic acid in 300mL phosphate buffer (pH 5.6).

3. *Enzyme extract* : Macerate one part of plant tissue with five parts (w/v) of 0.1M phosphate buffer (pH 6.5) in a homogenizer. Centrifuge the homogenate at 3,000g for 15 min. Use the supernatant as enzyme source. All procedures are carried out at 0-5°C.

4. Spectrophotometer with cuvettes.

5. Centrifuge refrigerated.

Procedure

1. Pipette out 3ml of substrate solution to each of sample and reference cuvettes of a spectrophotometer.
2. Add 0.1 ml of enzyme extract to the reference cuvette. (Enzyme is added so as to get a positive increase in absorbance values).
3. Measure the absorbance change at 265 nm in 30 sec intervals for 5min.

Calculation

From the linear phase of reaction, compute the change in absorbance per min.

Alternatively activity may be expressed in enzyme units. One enzyme unit is equivalent to 0.405 mole oxygen per min. Ascorbic acid requires only one atom of oxygen per mole oxidized. Therefore, 0.81mole of 1/2 oxygen per min is utilized per enzyme unit. Ascorbic acid has an $E_{1cm}^{1\%}$ of 760 at 265nm. From this it is calculated that ascorbic acid has an absorbance of 4.4 per mole in a 3 ml volume. Hence, one enzyme unit (0.81mole 1/2 oxygen per min) would be equivalent to an absorbance change of 3.58 per min.

Reference

Oberbacher, M. F. and Vines, H. M. (1963) Nature 197: 1203.

Malate dehydrogenase activity

Principle

Malate dehydrogenase (MDH), an oxidoreductase, is an enzyme of TCA cycle. It catalyses the following reaction.

$$\text{Oxaloacetate} + \text{NADH} + \text{H}^+ \xrightarrow{\text{Mg}+} \text{Malate} + \text{NAD}^+$$

During this reaction NADH is oxidized to NAD. NADH absorbs at 340nm. The decrease in absorbance at 340nm is utilized to measure the activity of MDH.

Reagents and equipment

1. Sample material: Freshly harvested plant material.
2. Acid washed river sand.
3. Extraction buffer: 50m M imidazole-HCl buffer containing 10m M dithiothreitol, and 2m M EDTA.

4. Polyvinyl pyrrolidine (PVP).
5. NADH solution: 6m M NADH solution in 50m M Tris- HCl (p^H 8.0) store at 0°C.
6. Assay buffer: make 50m M Tris- HCl (p^H 8.0) obtaining 1m M EDTA.
7. Oxaloacetate solution: 0.3m M solution of oxaloacetate in water is neutralized with solid Na HCO_3 till evolution of gas bubbles stop.
8. Refrigerated centrifuge.
9. Spectrophotometer.
10. Mortar and pestle.

Procedure

1. Homogenize 0.5g sample at 0-4°C in a chilled mortar and pestle using acid washed sand in the presence of 1% PVP and slow addition of 10ml of ice cold extraction buffer.
2. Centrifuge the homogenate at 10,000g for 20min in a refrigerated centrifuge at 0-4°C.
3. Decant the supernatant and preserve at 0°C till analysis.
4. Set the spectrophotometer to zero initially at 340nm. Then keep 1.5ml assay buffer and 0.1ml enzyme extract in a quartz cuvette and adjust to 100% transmission.
5. Take out the cuvette and keep in a water bath at 37°C for 2-3min. Add 0.05ml of NADH and mix completely and gently. Place it immediately in the spectrophotometer and note the reading.
6. Note the decrease in absorbance at intervals of 1.5sec up to 3min. This serves as blank.
7. Start the reaction by adding 0.05ml oxaloacetate solution and record the decrease in absorbance up to 3min at intervals of 15 sec.
8. Prepare a standard curve of NADH (0.0.3 μ moles range) prepared in assay buffer (the volume should be 1.7ml each time).

Calculation

The enzyme activity is expressed as μmoles of malate formed /g tissue/min. the calculation is as follows:

Weight of tissue taken = 0.5g, Final volume of enzyme extract = 10ml

Enzyme extract taken for assay = 0.1ml

Decrease in A_{340} in blank = 0.01 O.D. units/min, Decrease in A_{340} in sample = 0.06 units/min, Decrease in A_{340} due to MDH activity = 0.06- 0.01 = 0.05 O.D. units/min

Let it be assumed that from the standard curve O.D. of 0.12 corresponds to 40µmoles of NADH, then 0.12 O.D. units = 40 µmoles of NADH

0.05 O.D.units = 40/0.12 x0.05 = 16.66 µmoles of NADH

Total enzyme activity as µmoles/min in 10ml of enzyme preparation = 16.66/ 0.1x10

Total enzyme activity in 1g of tissue = 16.66/0.1x10x(1.0/0.5) = 3332 µmoles/ min

References

Gnanam, A. and Francis, K. (1976) Plant Biochem J 3: 11.

Sadasivam, S. and Gowri, G. (1981) Photosynthetica 15: 453.

Purification of alcohol dehydrogenase

Principle

Alcohol dehydrogenase catalyzes the oxidation of ethanol by DPN (coenzyme I). The reaction is conveniently assayed by measuring the rate of appearance of reduced DPN (DPNH) by means of change in optical density at 340mµ.

Reagents and equipment

1. Stirrer.
2. Refrigerated centrifuge.
3. Bakers' yeast.
4. 0.66MNa_2PO_4.
5. Acetone.
6. Dialysis tubing.
7. 0.001M phosphate buffer pH 7.5.
8. 3M ethanol (2ml 95% ethanol+ 8ml water).

9. 0.015M DPN (mol wt of DPN= 735).
10. 0.1% bovine serum albumin.
11. Water bath.
12. 0.01% versenate (EDTA).
13. 0.001M cysteine.

Procedure

1. 100g of dry yeast powder is extracted with 300 ml of 0.66M Na_2PO_4 for 2hr at 37°C with continuous stirring followed by extraction for 3hr at room temperature.
2. The yeast residue is removed by centrifugation at 13,000 rpm for 15 min.
3. The supernatant solution is quickly brought to 55°C. and maintained at this temperature in a water bath. After cooling the mixture is centrifuged.
4. To 100 ml of yeast extract (supernatant) at -10°C is added slowly (20 min) 50 ml of cold acetone (at -6°C). Do not allow freezing to occur.
5. The resulting precipitate is separated by centrifugation for 10 min at 2500 rpm at 0°C.
6. To the supernatant, 55ml acetone at -10°C is added (for each 100ml of yeast extract).
7. The mixture is centrifuged and the supernatant discarded.
8. The precipitate is dissolved in 0.01% versenate (EDTA) and 0.001 M cysteine and dialyzed for 3hr against 33litre batches of 0.001M phosphate buffer p^H 7.5
9. The precipitate is removed by centrifugation.
10. To the clear supernatant 36 g of ammonium sulphate/100 ml solution is added.
11. After standing at 0°C for 30 min, the mixture is centrifuged at 15,000 rpm for 20 min at 0°C.
12. The precipitate is dissolved in 20ml distilled water (D.W.) and 4 g of ammonium sulphate is added.
13. The precipitate is centrifuged off. From the supernatant the alcohol dehydrogenase starts to crystallize (further addition of 4 g ammonium sulphate).

14. The crystals are collected after centrifugation and re-suspended in minimum D.W. The crystals start forming immediately.
15. The enzyme can be stored in 50% saturated ammonium sulphate.
16. Dilute stock solution of enzyme in 0.1% bovine serum albumin in 0.01M potassium phosphate, p^H7.5 (200-1000 units of enzyme).
17. Place 2.2 ml D.W., 0.1ml ethanol and 0.1 ml pyrophosphate buffer into 1 cm quartz cell.
18. Start the reaction by adding 0.1 ml of enzyme.
19. The control cell contains all reagents except the substrate.
20. The optical density reading is taken at 340 mμ at 15min after addition of substrate.
21. Further readings are taken at 15s intervals after addition of substrate.
22. The increment in density between the 15-45s intervals into 2 = enzyme activity/min.

Isozyme analysis in wheat seedlings (Peroxidase and esterase)

Principle

Isozyme refers to those forms of an enzyme with similar enzymatic activity within a single species. Isozymes are efficient genetic markers. Hence their measurement is important. The assay is carried out as per standard methods.

Reagents and equipment

1. Tris-Cl (p^H 7.6) 50 m M.
2. EDTA 5m M.
3. β-ME 5m M.
4. Tris –glycine electrode buffer stock solution, p^H 8.3: Dissolve 6 g tris and 28.8 g glycine in water and make up to 1000 ml.
5. Electrode buffer: Dissolve Tris –glycine electrode buffer stock solution in 1:9 ratio.
6. Tris-chloride buffer stock solution p^H 8.9 : Take 48 ml of 1N HCl, 36.6 g tris and 0.23 ml TEMED dissolve in water and make up to 1000 ml.
7. Tris –chloride buffer stock solution p^H 6.7 : Take 48 ml of 1N HCl, 5.98 g tris and 0.46 ml TEMED dissolve in water and make up to 1000 ml.

8. Revolving gel acrylamide stock solution: take 28 g acrylamide and 0.74 g N'N'- methylene bisacrylamide and make up to 100 ml with distilled water. Store in dark at about 4°C.
9. Ammonium persulphate solution: Dissolve 0.1 g salt in 1ml distilled water. Prepare afresh.
10. Bromophenol blue: Take 25 g indicator and add 10 ml tris chloride buffer p^H6.7.
11. Revolving gel solution: Take 5 ml tris-chloride buffer stock solution p^H8.9, 10ml revolving gel acrylamide, 25 ml water and 300 µl ammonium persulphate.
12. Mortar and pestle.
13. Micro pipette.

Procedure

1. Extraction: a). germinate the overnight soaked wheat seeds at 25°C in dark for 7 days. b). esterase : grind the seedlings in a pre-chilled pestle and mortar in 50 mM tris-Cl p^H7.6 containig 5mM β-ME and 5 m M EEDTA at a ratio of 1:2 (w/v) c). peroxidase: grind the seedlings in 50 m M tris-Cl p^H7.6 at a ratio of 1:2 (w/v). Carry out all operations at 4°C.
2. Sample preparation and running the gel: to 100µl of the extract of the sample and 10µl of glycerol, 4µl bromophenol blue and mix thoroughly. Load appropriate quantity of samples into the wells using micropipette. Maintain 4°C and constant current 50 m A for 2x 1.0 mm gel during the run.
3. Staining: Once the tracking dye reaches the bottom of the gel, stop the electrophoresis and stain as follows.
   a) Esterase: incubate the gel in 50mM phosphate buffer (p^H6.0) containing 2ml of 1% a- naphthyl acetate in 60% acetone and 50mg fast blue RR at room temperature for 10-30 min.
   b) Peroxidase: incubate the gel in 0.6M sodium acetate buffer p^H 5.4 containing 0.5% 0- dianisidine HCl for 30 min at room temperature. Transfer the gels to 0.1M hydrogen peroxide till visible bands appear.

Phosphatase and peroxidase

Principle

Anionic system of disc acrylamide gel electrophoresis (Ornstein, 1964; Davis, 1964) is used for separating the multiple molecular forms of acid phosphatase and peroxidase.

Reagents

A.	1 N Hydrochloric acid	48.00 ml
	Tris- (hydroxy methyl)-amino methane	36. 60 g
	THEMED (N, N, N', tetramethyl Ethlene diamine)	0.23 ml
	Water to	100.00 ml
	pH	8.9
B.	1N Hydrochloric acid	48.00 ml
	Tris	5.98 g
	THEMED	0.46 ml
	Water to	100.00 ml
	pH	6.7
C.	Acrylamide	28.00 g
	N,N-methylen bis-acrylamide	36.00 mg
	Water to	100.00 ml
D.	acrylamide	10.00 g
	N,N-methylen bis-acrylamide	2.50 g
	Water to	100.00 ml
E.	Riboflavin	4.00 mg
	Water to	100.00 ml
F.	Ammmonium persulphate	0.14 g
	Water to	100.00 ml
G.	Sucrose	40.00 g
	Water to	100.00 ml

All these solutions should be stored in dark bottles in a refrigerator.

Buffer solution

Tris 0.	60 g
Glycine	2.88 g
Water to	1.00 lit
pH	8.30

Tracking dye

0.5% bromephonol blue in 100 ml of 1% acetic acid (v/v).

Procedure

1. A suitable aliquot of enzyme extract is applied to the acrylamide gel, which has been prepared in 0.5 cm x 9.0 cm glass columns with 7.5 cm of 7.5% running gel and 0.5 cm of 2.5% spacer gel.
2. Running and spacer gels are prepared by mixing the reagents A,C, water and G; and B.D.E. and F in the ratio 1:2 and 1:4 respectively.
3. Electrophorosis is conducted in the cold in tris glycine buffer at 1.5 ml/ tube.
4. After completion of electrophorosis which is indicated by the movement of tracking dye, the gels are used for locating the multiple molecular forms.
5. Rf value for each activity band is calculated with respect to distance travelled by the marker dye.

Stanining

i) Acid phosphatase

Acid phosphatase activity is located in the gels using alpha-naphthyl phosphate as the substrate, and Fast Garnet GBC salt as coupler. Gels are immersed in the staining mixture in small tubes and incubated at 37° C for 1 hr to get visible bands.

ii) Peroxidase

Activity bands of peroxidase are located as follows. The gels are incubated in freshly prepared saturated solution of bromidine (prepared in 25% acetic acid) mixed with equal volume of 0.05% H_2O_2. The gels are fully dipped in the staining mixture for 2-3 min for visible bands to appear.

6

NUCLEIC ACIDS

Nucleic acids were first discovered in 1868 by Friedrich Meischer. Nucleic acids are high molecular weight polymers that are present in our cells. They store and transfer genetic material from generation to generation. Knowledge of how genes are expressed and how they can be manipulated is becoming increasingly important for understanding of nearly every aspect of biochemistry. These macromolecules p ass on information for cellular growth and reproduction. All the genetic information in the cells is called the genome. Every time a cell divides, the information is copied and passed on to the new cells.

Nucleic acids fall into two main classes according to the type of sugar they contain: the Deoxyribonucleic acids (DNA) and Ribonucleic acids (RNA). Some sections of the DNA called genes contain the information to make a protein. RNA translates the genetic information in the DNA to the ribosomes, where the synthesis of proteins takes place. Nucleic acids have the following functions. a) DNA stores and transmits genetic information b) DNA expresses its encoded genetic information for the synthesis of RNA and protein for metabolic function. c) DNA controls all cellular activities. d) RNA is necessary for protein biosynthesis.

Nucleic acids are polymers of repeating units called nucleotides. Nucleotides are composed of nitrogenous base, sugar and phosphoric acid. Nucleotides in nucleic acids are linked by 3'5'phospho diester linkages.

Nucleic acid assay

Principle

The nucleic acids are extracted from defatted plant material and estimated colorimetrically.

Reagents

1. Orcinol reagent: The reagent is freshly prepared by mixing 10 ml of 1% orcinol, 40 ml of concentrated HCl and 1 ml of 10% ferric chloride in concentrated HCl
2. Diphenylamine reagent : 1.5 g diphenylamine is dissolved in 100 ml of glacial acetic acid and 1.5 ml of concentrated H_2SO_4 is then added. Just before use, 0.1 ml of 1.6% aqueous acetaldehyde is added per 20 ml of the reagent.
3. Water bath.
4. Incubator.
5. Centrifuge.
6. Spectrophotometer with cuvettes.

Procedure

Extraction of nucleic acids

Method 1: For nucleic acid extraction, the following procedure is used. The residue left after lipid extraction is used for extraction of nucleic acid. The residue is successively treated with 8 ml each of methanol, 0-2 M $HClO_4$ and ethanol to remove methanol, acid and ethanol soluble phosphates respectively. The residue obtained is extracted with 10-15 ml of 5% $HClO_4$ at 7^{0}C for 40 min. After centrifugation, the residue is discarded and the supernatant saved for the estimation RNA.

Method 2: A 5% homogenate of the tissue is made in 0.9% saline. This is filtered through four layers of muslin or cheese cloth. The filtrate is used for further analysis. A known volume of the extract (200 mg wet tissue/ml) is treated with an equal volume of 10% TCA. The solution is shaken well, centrifuged and the supernatant discarded. The residue is washed twice with 10% TCA followed by 5 ml of absolute alcohol twice. The residue is extracted with ether twice to remove phospholipids.

RNA estimation

1. One ml of nucleic acid solution is mixed with 2 ml of orcinol reagent 1 in a test tube which is heated on a boiling water bath for 8 min. It is then cooled and 7ml n-butanol is added.
2. The bluish green colour obtained is read at 665nm. RNA content in the samples is read from the standard curve (100 – 500 μg) prepared using yeast RNA.

DNA estimation

1. The nucleic acid extract is diluted with 0.5N $HClO_4$ and two ml of this extract is mixed with 4 ml of diphenylamine reagent in a test tube.
2. All the tubes are incubated for 15-17 hr at 28 $\pm$ 1°C and the absorbance read at 600 nm.
3. DNA content of the sample is read from a standard curve (15-75 μg) prepared, using calf thymus DNA.

Isolation of plant total DNA (Rapid method)

Principle

The DNA is extracted with a suitable medium. The sugars are removed by adding potassium acetate. The DNA is precipitated with isopropyl alcohol.

Reagents and equipment

1. Extraction buffer: 50m M Tris-HCl p^H8.0, 50m M EDTA Na_2^+, 250m M NaCl, 15% sucrose.
2. Ethanol.
3. Isopropanol.
4. 5m M Potassium acetate solution.
5. Porcelain mortar and pestle.
6. Incubator.
7. Centrifuge.

Procedure

1. Weigh out 3-5g of clean and sterile plant material into a porcelain mortar and pestle. Grind to a fine powder under liquid nitrogen.

2. Transfer the frozen powder to a flask containing 15ml extraction buffer at 65⁰C. Incubate at 65°C. for 15 min with gentle shaking.
3. Add 5ml of 5M potassium actate, shake well and incubate in ice for 20 min.
4. Centrifuge contents at 4000 rpm for 20 min.
5. Filter the supernatant (with DNA). Collect the filtrate in a test tube.
6. Add 2/3 volume of isopropanol to the filtrate (2ml to 3ml filtrate). Shake and incubate at -20°C overnight.
7. Pellet DNA by centrifuging at 10,000 rpm for 15 min. Wash the pellet with ice cold 70% ethanol followed by absolute alcohol. Dry the pellet under vacuum.

Reference

Dellaporta, S.L., Wood,J. and Hicks, J.B. (1983) A plant DNA mini preparation: version II. Plant Molecular Biology Reporter 1(4): 19-21.

Estimation of DNA

Principle

The deoxyribose moiety of DNA forms ù-hydroxy laevulaldehyde in TCA solution. This reacts with diphenylamine to give a blue colour which can be read at 565 nm.

Reagents and equipment

1. Dische's reagent: Dissolve 1.5 g of crystalline diphenyl amine in 100ml of glacial acetic acid and 1.5 ml of H_2SO_4. The reagent is stored in the dark. Before use add 0.1 ml of aqueous acetaldehyde (16mg/ml) to 20 ml reagent.
2. 1N perchloric acid.
3. 5mM NaOH.
4. Standard DNA solution: Prepare a stock solution (0.4mg/ml) in 5mM NaOH. Working solutions are made once every 20 days by mixing with an equal volume of 1N perchloric acid and heating at 70°C for 15min.
5. Water bath.
6. Glass marbles.
7. Spectrophotometer with cuvettes.

Procedure

1. Standard curve: Pipette different concentrations of DNA (50,100,200, 500μg etc.) into test tubes. Add 2ml of freshly prepared solution of Dische's reagent and make up to 5ml with glass distilled water. The tubes are covered with glass marbles and heated in a boiling water bath for 10-15min. Cool and measure the blue colour at 565nm in a spectrophotometer.
2. Take 0.1 ml or 0.5 or 1.0 ml of the nucleic acid extract into test tube. Add 2 ml of freshly prepared Dische's reagent and make up to 5 ml with glass distilled water. The tubes are covered with glass marbles and heated in a boiling water bath for 10-15 min. Cool and measure the blue colour at 565 nm in a spectrophotometer.
3. The control contains only water and Dische's reagent.

Calculation

From the standard curve, the DNA in the sample is read. Further calculations are made taking into account the weight of the sample and the dilutions made etc. Final values are presented as percent of weight of sample.

Isolation and estimation of RNA (U.V. Method)

Principle

The ribonucleoprotein complex is broken by SDS into RNA and protein. After deproteination the RNA is precipitated in the cold by adding alcohol.

Reagents and equipment

1. Phenol freshly distilled.
2. Extraction buffer (p^H 9.0): Tris-HCl (0.1M) =1.21g, NaCl (0.075M) = 0.44g, EDTA Na_2^+, (0.005M) = 0.19g make up to 100ml with water.
3. Ethanol.
4. SDS 10% w/v in water.
5. Ether.
6. Mortar and pestle.
7. Centrifuge.
8. Spectrophotometer.

Procedure

All operations should be carried out at 0-4°C.

1. Freeze 0.5-5 g plant material in a mortar and pestle with liquid nitrogen. Grind to a fine powder and extract with 10 volumes of extraction buffer.
2. Centrifuge at 2000 g for 3min.
3. Transfer supernatant to a volumetric flask and stir with 10% SDS for 2-3 min.
4. Add equal volume of buffered phenol (phenol saturated with 100 m M Tris- HCl p^H 8.5).
5. Partition contents by centrifuging at 2000 g for 5 min and collect the upper aqueous phase into a separating flask.
6. Stir the aqueous phase with equal volume of buffered phenol for 5 min.
7. Collect the upper aqueous phase containing RNA, dissolve it in 250 mg NaCl, add 2 volumes of cold phenol (96%). Leave the flask overnight at -20°C for RNA precipitation.
8. Collect RNA by centrifuging at 2000 g for 10 min. Wash the pellet with 70% ethanol, ethanol:ether (1:1v/v) and finally with ether. Dry under vacuum.
9. Dissolve RNA in elution buffer (0.01M Tris, 0.12g; 0.001M EDTA Na_2, 0.04 g; 10% SDS, 0.5 ml; water, 100 ml).
10. Dilute 20 µl aliquot to about 2 ml with buffer and read absorbance at 260 nm. One absorbance unit at 260 nm is equal to 50 µg RNA/ml.

Estimation of RNA (Colorimetric method)

Principle

The ribose moiety of RNA is converted to furfural which condenses with orcinol to yield a green colour. It is measured at 675 nm.

Reagents and equipment

1. 0.1% $FeCl_3$ solution in concentrated HCl.
2. 0.2% orcinol solution in absolute alcohol.
3. 0.1M NaCl.
4. Standard RNA solution: Dissolve crystalline yeast RNA in 0.1M NaCl

(5 mg/ml). if cloudy add a few drops of 0.1MNaOH, filter and store in cold.

5. Spectrophotometer with cuvettes.

Procedure

1. Standard curve: Take 100,200, 400, 500, 800 μg RNA solutions into different test tubes. Add 3ml 0.1% $FeCl_3$ solution and make up to 5.5 ml with glass distilled water. Then add 0.3 ml of orcinol reagent. Keep in a boiling water bath for 20 min. The green colour developed (when solution is clear) is read at 675 nm against a blank.
2. Take an aliquot of the RNA extract. Add 3ml 0.1% $FeCl_3$ solution and make up to 5.5 ml with glass distilled water. Then add 0.3 ml of orcinol reagent. Keep in a boiling water bath for 20 min. The green colour developed (when solution clear) is read at 675 nm against a blank.
3. The blank contains water and the reagents as above.

Calculation

From the standard curve sample RNA value is read. Further calculations are made taking into account the weight of the sample and the dilutions made etc. Final values are presented as percent of weight of sample.

Note

The interference from DNA is about 0.85%.

Estimation of ATP

Principle

Aqueous extracts of firefly lanterns which are not luminous emit light when ATP is added. The other requirements for the reaction are Mg^{++} and the fluorescent luciferin.

$$\text{ATP + Luciferin} \xrightarrow[\text{Mg++}]{\text{luciferase}} \text{adenyl luciferin + pyrophosphate}$$

$$\text{Adenyl-luciferin} \xrightarrow{\text{Oxygen}} \text{adenyl oxyluciferin} + H_2O + \text{light}$$

The light emitted is directly proportional to ATP concentration.

Reagents and equipment

1. 0.16M ATP: Dissolve 25 mg ATP (disodium salt) in 250 ml water. Store at -20°C. prepare afresh, whenever, needed. Dilute 1:100 prior to use.
2. 0.1M Arsenate- 50 mM Mg buffer p^H7.4: Dissolve 42.5g Na_3AsO_4. $12H_2O$ and 10g $MgCl_2$ $6H_2O$ in glass distilled water. Adjust the p^H 7.4 with 1N HCl and dilute to one litre.
3. Luciferase: Grind 50 mg of vacuum dried firefly lanterns (10 or12 lanterns) with 5 ml of ice cold 0.1M arsenate buffer for 5-10 min. Filter the mixture into a test tube kept in an ice bath. Dissolve 50 mg $MgSO_4$,$7H_2O$ in the filtrate.
4. Phenazine methosulphate (PMS) 0.5 mM: Required only if the rate of ATP production is to be followed. Dissolve 0.145 g of PMS in one litre.
5. Water bath.
6. Fluorimeter or a scintillation counter

Procedure

1. Preparation of sample: Heat the sample at 100° C in boiling water bath for 5-10 min. To avoid the enzymatic degradation of substrates in the sample during the heat inactivation, bring the sample to 100° C quickly. This can be done by adding the sample quickly into 2 or 3 volumes of boiling water. Immediately transfer the same to 0° C.
2. Take 0.2 ml luciferase solution and 0.6 ml water. Just before the measurement add 2 ml sample, mix rapidly and read the light intensity up to 30 sec.
3. Measurement can be made in a fluorimeter or a scintillation counter.

Calculation

The luminescence is directly proportional to ATP concentration. The light intensity produced is compared to that produced by the standard. A standard curve can be prepared. The quantity of ATP can be computed.

7

VITAMINS

Vitamins are essential organic nutrients required by an organism in small quantities. They are essential since they cannot be synthesized in sufficient quantities by an organism, and must be obtained from the diet. The term *vitamin* does not include other essential nutrients like dietary minerals, essential fatty acids, or essential amino acids which are needed in larger quantities than vitamins. The term *vitamin* is derived from "vitamine," coined in 1912 by the Polish biochemist Funk. The name is from *vital* and *amine,* meaning amine of life. Thirteen vitamins are generally recognized presently. Vitamins are used in small quantities by an organism to keep it in a healthy condition.

Vitamins are given letters of alphabet (A,B, C, D, *etc.*) in the order in which they have been discovered. The letters do not bear any relation to their structures or their importance. Vitamins are classified by their biological and chemical activity and not by their structure. Thus, each "vitamin" refers to a number of *vitamer* compounds that all show the biological activity associated with a particular vitamin. Vitamin A includes the compounds retinal, retinol, and four known carotenoids. Vitamers by definition are convertible to the active form of the vitamin in the body, and are sometimes inter-convertible to one another, as well.

Vitamins have diverse biochemical functions. Some have hormone-like functions as regulators of mineral metabolism (*Ex.* vitamin D), or regulators of cell and tissue growth and differentiation (*Ex.* some forms of vitamin A). Others function like antioxidants (*Ex.* vitamin E and sometimes vitamin C). The largest number of vitamins (*Ex.* B complex vitamins) function as precursors for enzyme cofactors, that help enzymes in their work as catalysts in metabolism. As co-factors, vitamins may be tightly bound to enzymes as part of prosthetic groups. For example, biotin is part of enzymes involved in making fatty acids. Vitamins may also be less tightly bound to enzyme catalysts as coenzymes, detachable molecules that function to carry chemical groups or electrons between molecules. For example, folic acid carries various forms of carbon groups – methyl, formyl, and methylene – in the cell. These roles in assisting enzyme-substrate reactions are vitamins' best-known and most important functions.

Vitamins are classified into two groups based on their solubility as water-soluble or fat-soluble. Presently, there are 13 vitamins. Out o f them four are fat-soluble (A, D, E, and K) and the remaining nine are water-soluble (8 B vitamins,1, 2, 3, 5, 6, 7, 9 and 12 and vitamin C). Water-soluble vitamins dissolve easily in water and, in general, are readily excreted from the body. Since they are not readily stored in the body, they need to be regularly taken to replenish their requirement. Good intake of fresh fruits and vegetables will satisfy the needs of water soluble vitamins. They are easily destroyed by heat, oxygen and U.V. light. Hence care should be taken while processing and storage of foods. These vitamins serve as cofactors to enzymes.

Fat-soluble vitamins are absorbed in the intestinal tract with the help of lipids (fats). They get accumulated in the body when taken in excess. These are non polar compounds and get deposited in fat deposits and membranes in the body.

Vitamin A (HPLC method)

Principle

Vitamin A is extracted with a suitable organic solvent after precipitation of the protein with ethanol. An aliquot of the organic phase which is evaporated under nitrogen is injected into a revesed phase HPLC column, followed by an eluting solvent of suitable polarity. A reversed phase column has a non-polar stationary phase with a polar mobile phase. Retinal is detected at 326 nm using a sensitive UV detector. An internal standard (Retinyl acetate) is used to account for processing losses. Deficiency of vitamin A causes night blindness, reduces immune response and inhibits growth.

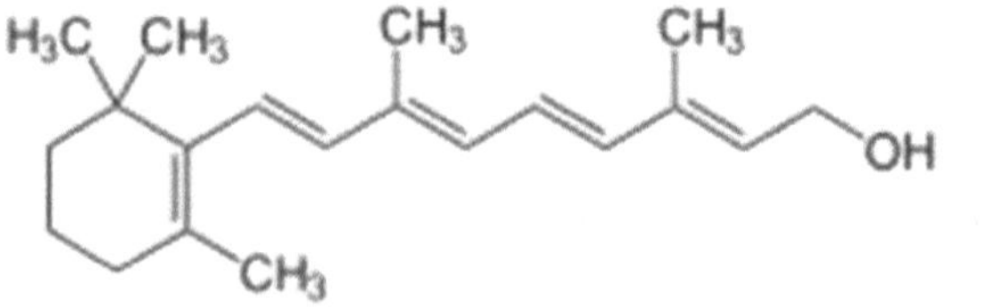

H_3C CH_3 CH_3 CH_3 OH CH_3

Reagents and equipment

1. Retinyl acetate and Retinol standards.
2. Absolute alcohol.
3. Hexane.
4. Methanol.
5. Water (all solvents HPLC grade).
6. Vortex mixer.
7. Centrifuge.
8. HPLC equipment with C-18 column, a UV-Visible spectrophotometric-detector and a data recorder.

HPLC chromatographic conditions

Column: C_{18} reverse phase, 5 mm particle size, 25 cm colum

(or 10µm, 15 cm column)

Mobile phase: Methanol-water (95:5 methanol : water)

Flow rate : 1ml/min

Detector : UV detector 326 nm, sensitivity 0.01 AUFS

Chart speed : 1 cm/min for recording

Temperature : Ambient.

Procedure

a). Sample preparation

1. Take 100 ml of oil sample and add 100 ml internal standard of retinyl acetate (1µg/ml), mix the contents for 1 min and then add 0.5 ml of n-hexane.
2. Vortex for 1 min and centrifuge at 3000 rpm for 5-min.
3. Remove the hexane layer into another tube.
4. Re-extract the sample with another 0.5 ml of n-hexane and centrifuge.
5. Remove the hexane layer and pool with the previous aliquot in the tube and evaporate the total contents to dryness under nitrogen.

6. To the dry tube, add 100 µl of HPLC grade methanol and mix thoroughly.
7. Inject 20 µl into HPLC instrument.

b) Preparation of standard retinol

1. A few crystals or small quantity of retinol (Sigma) is dissolved in absolute alcohol and the absorbance is determined by spectrophotometric method using the extinction coefficient of retinol(E 1%-1800) in alcohol. (An O.D. of 0.18 at 324 nm is equivalent to 1µg/ml).

c) Standard retinyl acetate

1. In the same way the activity in retinyl acetate standards is calibrated by the extinction value (E1% = 1550). An O.D. of 0.155 at 326nm is equivalent to 1µg/ml.

HPLC procedure

1. By use of an internal standard losses due to incomplete extraction, inaccurate aliquots or oxidation are automatically corrected, provided that the internal standard has physical and chemical properties sufficiently similar to retinol, is suitably separated from retinal on HPLC, does not coincide with other 326 nm absorbing material in the sample and not converted to retinol under the assay conditions.
2. A precisely known amount of internal standard added to the aliquot of sample is analyzed.
3. By determining the relative extraction efficiency and detector response of retinol and the internal standard, a standard curve is fashioned in which the ratio of peak heights (or areas) are plotted against the retinol concentration in the sample.
4. A standard curve is prepared by adding varying amounts of retinol to a fixed amount (ie 50mg) of internal standard on a final volume of 50 µl of eluting solvent.
5. The solvent is injected into HPLC under conditions mentioned above, the peak height is measured and peak height ratio is calculated. The peak height ratio is then plotted against retinol concentration.

Calculation

From the concentration of retinol in the standard and its relationship with Area-Under the Curve (AUC) of the peak, the concentration in the unknown is obtained from its corresponding AUC.

$$\text{Concentration in } \mu g/\ 100\ ml = \frac{\text{AUC of sample x conc. of std x 5000}}{\text{AUC of standard x 1000}}$$

Vitamin A (Spectrophotometric method)

Principle

Vitamin A in the oil sample is extracted into chloroform and estimated colorimetrically using thiobarbituric acid reagent.

Reagents and equipment

1. Solvent: Acetone: water (9:1 v/v) or 90% ethanol.
2. Thiourea reagent: Dissolve 4.0 g of thiourea in 100 ml of glacial acetic acid and filter the solution through glass wool or centifuge the solution.
3. Thio barbituric acid reagnt: Dissolve 600 ml of 2-thiobarbituric acid in 100 ml of absolute ethanol, filter and store the solution at 4°C.
4. Vitamin A aldehyde standard: Dissolve 10 mg all-trans retinal in 100 ml absolute ethanol (stock solution, store in dark at 4°C). Before analysis dilute 2 ml of stock solution to 25 ml with 90% ethanol to give a working solution of 8 µg ratinal/ml.
5. Spectrophotometer with quartz cuvettes.

Procedure

1. Place the sample of chloroform solution of lipids (containing 1-10 µg retinal) in a 15 ml glass stoppered tube and remove solvent in a stream of N_2 below 25°C. Immediately dissolve residue in 3 ml of 90% ethanol.
2. Add 1.5ml each of thiourea and thiobarbituric acid reagents, mix and leave at room temperature in the dark for 30 min. Centifuge the mixture to make it clear and measure absorbance of the red solution at 530nm in a 1cm stoppered cell against a reagent blank.
3. Standardize the procedure using 1-10 µg standard all trans- retinal (molar extinction coefficient = 58,000 at 530 nm).

Calculation

$$\mu g \text{ vitamin A aldehyde} = \frac{\text{absorbence of sample x } \mu g \text{ standard}}{\text{absorbence of standard}}$$

Vitamin A (Carr-Price method)

Principle

Vitamin A gives a blue coloured complex with antimony chloride with an absorption maximum at 620nm. β-carotene also gives the coloured complex and hence a correction is needed to the intensity of the blue colour measured.

Reagents and equipment

1. Antimony trichloride: 25% w/v of the reagent in chloroform containing 1%v/v acetic anhydride is prepared.
2. Spectrophotometer with cuvettes.

Procedure

1. An aliquot of the vitamin A solution in chloroform is taken in a cuvette and the volume made up to 1ml. Antimony trichloride (2ml) is added and the colour intensity is measured at 620nm within 15sec.
2. A standard curve is prepared using different concentrations of vitamin A. Correction for β-carotene is made based on its OD.
3. The correct OD for VitaminA = OD observed – OD for β-carotene.

Beta(β) Carotene

Principle

Carotenoids are separated on a column of calcium hydroxide and estimated colorimetrically. Carotene is a precursor of vitamin A. It is converted to vitamin A in the liver.

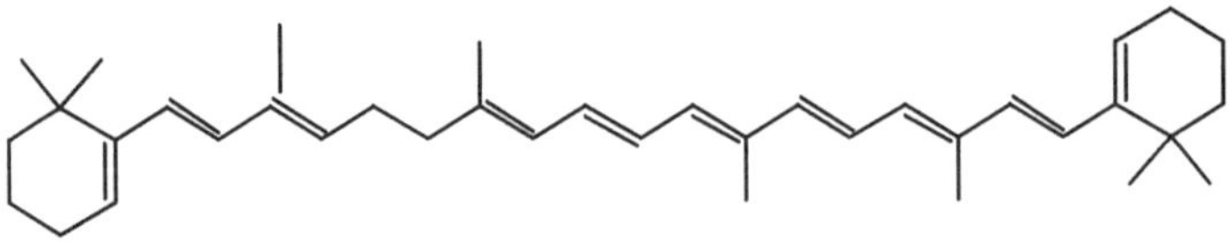

β Carotene

Reagents and equipment

1. Petroleum ether (40-60° C).
2. Acetone.
3. Calcium hydroxide.

4. Anhydrous sodium sulphate.
5. Rotary flash evaporator.
6. Glass columns (1.5X30 cm).
7. Spectrophotometer.
8. Vacuum pump.

Procedure

1. About 20 g of fresh oil is allowed to stand overnight in a mixture of 100 ml of petroleum ether: acetone (1:1).
2. The extract is filtered and the residue on the filter washed twice to remove the entire yellow colour.
3. The pooled filtrate is shaken with 50 ml portions of water twice and then discarded.
4. The organic layer is then dried over anhydrous sodium sulphate and conentrated under reduced pressure to a final voume of 5–6 ml.
5. A column (1.5 x 30 cm) is packed with alumina containing 3% anhydrous sodium sulphate. An aliquot of the sample is put on the colum eluted with 3% acetone in petroleum ether.
6. The eluate is concentrated and made upto 5 ml.
7. The O.D. of the sample is determined spectophotometrically at 460 nm.
8. The concentration is read from a graph prepared with standard beta carotene.

Calculation

1 OD=4μg/ml of beta carotene in a 1 cm cell.

Thiamine -Vitamin B_1 content

Principle

Thiamine helps carbohydrate to release energy. Deficiency causes the disease called "beriberi". The thiamine is oxidized to thiochrome which fluoresces under U.V. light. Under standard conditions and in the absence of any other fluorescing substance the fluorescence is proportional to the thiamine concentration.

Vitamin B_1 Thiamine

Reagents and equipment

1. 1. Acetate buffer (pH 4-4.2): 0.1N acetic acid (66.05 ml glacial acetic acid in one litre water), 0.1 M sodium acetate (136 g sodium acetate $3H_2O$ in 1litre), Mix 70 ml of 0.1N acetic acid and 30 ml of 0.1M sodium acetate and make up to 500 ml.
2. Enzyme suspension: 150 mg taka diastase and 75 mg papain in 5 ml acetate buffer.
3. Standard thiamine : 25 mg thiamine dissolved in 250 ml 0.01 N HCl.
4. Working standard: 1 ml of stock standard thiamine is diluted to 100 ml.
5. 3% H_2O_2.
6. 1N NaOH.
7. 2% Potassium ferricyanide.
8. Oxidising agent: 2 parts of 40% NaOH + 1 part of 2% potassium ferricyanide.
9. Isobutyl alcohol.
10. Water bath.

Procedure

1. Take 5-10 g sample, add 100ml acetate buffer (pH 4-4.2) and 5ml enzyme suspension. Keep the contents overnight at 37^{0}C. Heat on a waterbath at 80^{0}C for 10 min to inactivate the enzymes. Filter and take 10ml of filtrate into three separating funnels (SF).

Blank (SF)	Experimental (SF)	Recovery (SF)
10 ml filtrate+ 1 ml of D.W.+ 1 ml 1N NaOH	10 ml filtrate+ 2 ml distilled H_2O	10 ml filtrate+ 1 ml standard B_1

2. Shake the contents of all the SFs. Then add 2.5 ml of oxidising agent to each of the separating funnels and again shake. Add 3-4 drops of 3% H_2O_2 to the three funnels. This discolourises the solution.
3. Keep for for 2 min. Add a mixture of 7.5 ml H_2O and 7.5 ml of isobutyl alcohol to each separating funnel. Shake vigourously.
4. Throw away the aqueous layer and take the alcohol layer into a test tube.
5. Read fluorescence in a fluorimeter or a spectrofluorimeter.
6. Note the readings separately for the sample (S), blank (B) and recovery sample (R) and calculate the thiamine content.

Calculation

$$\text{Thiamine, mg /100g sample} = \frac{S-B}{R-S} \text{ X } \frac{\text{Dilution x 100}}{\text{Weight of sample}}$$

Riboflavin – vitamin B_2 content

Principle

Riboflavin is a coenzyme useful in the breakdown of fats, lipids and proteins. Deficiency occurs mostly among the vegetarians. The native fluorescence of riboflavin in neutral p^H is used for the estimation of riboflavin.

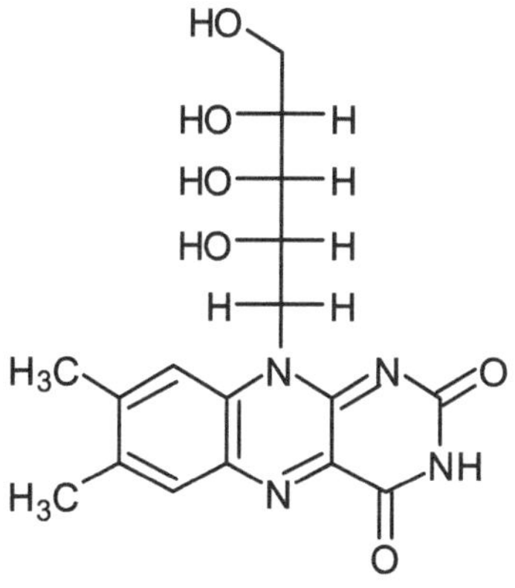

Vitamin B_2 Riboflavin

Reagents and equipment

1. Riboflavin standard: 25 mg of riboflavin is dissolved in 300-400ml water, add 1.2 ml glacial acetic acid. Warm at low temperature to aid dissolution. The contents are then made up to one litre. This solution has 25μg

riboflavin/ml (25 ppm). 2ml of this stock solution is diluted to 50 ml to obtain 1µg/ml or 1 ppm.

2. 4% potassium permanganate
3. 1:1 H_2O_2 –water mixture.
4. $NaHSO_3$.
5. Fluorimeter.

Procedure

1. To about 25 ml vitamin extract, add 1-2 drops of acrylic alcohol are added. Then add 3 ml freshly prepared $KMnO_4$ solution. Stir well and within 2 min add 1:1 H_2O_2 –water mixture to discharge the $KMnO_4$ colour.
2. The p^H is adjusted to 7.0 with NaOH. The volume is made up to 35ml and the solution is filtered.
3. Measure the fluorescence of the filtrate in a fluorimeter.

Calculation

1. The fluorescence of the known aliquot is = A
2. One µg of riboflavin is added and the reading noted = B
3. A small pinch of $NaHSO_3$ is added to destroy riboflavin, the reading noted = C
4. The quantity of riboflavin = A-C/B-C X 1 µg
5. The riboflavin content is calculated taking into account dilution and enzyme blank.

Niacin-vitamin B_3 content

Principle

Niacin has a role in the release of energy from food. Deficiency results in a disease called "pellagra". Nicotinic acid reacts with cyanogen bromide to give a yellow coloured compound which is measured colorimetrically.

Reagents

1. Ca $(OH)_2$ suspension: 250 g $Ca(OH)_2$ in one litre of water. This is diluted to get 2N $Ca(OH)_2$. Check normality through titration.
2. $ZnSO_4$ solution: 80g $ZnSO_4$,$7H_2O$ dissolved in 100 ml water.
3. Phosphate solution: 5g KH_2PO_4 in 100 ml water.
4. Niacin standard: 500 mg dissolved in water, add 5 ml 10N H_2SO_4 and make up to 500 ml.
5. Cyanogen bromide solution: 27 ml bromine in a conical flask is immersed in ice. Neutralize with fresh 10% NaCN using a burette till the solution becomes colourless. Ten extra drops of NaCN are added and the solution is made up to one litre.
6. Buffered CNBr: 500 ml CNBr solution (above) are added to 25 g KH_2PO_4 in a 500 ml volumetric flask. Make up to 500 ml after the contents are dissolved.
7. 8% metol in 0.5N HCl.
8. Caprylic alcohol
9. Water bath.
10. Centrifuge.

Procedure

1. Digestion of sample: 2 g powdered sample is suspended in 10 ml $Ca(OH)_2$ and the volume made up to 50 ml.
2. Sample and blank tubes are simultaneously prepared and all of them are heated in a boiling water bath for 90 min. The contents are stirred frequently.
3. The contents are cooled in a cold water bath and the solution made up to 50ml and centrifuged at 2000 rpm for 20 min.
4. A 25 ml of the supernatant is transferred to a calibrated tube and acidified with 0.5 ml 8N HCl. Then add 2 ml $ZnSO_4$, 2 drops of caprylic alcohol and 2 drops of phenolphthalein reagent.

5. Add 4N NaOH with cooling till $Zn(OH)_2$ is formed. Then add 1NNaOH till the mixture turns pink.
6. Then add $5NH_2SO_4$ till the pink colour disappears. The solution is made up to 50ml. The contents are centrifuged.
7. The supernatants are used for development of colour.
8. Colour development: Each solution is taken in 4 tubes (A,B,C,D) to allow for correction to amine, CNBr and others.

Sample/Reagents	A ml	B ml	C ml	D ml
Filtrate	5	5	5	5
CNBr reagent	2	-	2	-
5% KH_2PO_4	-	2	-	2
8% metol in 0.5N HCl	3	3	-	-
0.5 N HCl	-	-	3	3

The yellow colour is read at 400 nm in a spectrophotometer.

Calculation

The colour due to 5 ml filtrate (0.1 g sample) = A- (B+ C-D) = a

The colour in the recovery tube

(0.1 g sample + 5μg niacin) = A_R-(B_R+C_R-D_R) = b

Therefore b-a μg of niacin = b-a/5

0.1g of substance contains 5a/ b-a

Therefore $\% \, niacin = \frac{5a}{b-a} X \frac{100}{0.1}$

Ascorbic acid (Vitamin C) in fruit juice

Principle

Many fruits contain ascorbic acid (vitamin C). The fruit juice can be diluted to reduce interferences. Use of DPP (differential pulse polarography) in the anodic mode helps in separation and quantification of ascorbic acid. it is essential for skin and general health. Deficiency causes a disease called "scurvy".

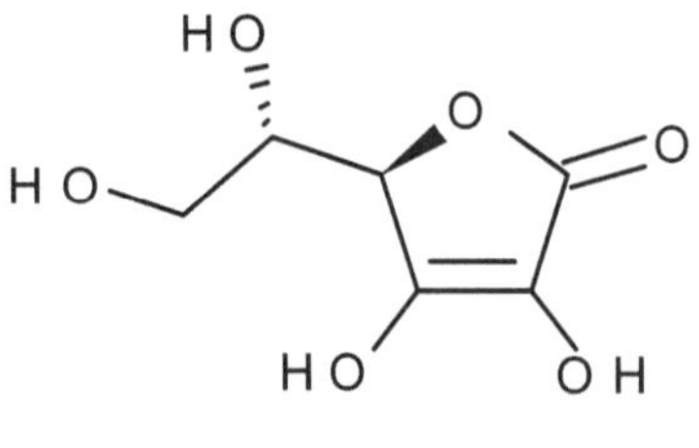

Vitamin C Ascorbic acid

Reagents and equipment

1. Stock solution of ascorbic acid: 30 mg ascorbic acid in 100 ml of electrolyte (300μg/g).
2. Electrolyte solution: Mix equal volumes of 0.05M ethanoic acid and 0.01M $NaNO_3$, p^H = 3.0. This solution should be degassed to remove dissolved O_2.
3. Polarography equipment (DPP).
4. Microsyringe.

Procedure

1. Calibration curve: 20 μl of stock solution + 10 ml degassed electrolyte in the poalrographic cell are utilized for differential pulse polarography under the following conditions.

Drop time	1s
Starting potential	-0.1V
Scan range	0.4V/s
Scan rate	2Mv/s
Pulse modulation amplitude	25Mv
Current range	1-2μA f.s.d.

2. A 20-40 μl aliquot of fruit juice is added to a fresh, degassed 10 ml of electrolyte in the cell and the polarographic trace is determined as above.
3. A calibration curve prepared from standard solutions provides a graph.
4. The amount of ascorbic acid in the fruit juice can easily be extrapolated.

Ascorbic acid (Titrimetric method)

Principle

Ascorbic acid is a water soluble and heat labile vitamin. It is sufficiently present in citrus fruits and fresh vegetables. Ascorbic acid reduces 2-6 dichlorophenol indophenols to a colorless leucobase. The blue dye 2-6 dichlorophenol indophenol is red in acid solution. On titration, the ascorbic acid reduces the dye to a colorless leucobase and the ascorbic acid gets reduced to dehydro ascorbic acid.

Reagents and equipment

1. Glacial acetic acid.
2. 2-6 dichlorophenol indophenol: Dissolve 40 mg of the dye in 100 ml distilled water. 1 ml of this solution = 0.2 mg of ascorbic acid.
3. Burette and related titration assembly.

Procedure

1. Pipette 0.5 ml of the dye into a test tube and add 1 ml of glacial acetic acid.
2. The dye solution is titrated with a sample solution of ascorbic acid with constant stirring till the red color gets discharged.
3. Note the burette reading exactly when color gets discharged.

Calculations

Amount of ascorbic acid = 0.1/ titer value X 100

Tocopherols - vitamin E (HPLC Method)

Principle

Vitamin E acts as an antioxidant and prevents the formation of free radicals. Deficiency causes hemolysis and anemia. Tocopherols are extracted and analysed as per the method of Indyke, (1990).

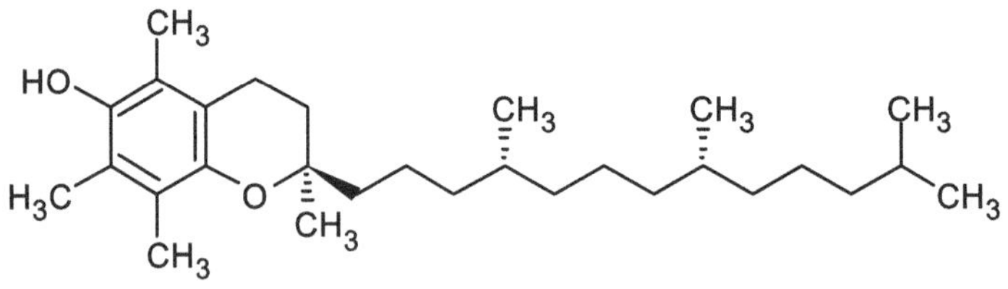

Reagents and equipment

1. Ethanol.
2. Potassium hydroxide 0.5.
3. Hexane : diethyl ether(3:1).
4. Mobile phase: Methanol (HPLC grade).

5. HPLC equipment with C-18 column, a UV-Visible spectrophotometric detector and a data recorder.
6. Centrifuge.

Procedure

a). Extraction

1. 0.1 g of the oil sample is accurately weighed into a test tube. 10 ml of ethanol is added to the sample and the mixture is agitated to avoid agglomeration.
2. 0.5N KOH (2 ml) is added quickly and the tubes are stoppered and incubated for 10 min at 70°C to complete the saponification process.
3. Later the product is cooled to room temperatre and extracted with hexane/ diethyl ether (3:1) and kept for shaking for 5-10 min.
4. 30 ml of water is added into the mixture and inverted 15-20 times and centrifuged at 1000 rpm (180 g) for 10 minutes.
5. An aliquot (10 ml) of the upper layer is evaporatd to dryness under vacuum at 45°C.
6. The residue is dissolved in ethanol (1.0 ml), filtered through 0.45 mm nylon membrane (Filtech Pharmalab, USA).

b). HPLC separation of tocopherols and their quantification

1. The sample is separated using HPLC on a C-18 column, 4.6 x 25 cm.
2. 2ml of the sample is injected into the column. An isocratic mobile phase of methanol (100%) is used at a flow rate of 1.0 ml/min at ambient temperatures after filtration and degassing.
3. Detection is done at 210 nm using a spectrophotometric detector. Identification and quantification are based on comparison of RT values of standard tocopherols.

Reference

Indyke, H.E. 1990. Simultaneous Liquid Chromatographic Determination of Cholesterol, Phytosterols and Tocopherols. Analyst. 91: 115.

Total tocopherols (Colorimetric method)

Principle

By using a preliminary heat-bleach at 250°C the Emmere-Engel method has been adapted for determination of total tocopherols (including tocotrienols) in crude as well as refined palm oil, olein and stearin. Tocopherols and tocotrienols reduce ferric ions to ferrous, which reacts to form a coloured complex with dipyridyl.

Reagents and equipment

1. Toluene.
2. Ethanol.
3. Bipyridine: 0.07% 2-2' bipyridne in 95% of ethanol.
4. $FeCl_3$: 0.2% $FeCl_3$ 6 H_2O in 95% ethanol.
5. Spectrophotometer with quartz cuvettes.

Procedure (refined oils)

1. 200 mg of oil sample is weighed accurately into a 10 ml volumetric flask.
2. 5ml of toluene is added by a pipette and the oil taken into solution.
3. 3.5ml of 2,2'-bipyridine- 0.07% w/v in 95% aqueous ethanol and 0.5ml of $FeCl_3$ 6 H_2O (0.2% w/v in 95% aqueous ethanol) are added in that order.
4. Make up the solution to 10 ml with 95% aqueous ethanol.
5. Let stand for one minute and read absorbence at 520 nm against a blank (prepared as above without the oil). Protect solutions from strong light during colour development.
6. Calibrate the method by preparing standards containing 0-240 mg of pure α-tocopherol in 10 ml of toluene and analyse as above.

	Toluene ml	Bipyridine ml	$FeCl_3$ $6H_2O$,ml	Toluene ml	Read at
Sample 200 mg	5ml	3.5	0.5	1	520 nm.
Blank -	5ml	3.5	0.5	1	520 nm.

Calculations

Total tocopherols (ppm) = A-B/ MxW.

Where, A = Absorbence of sample in 10mm cell.

B = Absorbance of blank in 10 mm cell.

M = Gradient of absorbance, it is approximately 8.0 x 10^{-3} absorbance

per µg β- tocopherol.

W = weight of sample in gms.

Reference

Wong, M.L., Timms, R.F. and Goh. Kempas, E.M. Colorimetric Determination of Total Tocopherols in Palm Oil, Olein and Stearin. JAOCS, 65(2): 1988

Folic Acid (Vitamin B_9) Assay

Principle

Folic acid is estimated based on the absorbance at 595nm. Folic acid is needed for the formation of hemoglobin. Deficiency leads to abnormal blood cells, anemia and loss of hair.

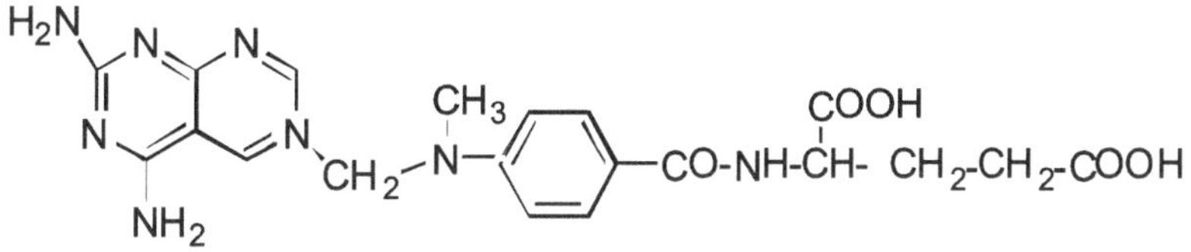

Vitamin B_9 Folic acid

Reagents

1. 0.9% saline.
2. *L. casei* media.
3. 0.5 M Manganous Sulfate: Dissolve 0.845 g/10 ml water and autoclave.
4. *L. casei* media: Dissolve 47 g of media in 500 ml water. Carefully bring to a boil and allow media to cool to the touch. Add 0.25 g ascorbic acid. Sterile filter and save at 4° C for one month.
5. Assay buffer (potassium phosphate buffer pH 6.3 with ascorbate). Dissolve 5.25 g KH_2PO_4, 2.0 g K_2HPO_4 and 0.5 g sodium ascorbate in 300 ml

water. Adjust pH to 6.8. Bring to a final volume of 500 ml. Filter, sterilize and use within one week. Store at 4°C.

6. Folic acid standard (use new glassware each time):

   Dissolve 0.10 g of folic acid in 10 ml 5 mM NaOH (solution A).

   Add 1 ml of the above solution to 19 ml of methanol and store 1 ml aliquots at -70° C (solution B).

   On assay day, remove 100 ml of solution B and add to 25 ml of assay buffer (solution C).

   Remove 50 ml of solution C and add to 50 ml of assay buffer (solution D).

   Remove 50 ml of solution D and add to 50 ml of assay buffer (solution E).

7. The solution is now 2 ng/ml. Use this for all assays.
8. Autoclave.

Procedure

Part I - Preparation of cell suspension - use sterile glassware

1. Thaw 1 ml of frozen cells and add to 10 ml of 0.9% saline.
2. Add 1 ml of the above dilution to 50 ml of *L. casei* media.
3. Add 1 ml of 0.5 M Manganous sulfate (0.845 g /10 ml water, autoclave).

Part II - Preparation of standard curve

1. Pipette 0.5 ml of phosphate buffer into 12 wells.
2. Into well 1 add 0.5 ml of solution E (folic acid standard).
3. Mix and remove 0.5 ml and place into well 2.
4. Mix and remove 0.5 ml and place into well 3.
5. Repeat for well 4, 5 and 6.
6. Perform in duplicate for wells 7 through 12.
7. Remove and discard 0.5 ml from wells 6 and 12.
8. Add 0.5 ml of cell suspension from part I.

Part III - Determination of tissue folate

1. Pipette 0.4 ml of phosphate buffer into a few (3-4) wells.
2. Pipette 0.1 ml of tissue sample into each well.
3. Add 0.5 ml of cell suspension from part I.
4. Incubate in a plastic bag overnight at 37°C.
5. Read absorbance at 595 nm.

Calculation

1. Make calculations based on absorbances of sample and the standards.

Choline determination (Reineckate procedure)

Principle

When methionine intake is insufficient or deficient, choline is utilized to make it up. Thus choline serves as a supplementary vitamin. Choline reacts with ammonium reineckate to form choline reineckate and the colour developed is measured at 526nm spectrophotometrically to estimate the choline content.

$$HOCH_2\,CH_2\,N^{+}(CH_3)_3 \xrightarrow[NaOH//MeOH\ /H_2O]{NH_4\,[Cr\,(NH_3)_2\,(SCN)_4]} HO\,CH_2\,CH_2\,N^{+}\,(CH_3)_3\,[Cr(NH_3)_2(SCN)_4]$$

Choline — Choline reineckate (crystalline ppt)

Reagents and equipment

1. NaOH (2N),
2. Choline standard (8.25 mM) : 1.150 g (dried over H_2SO_4) choline chloride in 1 litre water (1ml = 1mg choline),
3. Phenolphthalein indicator (1%) : 0.5 g phenolphtholein in 50 ml ethanol -H_2O (1:1 v/v).
4. Ammonium reineckate: MeOH solution (2%) : 1g $NH_4[Cr(NH_3)_2(SCN)_4]$ in 50 ml MeOH, centrifuge to obtain clear solution, if neccesary.
5. Vortex mixer.
6. Centrifuge.

Procedure

1. Place an aliquot (containing 0.5-2.0 mg or 4-16 m moles choline) of MeOH-water phase of the methanolic – HCl or methanolic – NaOH hydrolysate or of an aqueous solution of choline in a 15 ml graduated centrifuge tube.
2. Add two drops of phenolphthalein indicator and 2N NaOH untill distinctly alkaline and dilute with water to 5 ml.
3. Add 2.5 ml of of MeOH reineckate solution. Mix on vortex mixer and keep at 4^0C for 2 hr. Centrifuge pink crystals at 3000 rpm and decant supernatant carefully.
4. Wash choline reineckate crystals twice with 1 ml portions of n-propanol and centrifuge. Remove washings by Pasteur pipette.
5. Dissolve residue in 5.0 ml acetone. Centrifuge the solution and transfer to a 1 cm stoppered cell or cuvette and read the absorbence at 526 nm against a blank in which substrate is omitted. The procedure is standardized against known choline (4-20 μ moles or 0.5-2.5 mg).

Calculations

$$\mu \text{ moles choline} = \frac{\text{absorbence of sample} \times \mu \text{ moles of standard}}{\text{absorbence of standard .}}$$

$$\% \text{ choline} = \frac{\text{absorbence of sample} \times \text{mg std. X 100}}{\text{weight of parent lipid(mg)} \times \text{absorbence of standard}}$$

Reference

Glick, D., 1944. J. Biol. Chem 156, 643.

Indole acetic acid (Plant growth harmone)

Principle

Indole-3-acetic acid (IAA) is an auxin and a plant growth hormone. Auxins are compounds that positively influence cell enlargement, bud formation and root initiation. IAA reacts with trifluoroacetic acid in the presence of acetic anhydride to form indole - α-pyrone which is measured fluorometrically.

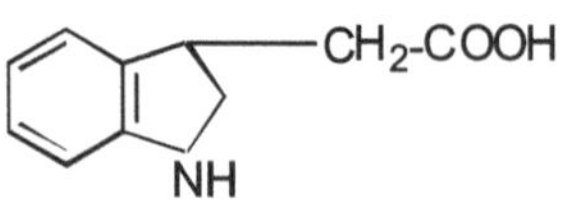

CH_2-COOH

NH

INDOLE 3- ACETIC ACID

Reagents

1. Methanol (re-distilled).
2. Trifluoroacetic acid- acetic anhydride reagent: mix equal volumes of the chemicals. Prepare afresh.
3. Spectrofluorimeter: primary filter 440 nm and secondary filter 490 nm.
4. Liquid nitrogen.
5. 0.5M K_2HPO_4.
6. Petroleum ether.
7. 2.8M Phosphoric acid.
8. Diethyl ether.

Procedure

1. Extraction of IAA: Take 5 g of plant material and freeze under liquid nitrogen. Grind to a fine powder. Add 10 ml methanol and grind further. Filter under suction. Wash the material with 10 ml methanol. Evaporate filtrate at 30°C in a flash evaporator. To the residue add 10 ml cold 0.5 M K_2HPO_4 solution so that the p^H reaches about 8.5. Extract the contents with 10ml petroleum ether. Repeat extraction and discard the lipid fraction. Adjust the aqueous fraction to p^H 3 by adding 3 ml 2.8 M phosphoric acid. Extract IAA with 10 ml diethyl ether. Evaporate the ether and dissolve the residue in 5 ml redistilled methanol.
2. Pipet 1 ml methanolic extract into four different test tubes.
3. To the test tubes add 1 ml methanol containing 0,10,20 and 30 ng IAA.
4. Dry the contents under reduced pressure and cool to 0°C.
5. To each flask add 0.2 ml ice cold trifluoroacetic acid- acetic anhydride reagent and mix.
6. Place in ice for 15 min, stop reaction by adding water.
7. A blank may be prepared by adding 3 ml water to one test tube and 0.2 ml reagent after 15 min.
8. Take the readings in a spectrofluorimeter using excitation at 440 nm and emission at 490 nm.
9. Calculate the amount in the sample.

References

Stroessl, A. and Venis, M. A. (1970) Anal. Biochem 34: 344.

Knegt, E. and Bruinsma, J. (1973) Phytochem 12: 753.

8

PHENOLICS AND ANTINUTRIENTS

Phenol is normally a benzene ring containing an –OH group, and hence an aromatic alcohol. The term "phenolic" refers to similar aromatic compounds with one or more -OH groups. Phenolics constitute one of the most widespread classes of secondary metabolites. Commonly talked about class of compounds like anthocyanins, flavonoids, lignins, tannins, coumarins and chalcones are all phenolic substances. Phenolics are water insoluble substances. However, the water solubility increases with the increase in the number of -OH groups. Phenolics absorb electromagnetic radiation strongly. Many of the phenolics absorb in the visible part of the spectrum (*Ex*. anthocyanins) and are coloured. Phenolics are produced in plants by 1. Shikimate pathway through phenylalanine (C_6-C_3) and 2. Acetate/malonate pathway.

Plants synthesize many chemicals that cause toxic reactions when consumed by animals or men. Plant foods contain protease inhibitors, lathyrogens, flavism agents, cyanogens, haemagglutinins and saponins. Protease inhibitors are present in cereals and pulses. Saponins are present in spinach and asparagus. Goitrogens (cause hypothyroidism) are present in cabbage and rapeseed. Oxalic and phytic acids are present in oilseeds. Cooking is one way of reducing the toxicity levels. Analysis of these constituents is essential to assess their safety status.

Some important phenolic compounds

Phenolic compounds occur exclusively in the form of glycosides and esters. Polyphenols like tannins, coumarins, flavonols, anthocyanins and hydroxylated cyclohexanes are present in plants. Chlorogenic acid, caffeic acid, ferulic acid, quinic acid, scopoletin, rutin, coumaric acid, eugenol, guaicol, vanillic acid etc.are some important phenolics in plants.

Lignin, a cell wall strengthening material, is a C_6-C_3 polymer. It is more abundant in wood, corky tissues and straw. Lignin imparts resistance to degradation by water and microorganisms. In plant cell walls where lignin is more common, lignin remains embedded in cellulose microfibrils in a dense cross connected network. Lignin is made up of three aromatic alcohols of which the most important is p-coumaryl alcohol.

Flavonoids are 15 carbon phenolics with C_6-C_3-C_6 carbon skeleton. Flavonoids are widely distributed in the plant kingdom. They absorb radiation very strongly in the visible part of the electromagnetic spectrum and are differently coloured. Anthocyanins which impart colour to flowers, leaves, and fruits of many plants are flavonoids. In flowers, they help attract the pollinating insects. A specific example of flavonoid is delphinidin, the substance that gives blue color to grapes. Flavonols and flavones are also important flavonoids. Flavonoids occur as glycosides attached to sugars, normally glucose and occasionally galactose.

Tannins are C_6-C_3-C_6 ploymeric phenolics. They are astringent phenols and tan animal skin to leather. Tannins cross link proteins, thus denaturing them. They thus keep the microorganisms away and impart a protective function. Coumarins form the starting metabolite of a number of important phenolics. Aroma of freshly cut grass is due to the coumarins.

Phytic acid

Phytic acid is the main component of phosphorus storage in plant. It also is a source of energy and cations, a natural antioxidant for plants, and can be a source of myoinositol which is one of the preliminary pieces for cell walls.

Phytic acid is also known to bond with many different minerals, and by doing so prevents those minerals from being absorbed; making phytic acid an anti-nutrient. There is a lot of concern with phytic acids in nuts and seeds because of its anti-nutrient characteristics. Cooking can reduce the amount of phytic acid in food but soaking is much more effective. Phytic acid is an antioxidant found in plant cells and serves the purpose of preservation. This preservation is removed when soaked, reducing the phytic acid and allowing the germination and growth of the seed. When added to foods it can help prevent discoloration by inhibiting lipid peroxidation. There is also some belief that the chelating property of phytic acid may have potential use in the treatment of cancer.

Gossypol

Gossypol ($C_{30}H_{30}O_8$), an yellow pigment, is found in cotton plants. It occurs mainly in the root and/or seeds of different species of cotton plants. Gossypol exists in three forms: gossypol, gossypol acetic acid, and gossypol formic acid. All of them have similar biological properties. Gossypol is a polyphenolic aldehyde, and has a formyl group. The formation of gossypol occurs through an isoprenoid pathway. Gossypol's main function in the cotton plant is to act as an enzyme inhibitor. An example of gossypol's enzyme inhibition is its ability to inhibit nicotinamide adenine dinucleotide-linked enzymes of *Trypanosoma cruzi* which is a parasite causing Chaga's disease. It is an effective protein kinase C inhibitor. It also inhibits replication of HIV-1 virus. It is an inhibitor of several dehydrogenase enzymes.

Gossypol has been considered as a waste product produced during the processing of cottonseed. Recent studies have shown that gossypol has other functions. It can act as a male contraceptive. Gossypol has also been linked to causing hypokalemic paralysis, a disease characterized by muscle weakness. Gossypol induced hypokalemic paralysis is easily treatable with potassium repletion. Cotton seeds contain gossypol in the range of 0.7 – 1.2 %.

Phenol content (Folin Ciocalteau method)

Principle

The estimation of phenols is based on the reaction with an oxidizing agent phosphomolybdate forming a blue complex. The intensity of the complex is measured colorimetrically.

Reagents and equipment

1. Chloroform: methanol : water : (25:60:15) or Alcohol.
2. 20% Na_2CO_3 in water
3. Diluent : Isopropanol : water (1:1)
4. F.C. (*Folin – Ciocalteau*) Reagent (F.C. reagent: water: 1:1, dilute ready made reagent with distilled water). If ready made solution is not available, prepare the solution as follows: Dissolve 100 g of sodium tungstate and 25 g of sodium molybdate in 700ml distilled water. Add 50 ml of 85%

orthophosphoric acid and 100 ml concentrated hydrochloric acid and 150g lithium sulphate dissolved in 50 ml water and 4-5 drops of liquid bromine. Boil the mixture without any condenser for about 15 minutes to remove excess bromine. Cool and dilute to one litre with water and filter. The reagent should be golden yellow in colour (store the solution in a brown bottle, preferably in a refrigerator). Just before use dilute 1 volume with 2 volumes of water. If the reagent acquires a green colour, it is not fit for use. However, it can be regenerated by boiling with a few drops of bromine. Boil out excess bromine.

5. Standard: Catechol or phenol or caffeic acid (10 ppm solution, 10 mg/ 1000ml).
6. Water bath.
7. Spectrophotometer with quartz cuvettes.
8. Centrifuge.

Procedure

Standard curve

1. Take 0.0 0.1, 0.2, 0.3, 0.4, 0.5, 0..6, 0.7, and 0.8, ml of standard catechol (10 ppm solution,) to a 25 ml volumetric flask and add 0.5 ml of F.C. reagent (1:1). Add 2.0 ml of Na_2CO_3 solution shake thoroughly and boil over a hot water bath for one minute.
2. Take out and cool under running tap water.
3. Make up to volume with diluent and measure the absorbance at 650 nm.
4. Plot a graph with absorbance versus concentration.

Extraction of phenols

1. Weigh 0.5g finely powdered plant sample in a centrifuge tube and add 5.0 ml of reagent No.1 and shake for 10 min.
2. Centrifuge and take the supernatant in a 10 ml volumetric flask.
3. Wash the residue twice with the solvent mixture (2ml each) and pool the supernatants and make upto 10 ml.

Estimation of phenols

1. Take 0.1 ml of the above extract, add 0.5 ml F.C. reagent (1:1) and 2 ml of 20% Na_2CO_3 into a 25 ml volumetric flask.

2. Shake thoroughly and keep the sample in a boiling water bath for 1 min.
3. Immediately take out and cool under running tap water.
4. Make up the volume to 25 ml by adding the diluent, and read the absorbance at 650 nm in a spectrophotometer.
5. Read the concentration using the absorbance value from the standard graph already prepared.

Calculations

% Phenols = ppm/ W x 100

ppm= Concentration in ppm from the graph,

W= Weight of sample

Chlorogenic acid content (HPLC method)

Principle

This method is useful in separating phenolic acids like chlorogenic acid through HPLC. Identification and quantification also can be carried out (Srinivas *et al.*, 2007).

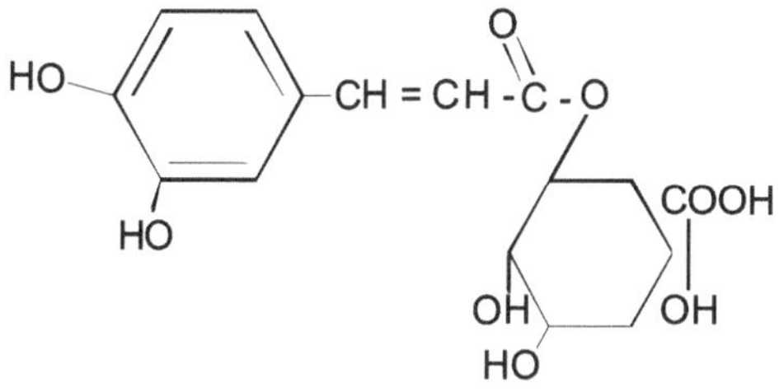

CHLOROGENIC ACID (3-O- CAFFEOYL-d-QUINIC ACID)

Reagents and equipment

1. HPLC equipment with C-18 column, a UV-Visible spectrophotometric detector(SPD) and a data recorder.
2. Methanol 90% aqueous.
3. Potassium hydroxide 5% aqueous.
4. Sulphuric acid 5% aqueous.
5. Diethyl ether.
6. Mobile phase: Methanol 100% ml (HPLC grade).

Procedure

Isolation of Phenols

1. The phenols can be isolated as per the method of Stedmann, (1957), Stedmann and Swain (1962).
2. 20g of defatted oil seed (castor sunflower)/ plant sample is extracted with 50 ml portions of 90% methanol thrice. The pooled extract is filtered. The methanol solution is concentrated to 50 ml and re-extracted with diethyl ether (50-100 ml).
3. The phenols are extracted with 5% KOH thrice (25, 20 and 20 ml portions).
4. The pooled aqueous solution is then acidified using 5% H_2SO_4. The acidic layer is saturated using sodium bicarbonate.
5. The aqueous solution is extracted thrice with diethyl ether (10, 5 and 5 ml portions). Finally the combined ether layers are concentrated to 1 ml..

Separation and characterization (HPLC)

1. 20µl of the concentrated phenolic extract is injected into the pump and separated on a C-18 column (2.5 X 30 cm) of the HPLC instrument. The flow rate is 0.5ml per minute under isocratic conditions with the mobile phase being 100% methanol.
2. Identification of phenolic acids is based on comparison of RT values of standard phenols (cinnamic acid, ferulic acid, syringic acid, coumaric acid, chlorogenic and delphic acids).

Reference

Srinivas, V.S. Chakravartula and Nagaraj Guttarla. 2007. Identification and characterization of phenolic compounds in castor seed. Natural Products Research. 21(12): 1073-1077.

Stedman, R.L. 1957. Polyphenolic constituents of Tobacco. Tobacco science.6:1.

Stedman, R.L., Swain, A.P. 1962. Tobacco hexane extractives. Tobacco Science. 1: 155

Oxalic acid content (Volumetric procedure)

Principle

Oxalic acid is extracted with dilute acid, precipitated as calcium oxalate and the same estimated by titration with $KMnO_4$ solution.

Reagents

1. Hydrochloric acid (2N) : To 82.91 ml distilled water 17.09 ml concentrated HCl is added.
2. Phospho tungstate reagent : Dissolve 24 g sodium tungstate in water. To this 40 ml of phosphoric acid is added and the solution made upto 1 litre with water.
3. Methyl red (0.1%) in 95% ethanol.
4. Calcium chloride buffer: Dissolve 25g $CaCl_2$ in 50% glacial acetic acid. Dissolve 330g sodium acetate in 500 ml water. Both the solutions are mixed well to obtain the buffer.
5. $KMnO_4$ (0.01N): Dissolve 0.0316 g $KMnO_4$ in distilled water and make upto 100 ml.
6. A mechanical shaker.

Procedure

1. To 5 g of the powdered sample, 100 ml of 2N HCl is added and the mixture shaken well for about 22hrs in a mechanical shaker.
2. It is then centrifuged and filtered. The filtrate is transferred into a beaker and boiled for 15 min. It is then cooled to room temperature. Some distilled water is added followed by 2N HCl to make up to 100 ml.
3. To 20ml of the filtrate, 2 or 3 drops of methyl red is added and neutralised with ammonia. Five ml of calcium chloride buffer is added and stirred well.
4. The mixture is allowed to stand overnight and later it is filtered through Whatman No. 40 filter paper and washed thoroughly to free the chlorides with distilled water.
5. The precipitate along with the filter paper is transferred to a beaker and some distilled water is added followed by the addition of 5ml of 2N sulphuric acid.
6. The mixture is heated to 80°C over a burner and titrated against N/100 potassium permagnate solution.

Calculations

1 ml of 0.01 N $KMnO_4$ = 0.45 mg oxalic acid.

mg oxalic acid/100g sample =

$$\frac{\text{Titre value} \times \text{Normality of } KMnO_4 \times \text{dilution factor}}{\text{Weight of the sample}}$$

Estimation of phytic acid

Principle

The phytic acid and phytic acid chelates react with ferric chloride and form ferric phytate. The available ferric ion after the reaction is determined by developing the blood red colour with KSCN.

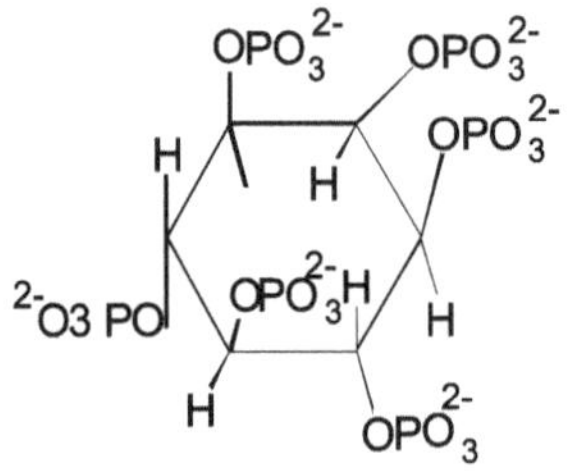

PHYTIC ACID - MYOINOSITOL HEXAPHOSPHATE

Reagents and equipment

1. $FeCl_3$ solution: Containing 2 mg ferric iron per ml in 3% TCA.
2. 5% TCA: Dissolve 5g TCA in 100ml water.
3. 1.5 M KSCN solution.
4. Centrifuge
5. Incubator
6. Spectrophotometer with quartz cuvettes

Procedure

1. 250 mg of defatted sample is taken in a 15 ml centrifuge tube. 7.5 ml of 5% TCA (trichloroacetic acid) is added while shaking.
2. The mixture is incubated for 10 min at 60°C. The supernatant is collected after centrifuging at 5000 rpm and kept for 10 min.

3. The supernatant is then transferred into 25 ml standard flask and made up to the volume by using distilled water.
4. 20ml is pipetted from the flask, to which 5ml of ferric chloride solution is added and heated in a water bath at 95°C for 45 min.
5. The reaction mixture is cooled and filtered using Whatan 42-filter paper.'
6. 2ml of KSCN is added and the colour developed is measured at 485 nm.

Calculation

Calculations are made based on values obtained from calibration graph prepared with standard phytic acid (10-50 or 100 ppm concentrations).

Reference

Wheeler, E.L. and Ferrel, R.E. 1971. A Method for Phytic Acid Determination in Wheat Fractions. Cereal Chemistry. 48: 312- 315

Estimation of polyphenols

Principle

The phenolics comprise of a very complex tannin compounds. Polyphenolics are estimated by Folin Denis method (AOAC, 1984) spectrophotometrically at 760 nm.

Reagents equipment

1. Folin – Denis reagent: To 750 ml water, add 100 g sodium tungstate ($Na_2WO_4.2H_2O$), 20 g phosphomolybdic acid and 50 ml phosphoric acid. Reflux for 2 hr, cool and make upto 1 litre with water.
2. Saturated sodium carbonate solution: Dissolve 45 g of Na_2CO_3 in 100 ml distilled water at 70-80°C; and cool overnight. The supersaturated solution is then filtered through glass wool.
3. Tannic acid standard solution: Dissolve 100 mg tannic acid in 1 litre water. Fresh solution is prepared for each determination.
4. Methanol – HCl: Mix 10 ml concentrated HCl with methanol and make up to 1 litre with methanol.
5. Vortex mixer/cyclomixer.
6. Water bath.
7. Spectrophotometer with quartz cuvettes.

Procedure

1. Weigh 200 mg powdered defatted plant or seed sample into a 250 ml round bottomed flask.
2. To this sample 100 ml of methanol – HCl is added and the contents are refluxed for 2 hrs and allowed to cool.
3. The extract is filtered through Whatman No. 40 filter paper into a 100 ml volumetric flask and volume made up with methanol-HCl, after a few washings. The extract is taken for estimation of polyphenols.

Standard Curve

1. Pipette 0.2, 0.4, 0.6, 0.8, and 1 ml of standard tannic acid solution into 20ml test tubes. Volume is made upto 5ml with water. Add 0.5 ml of Folin- Denis reagent and vortex it.
2. To all the tubes 1 ml of saturated sodium carbonate solution is added. Make up the final volume to 10 ml and mix well using cyclomixer.
3. Allow to stand for 30 min and read the absorbence at 760 nm.

Sample analysis

1. One ml of sample extract is taken in triplicate in 20ml test tubes. Volume is made upto 5ml with water. Add 0.5 ml of Folin- Denis reagent and vortex it.
2. To all the tubes 1 ml of saturated sodium carbonate solution is added. Make up the final volume to 10 ml and mix well using cyclomixer.
3. Allow to stand for 30 min and read the absorbence at 760 nm.
4. Using the standard curve intercept the concentration and calculate the polyphenols in the experimental material taking into account sample weight and dilutions made, if any.

Reference

Official Methods of Analysis of the Association of Official Analytical Chemists (A.O.A.C.). 1984. 14th Edition. Association of Official Analytical Chemists, Inc. 1111, North Nineteenth Street, Suite 210, Arlington, VA 22209, U.S.A. p.187- 188.

Estimation of tannins (Vanillin-HCl method)

Principle

Tannins are estimated by Vanillin Hydrochloric acid method (Price *et al.,* 1978).

The colour complex formed absorbs at 500 nm.

Reagents and equipment

1. Hydrochloric acid:methanol (8%): Mix 8 ml concentrated HCl with 92 ml methanol.
2. Vanillin (1%): Dissolve 1g vanillin in methanol and make upto 100ml.
3. Vanillin – HCl reagent: Mix equal volumes of solutions 1 and 2 before use.
4. HCl in methanol (4%): Mix 4 ml concentrated HCl with 96 ml methanol.
5. HCl in methanol (1%): Mix 1 ml concentrated HCl with 99 ml methanol.
6. Standard solutions: A stock solution containing 1mg/ml catechin is prepared in methanol (0.1ml of stock solution is equivalent to 100 µg catechin).
7. Centrifuge.
8. Spectrophotometer with quartz cuvettes.

Procedure

Tannin extraction

1. Five hundred mg defatted sample is weighed and transferred to a centrifuge tube. To this 10 ml acidic- methanol (1%) is added and shaken for 20 min on a shaker.
2. The tubes are centrifuged for 10 min and the supernatants transferred to a 25ml volumetric flask.
3. To the residue in the centrifuge tubes 5ml acidic methanol (1%) is added and shaken for 20 min.
4. The tubes are centrifuged for 10 min and the supernatants are transferred to the first extraction. The volume is made upto 25 ml and mixed well.

Standard curve

1. The standard solution is taken at levels of 0.1,0.2, 0.3, 0.4,0.5 0.6, 0.7, and 0.8 ml in different test tubes. The volume is made upto 1 ml in all the tubes.
2. Five ml of vanillin – HCl reagent is added and the intensity of the colour developed is read in the spectrophotometer at 500 nm against a sample blank.
3. Sample blank is prepared by the addition of 5 ml of 4% HCl in methanol to 1ml of the sample solution.
4. The absorbance is plotted against different concentrations of standard.

Estimation

1. One ml extract is taken in triplicate into three 20ml test tubes.
2. To this 5 ml of freshly prepared Vanillin – HCl reagent is added slowly and the intensity of the colour developed is read in the spectrophotometer at 500 nm.
3. The concentration of the sample is intercepted from the standard graph using the absorbance value of the sample.

Calculation

Catechin equivalents (CE%) =

$$\frac{\text{(mg catechin/ml) x volume made up (25 ml) x 100}}{\text{Volume of extract taken (1 ml) x weight of the sample (500 mg)}}$$

Reference

Price, M.L. Van Scoyoc, S. and Butler, L.G. 1978. A critical Evaluation of the Vanillin Reaction as an Assay for Tannins in Sorghum Grain. Journal of Agricultural and Food Chemistry.

Tannins (Folin-Denis Method)

Principle

Tannin-like compounds reduce phosphotungstomolybdic acid in alkaline solution to produce a highly colored blue solution, the intensity of which is proportional to the amount of tannins. The intensity is measured in a spectrophotometer at 700 nm.

Reagents

1. *Folin-Denis Reagent*: Dissolve 100g sodium tungstate and 20g phosphomolybdic acid in 750ml distilled water in a suitable flask and add 50ml phosphoric acid. Reflux the mixture for 2h and make up to one liter with water. Protect the reagent from exposure to light.
2. *Sodium Carbonate Solution:* Dissolve 350g sodium carbonate in one liter of water at 70-80°C. Filter through glasswool after allowing it to stand overnight.
3. *Standard Tannic Acid Solution* : Dissolve 100mg tannic acid in 100ml of distilled water. Dilute 1: 10 for development of colour.
4. *Working Standard Solution* : Dilute 5ml of the stock solution to 100ml with distilled water. One ml contains 50g tannic acid.
5. Centrifuge.
6. Water bath.
7. Spectrophotometer with quartz cuvettes.
8. Vortex/cyclomixer.

Procedure

1. *Extraction of Tannins*: Weigh 0.5 g of the powdered sample and transfer to a 250 ml conical flask. Add 75ml water. Heat the flask gently and boil for 30 min. Centrifuge at 2,000 rpm for 20 min and collect the supernatant in 100ml volumetric flask and make up the volume.
2. Transfer 1 ml of the sample extract to a 100ml volumetric flask containing 75 ml water.
3. Add 5 ml of Folin-Denis reagent, 10 ml of sodium carbonate solution and dilute to 100 ml with water.
4. Shake well and read the absorbance at 700 nm after 30 min.
5. If absorbance is greater than 0.7, make a 1 + 4 dilution of the sample.
6. Prepare a blank with water instead of the sample
7. Prepare a standard graph by using 0-100μg tannic acid.

Calculation

Calculate the tannin content of the samples as tannic acid equivalents from the concentration value read from the standard graph.

Reference

Schanderl, S H (1970) In: Method in Food Analysis. Academic Press. New York, p 709.

Phytate P Determination

Principle

The phytate is extracted with trichloroacetic acid (TCA) and precipitated as ferric phytate. Iron content is estimated spectrophotometrically. The quantity of phytic acid is calculated based on the phosphorus content using the ratio of 4: 6 for Fe and P.

Reagents and equipment

1. 3% TCA.
2. $FeCl_3$ solution (containing 2 mg ferric iron per ml in 3% TCA.
3. 3% Na_2SO_4 in 3% TCA.
4. 1.5N NaOH.
5. 3.2N HNO_3.
6. Fe $(NO_3)_3$.
7. Spectrophotometer with cuvettes.
8. Centrifuge.
9. Water bath.
10. 1.5 M KSCN

Procedure

1. Weigh a finely ground sample(40 mesh) estimated to contain 5 to 30 mg Phytate P into a 125 ml Erlenmeyer flask.
2. Extract with 50 ml 3% TCA for 30 min with mechanical shaking or 45 min with occasional swirling by hand.
3. Centrifuge the suspension and transfer a 10 ml aliquot of the supernatant into a 40 ml conical centrifuge tube.
4. Add 4 ml $FeCl_3$ solution (containing 2 mg ferric iron per ml in 3% (TCA) to the aliquot by blowing rapidly from the pipet.

5. Heat the contents in a boiling-water bath for 45 min. If the supernatant is not clear after 30 min, add 1 or 2 drops of 3% Na_2SO_4 in 3% TCA and continue heating.
6. Centrifuge for 10-15 min and carefully decant the clear supernatant.
7. Wash precipitate twice by dispersing well in 20 to 25 ml 3% TCA, heating in boiling-water bath 5 to 10 min, and centrifuging.
8. Repeat wash once with water.
9. Disperse the precipitate in a few ml water and add 3 ml 1.5N NaOH with mixing.
10. Bring volume to about 30 ml with water and heat in boiling-water bath for 30 min.
11. Filter hot (quantitatively) through a moderately retentive paper (Whatman No. 2).
12. Wash precipitate with 60 to 70 ml hot water and discard filtrate. If phosphorus determination is desired, it may be made on this filtrate.
13. Dissolve the precipitate from the paper with 40 ml hot 3.2N HNO_3 into a 100-ml volumetric flask.
14. Wash paper with several portions of water, collecing the washing in the same flask.
15. Transfer a 5 ml of 1.5M KSCN, dilute to volume, and read color immediately (within 1 min) at 480 nm.
16. Run a reagent blank with each set of samples.
17. Calculate iron content from a Fe $(NO_3)_3$ standard run at the same time or read from a previously prepared standard curve.
18. Calulate the phytate phosphorus from the iron results assuming a 4:6 iron: phosphorus molecular ratio.

Phytic acid content

Principle

Phytic acid is determined by the method of Davies and Reid (1979). The colour developed with ammonium thiocyanate is read spectrophotometrically.

Reagents and equipment

1. Nitric acid (0.5N) : To 484.1ml distilled water, 15.9ml 69.5% nitric acid (AR grade specific gravity 1.42) is added.
2. Ferric ammonium sulphate : Dissolve 216 mg ferric ammonium sulphate in water and make upto 500 ml.
3. Ammonium thiocyanate: Dissolve 10 g of ammonium thiocyanate in water and make upto 100 ml.
4. Iso-amylalcohol.
5. Sodium phytate: Dissolve 30.54 mg sodium phytate in 100 ml of 0.5N nitric acid to give 20 mg/100 ml or 200 μg/ml.
6. Vortex or other flask shaker.
7. Water bath.
8. Centrifuge.
9. Spectrophotometer with cuvettes.

Procedure

Extraction

1. One g finely ground sample is extracted with 25 ml of 0.5 N nitric acid for 3 hrs with continuous shaking in a shaker. After proper shaking it is filtered through Whatman No: 1 filter paper.

Estimation

1. Into 3 tubes 1.0 ml of the filtrate is taken and diluted with distilled water to a final volume of 1.4 ml.
2. To the standard and sample tubes 1 ml of ferric ammonium sulphate solution is added.
3. The tubes are mixed thoroughly and placed in a boiling water bath for 20 min.
4. After cooling to room temperature, under running water, 5 ml of isoamyl alcohol is added.
5. Immediately after mixing the tubes by inversion method, 0.1 ml of ammonium thiocyanate solution is added.

6. The tubes are shaken well and then the tube contents are centrifuged at 3000 rpm for 10 min. The alcoholic layer is separated with a pasteur pipette.
7. Finally the colour intensity of alcoholic layer is read at 465 nm against amyl alcohol blank exactly after 15 min. The colour in amyl alcohol layer is inversely proportional to phytate concentration.

Standard graph

1. For plotting a standard curve, different concentrations i.e. 0.4 to 1.0 ml of standard phytate solution containing 80 to 200 μg of phytic acid is taken and made up to 1.4 ml with water.
2. Follow steps 2 to 7 above.
3. The readings obtained are plotted against different concentration of the standard and the concentrations of the sample is intercepted from graph.

Calculation

1. Calculations are made based on the concentration read from the standard curve and the weight of sample taken for estimation etc.

Estimation of saponins

Principle

The saponins are extracted with H_2SO_4 and fractionated over active alumina column. The concentration is read at 530nm.

Reagents and equipment

1. Standard saponin solution: Dissolve 100 mg of saponin in acetic acid and make upto 100 ml.
2. 1N H_2SO_4 in Dioxane: Water (1:2 v/v): To 97.21 ml of 1:2 dioxane and water 2.79 ml of concentrated H_2SO_4 is added.
3. Acetic acid (10%): Mix 10ml of acetic acid with 90ml of distilled water.
4. Concentrated sulphuric acid.
5. Sodium sulphate.
6. Refluxion assembly.
7. Active alumina column.
8. Spectrophotometer with cuvettes.

Procedure

Extraction and determination

1. Five hundred mg sample is taken in an extraction flask, dispersed in 50ml of 1N H_2SO_4 in dioxane: water (1:2) and hydrolysed under reflux for 8 hrs.
2. The contents are cooled and diluted with addition of 50 ml water.
3. Sapogenins are extracted with 25ml and then with 3 successive portions of 15ml diethyl ether.
4. The combined ether extracts are washed with water and made moisture free by adding Na_2SO_4 and then dried.
5. After taking to dryness the residue is taken up in minimal amount of benzene and purified on a column of Alumina (Aluminium oxide active-acidic) suspended in 100ml benzene and 0.2ml of 10% aqueous acetic aid. It is stirred vigorously for 30 min, poured into a column of 15 mm diameter and washed with 250 ml benzene. Sapongenin extract (1-2ml) is loaded onto the column.
6. Various impurities are then removed by washing the column with 100ml of 3% benzene solution in metnanol.
7. This sapogenin solution is concentrated readily to dryness and the residue taken up in 10ml acetic acid.
8. To 2ml acetic acid solution of sapogenin 1ml glacial acetic acid followed by 2ml concentrated H_2SO_4 are added.
9. The tubes are cooled to room temperature. The absorbence is read at 530 nm against a suitable blank (containing 3ml glacial acetic acid and 2ml concentrated H_2SO_4).

Standard curve

1. A standard curve is prepared by taking 0.5ml to 3ml standard saponin solution (0.5 to 3mg saponin) and treated in the same way as that of the sample. The concentrations of unknowns are read from the standard graph.

Calculation

1. Calculations are made taking into account the concentration from the graph and the weight of the sample etc.

Caffeine content

Principle

Caffeine has a E max value of 272nm. The O.D. of a clear solution is measured and the amount of caffeine is calculated using a standard curve.

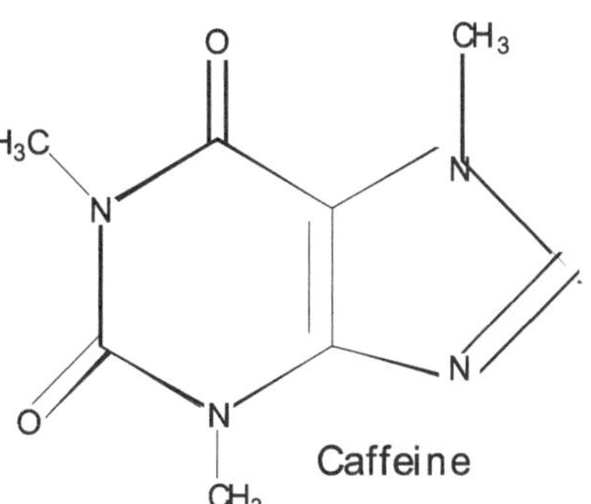

Caffeine

Reagents and equipment

1. 0.1N sulphuric acid.
2. Heavy MgO.
3. Pure caffeine (0.1mg/ml).
4. Spectrophotometer with cuvettes.

Procedure

1. Weigh accurately about 1g of roasted coffee.
2. Transfer to a one litre flask. Add 50 ml 0.1N sulphuric acid and 250 ml water carefully boil for 20 min.
3. Add 50 mg heavy MgO and boil further for 20min. Cool and make up to one litre and filter.
4. Dilute 50 ml aliquot to 100ml.
5. Measure OD at 272 nm.
6. Prepare a standard curve using pure caffeine (0.1mg/ml). Take 1-20ml aliquots and dilute to 100ml. Measure the OD of these solutions.
7. Using the standard curve measure the caffeine concentration of the sample based on its OD.
8. Make calculations based on the sample weight etc. Express caffeine as per cent.

Caffeine isolation and estimation

Principle

The caffeine is isolated from the aqueous extract with chloroform. The content is estimated spectrophotometrically.

Reagents and equipment

1. Chloroform.
2. Refluxion assembly.
3. Buchner funnel.
4. Silicagel (filter aid).
5. Dilute KOH solution.
6. Pb $(CH_3COO)_2$ solution (saturated).
7. Rotary flash evaporator.
8. HCl solution dilute.
9. Dilute H_2SO_4.
10. Spectrophotometer with quartz cuvettes

Procedure

1. Caffeine isolation with chloroform: 20 g of coffee/tea and 90 ml of distilled water are refluxed for 30 min, and filtered under vacuum.
2. The residue is again refluxed and filtered.
3. The filtrates are combined, 12.5 ml of Pb $(CH_3COO)_2$ solution is added, boiled (5 min), and filtered through a Buchner funnel with silica gel layer.
4. The filtrate is extracted four times with chloroform (40 ml).
5. Combined chloroform phases are washed with KOH solution and then with distilled water.
6. Chloroform is removed from extracts by rotary evaporator. After evaporation, extracted caffeine is weighed and expressed in mg/l.
7. Caffeine determination using the lead acetate solution: The caffeine extract is treated with HCl solution (5 ml), $Pb(CH_3COO)_2$ and H_2SO_4 solution.
8. Absorbance of obtained extracts is measured at 274 nm.

9. The content of caffeine (mg/l) is calculated using a standard curve derived from caffeine (0–250 mg/l).
10. All measurements are performed in triplicate.

Gossypol

Principle

Gossypol occurs in cotton plant and mainly its seeds. It reacts with phloroglucinol under acidic conditions to produce a reddish-brown product. The colour is measured to quatify the gossypol.

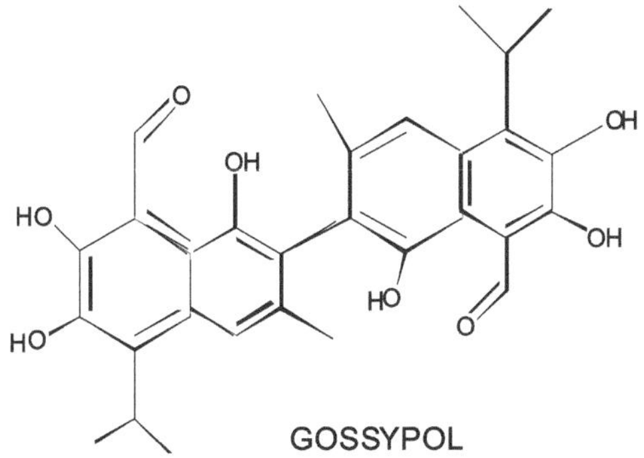

GOSSYPOL

Reagents and equipment

1. Ethanol (96%).
2. Phloroglucinol : 5mg phloroglucinol in 100ml 80% ethanol.
3. Diethyl ether.
4. Concentrated HCl.
5. Spectrophotometer.
6. Gossypol acetate.
7. Incubator.

Procedure

1. Extraction: Weigh 5g fresh tissue . Make into small pieces. Put in 20ml boiling ethanol (96%) for 5min. Filter and collect the extract. Re-extract the residue two or three more times. Combine the extracts. Dilute with 40% ethanol. Adjust p^H to 3.0. Mix contents with 1.5 volumes of diethyl ether at $10^{\circ}C$ in a separating funnel. Separate the ether phase. Re-extract

the alcoholic phase and pool the ether extracts. Evaporate ether and dissolve the ether extractives in 95% ethanol.

2. Pipet 1 or 2ml of the above alcoholic extract into test tubes. Add 0.5ml phloroglucinol reagent followed by 1m conc. HCl to the tubes.
3. Incubate for 30min with stirring at room temperature.
4. Make the volume to 10ml with 80% ethanol.
5. Read absorbance at 550nm against a reagent blank.
6. Prepare a standard graph using gossypol acetate.
7. Read the concentration of sample based on its absorbancy from the standard graph.
8. Make calculations taking into sample weight and diutions etc.

References

Bell, A A (1961) Phytopathol 57:159.
Sood, D R and Singh, A (1973) Cotton Dev 3:15.

Karanjin content (HPLC Method)

Principle

Karanjin is the active principle in karanj oil. Karanjin is a furanoflavonoid. The methanolic extract is separated and estimated using HPLC.

Reagents and equipment

1. Standard karanjin (1mg/ml).
2. Methanol.
3. Methanol: water(80:20).
4. Soxhlet apparatus.
5. HPLC equipment with spectrophotometric detector.

Procedure

1. Extraction of karanjin: 8g of karanjin cake is extracted with methanol for 12hr in a Soxhlet apparatus. The clear methanol extract is taken and the excess solvent evaporated under vacuum. The residue is dissolved 1-5 ml methanol. If necessary the extract can be diluted to get asuitable peak through the HPLC.

2. Use suitable aliquot (20µl) for injection into HPLC.
3. HPLC analysis:

Column oven	:	CTO-10 As vp
On line degasser	:	DGU-14A
Flow contol valve	:	FCV-10AL vp
Solvent delivery nodule	:	LC-10AT vp
Detector	:	SPD-M 10A vp
Column	:	C18 reverse phase
Mobile phase flow rate	:	1.0 ml/min
Detection	:	250nm UV detector
Sample size	:	20µl
Average pressure	:	2000psi
Oven temperature	:	40^0C

4. The mobile phase is methanol : water(80:20).
5. Standards and samples injected to the HPLC equipment (20µl)
6. The chromatograms are obtained from the instrument

Calculation

Area points are taken into consideration to calculate karanjin concentration in the sample based on area of the standard.

$$\text{Karanjin (mg/ml)} = \frac{\text{area point of sample}}{\text{area point of standard}}$$

$$\text{Karanjin in the sample} = \frac{A \times B \times 100}{W \times 100}$$

Where A = karanjin in mg/ml, B = methanol extract (%), W = weight of sample

Azadirachtin by HPLC

Principle

Azadirachtin is a low molecular weight bio-pesticide. It is a tetranortriterpenoid ($C_{35}H_{44}O_{16}$) from the neem tree. It is extracted from the deoiled neem seed kernel into methanol and analyzed by HPLC.

Reagents and equipment

1. Petroleum ether(40-60^{0}C) or hexane.
2. Methanol.
3. 0.2micron filter.
4. Soxhlet extraction apparatus.
5. HPLC with C-18 ODS column and a spectrophotometric detector (SPD).

Procedure

1. Weigh 10g of dry neem kernel powder and put it in a thimble.
2. Extract at 40-60^{0}C with petroleum ether or hexane for 3-4 hrs in a Soxhlet apparatus.
3. Collect the extract, air dry it and weigh the oil content.
4. Soak the thimble with the deoiled cake in methanol for 16-18 hrs in the Soxhlet extractor itself.
5. Then extract with methanol in the Soxhlet for 6-8 hrs.
6. Collect the extract in an amber coloured bottle and store in a fridge.
7. Concentrate the extract to 100ml.
8. Filter through 0.2 micron filter and use the filtrate for HPLC analysis.
9. Inject to the HPLC system 20 μl of methanolic extract (filtered through filter 0.2 μm) through syringe with a C-18 ODS column (250 x 4.60mm) and a spectrophotometric detector (a suitable aliquot - 10-20μl).
10. The chromatographic separations are accomplished using a mobile phase consisting of 40% acetonitrile. Mobile phase is pumped at a flow rate of 1 ml/min at room temperature. Absorbance at 217 nm is measured.
11. The retention time of Azadirachtin (9-12 minutes) is determined using HPLC.
12. For identification and quantification standard azadirachtin can be used with the HPLC.
13. Based on standard peak size and the quantity of sample injected and the initial sample powder and the dilution of the extracts, calculations are made.

Calculation

Azadirachtin % can be calculated by using the formula:

$$\frac{\text{Standard wt.}}{\text{standard sample weight}} \text{ X } \frac{\text{Standard volume}}{\text{sample volume}}$$

$$\frac{\text{Sample peak area}}{\text{standard peak area}} \text{ X purity of the standard}$$

Raffinose, Stachyose and Verbacose (T LC Method)

Principle

Mixture of sugars are separated on TLC plates depending upon their Rf value which is the ratio of component front to solvent front. The individual components can be qualitatively and quantitatively determined.

Reagents and equipment

1. Balance.
2. Thin layer applicator set.
3. Hot air oven.
4. Hot air blower.
5. TLC solvent tank.
6. Chromatography sprayer.
7. Mixture of chloroform, acetic acid, water (for sugars), (3:3.5:0.5, v/v) ratio.
8. Silica gel with 13% calcium sulphate-silica gel G-60.
9. Diphenylamine spray reagent: Dissolve 1g diphenylamine in 100ml acetone. Add 1ml of aniline, 8.5ml orthophasphoric acid and 1.5mlof water. Mix well before use.
10. Densitometer.

Procedure

1. Arrange 5 clean TLC plates (20 x 20cm) on the TLC applicator (There should be no gaps in between plates).

2. Weigh 40g of silica gel (or silica gel-G-60, reagent 8 above) into a conical flask.
3. Add 80-85 ml distilled water and close the flask with a rubber stopper.
4. Shake vigorously for 5 to 7 min. The slurry is ready for spreading when particles swell and become more viscous.
5. Pour the complete slurry into the spreader at once and move the spreader forward with uniform force (slow spreading causes uneven distribution).
6. Leave the coated plates undisturbed for 1h at room temperatue.
7. Activate the plates by keeping them in an oven at 100°C for 1h.
8. Remove the plates and store in desiccator when not in use. Leave an inch gap at the bottom of the coated plate and spot the required volume (e.g., 5mg equivalent of sugars in aqueous or alcohol medium) of the extract. Also spot standard sugars like glucose raffinose, stachyose, verbacose etc.)
9. During spotting use hot air blower for drying spots.
10. Saturate TLC chamber 2-3 hr or overnight with suitable solvent system (e.g., chloroform: acetic acid: water – 3.0:3.5:0.5, v/v).
11. Keep the spotted plates in the chamber for developing.
12. Remove the plates when the solvent front is one inch below the top of the plate.
13. Dry the plates with the help of a hot air blower.
14. Spray the dried plates with diphenylamine reagent.
15. Dry the plates in an oven until coloured spots develop.

Calculations

1. Calculate the Rf value using a scale by measuring the distance covered or eluted by the spot and the solvent front. (Rf = spot front/solvent front). Identify the unknown spots using Rf values of known sugars.
2. Measure the color of the spot on a densitometer using appropriate filter. Compare with densitometer readings of a known standard sugar developed on TLC plates as above.

Sesame Lignans

(Hemalatha and Ghafoorunnissa, 2004)

Principle

The lignans in the unsaponifiables of sesame oil are extracted and separated on a C-18 column of HPLC equipment. They are qualitatively and quantitatively estimated with the help of standards.

Reagents and equipment

1. Mobile phase: Methanol : water, 75:25 (HPLC grade).
2. Sesamol – authentic sample, available commercially.
3. Sesamin and sesamolin, extracted and prepared as described below.
4. HPLC equipment with C-18 column, a UV-Visible spectrophotometric detector and a data recorder.
5. Centrifuge.
6. Cryofreezer.

Procedure

Sesamin and sesamol extraction

1. Sesame oil (500g) in acetone (1:8 v/v) is cooled over night at –70°C. The glyceride crystals formed are separated by filtration.
2. The mixture is kept at –20°C. Removal of acetone from the filtrate gives a yellow oil, which is mixed with iso-octane and left at 4°C for 4-5 days.
3. Sesamin crystallises, which is separated through centrifugation. The filtrate after evaporation of iso-octane is saponified with ethanolic KOH.
4. The mixture is diluted and extracted with ether. The filtrate after removal of ether is dissolved in chloroform and petroleum ether (80-100°C). Upon cooling at 5°C, sesamolin crystallises.
5. Sesamin and sesamolin are recrystallised in ethanol. These are used as standards in HPLC separation.

Procedure

1. The oil at 1:100 dilution in hexane is directly injected using a reversed phase HPLC system into a C-18 column(4.6 mm x 250 mm) through an injector.

2. The peaks are recorded using the UV-Visible spectrophotometric detector. The mobile phase is methanol: water, 75:25 (HPLC grade). Flow rate is 0.8 ml/min.
3. Peaks can be identified and quantified by comparison with authentic lignans (sesamol, sesamin and sesamolin).
4. Sesamol can be used as an internal standard for calculating the percentage recovery of the lignans.

Reference

Hemalatha, S. and Ghafoorunissa. 2004. Lignans and Tocopherols in Indian Sesame Cultivars. Journal of the American Oil Chemists' Society. 81(5): 470.

Trypsin inhibitor activity

Principle

The casein digestion method is employed for the measurement of trypsin inhibitor activity (TIA) in grains. The method involves the use of 2% casein and determination of the breakdown products by trypsin in the presence and absence of the inhibitor. Later inhibition of this activity is measured to calculate the trypsin inhibitor activity.

Reagents and equipment

1. 0.1M sodium phosphate buffer,p^H 7.5.
2. 2% casein solution in phosphate buffer.
3. Trypsin solution 5mg/ml.
4. 0.001M HCl.
5. 5%TCA.
6. Centrifuge.
7. Incubator.
8. Spectrophotometer with cuvettes.

Procedure

1. Extraction of samples: 4g of powdered defatted plant material is treated with 40ml of 0.1M sodium phosphate buffer p^H 7.5 and 40ml distilled water. Shake samples for 3hr and then centrifuge at 700g for 30min at 15^{0}C. Use the supernatant as inhibitor extract.

2. Suitable dilutions can be made in such a way that inhibition is around 40-60% of enzyme activity.
3. The incubation mixture consists of 0.5 ml of trypsin solution, 2ml of 2% casein, 1.0ml of sodium phosphate buffer, 0.4 ml HCl (0.001M) and 1 ml extract of sample. The volume of total incubation mixture is kept at 4.0 ml. Incubations are carried out at 37^{0}C for 20 min. Later 6.0 ml of 5% TCA is added to stop the reaction. Blanks are run side by side.
4. In this method, one unit of trypsin activity is chosen as an increase of 0.01 absorbance unit at 280 nm for 10ml of reaction mixture.
5. The trypsin inhibitor activity is measured as the number of trypsin units inhibited.

Calculation

1. One trypsin unit (TU) is an increase of 0.01 absorbance units at 280 nm in 20min per 10 ml of the reaction mixture under the conditions mentioned above. TIA can be expressed as number of TU inhibited either per mg or per ml of extract.

Reference

J. Agi.Food Chem. 19: 257 (1971)

Chymotrypsin inhibitor activity

Principle

Casein digestion method is used for the measurement of Chymotrypsin inhibitor activity (CIA) in grains. The method involves the use of 1% casein and the spectrophotometric determination of the breakdown products by a given concentration of chymotrypsin in the presence of the inhibitor.

Reagents and equipment

1. 0.1M borate buffer: P^H 7.6: Dissolve 6.183g boric acid in water and make upto one litre. Dissolve 38.139g disodium tetraborate($NaB_4O_7.10H_2O$) in water and make upto one litre. Mix both solutions to get P^H7.6.
2. Casein solution: Mix1g casein in 80ml 0.1M borate buffer P^H 7.6 and dissolve by heating on a steam bath for 15 min. Cool and adjust to P^H7.6 and make upto 100 ml.
3. 0.08M $CaCl_2$ in 0.001M HCl: Dilute 0.083ml of conc. HCl in one litre water (0.001M HCl). Then dissolve 11.762g $CaCl_2.2H_2O$) in one litre 0.001M HCl.

4. Chymotrypsin solution: Dissolve 4mg chymotrypsin in 100 ml of 0.001M HCl containing 0.08M $CaCl_2$.
5. Trichloro actic acid (TCA) reagent: Dissolve 18 g anhydrous sodium acetate and 20 ml glacial acetic acid in water and make upto one litre.
6. Balance.
7. P^H meter.
8. Shaker.
9. Centrifuge.
10. Water bath.
11. Vortex mixer.
12. Spectrophotometer.

Procedure

Extraction

1. Weigh 200 mg defatted fluor into a screw cap centrifuge tube.
2. Add 10ml 0.1M borate buffer and shake contents for 1hr on a shaker.
3. Centrifuge the suspension for 5min at 3000g and filter through No. 42 filter paper. Use the filtrate as inhibitor extract.

Chymotrypsin assay

1. Pipet 0.2, 0.4, 0.6, 0.8 and 1.0ml stock chymotrypsin solution into a triplicate set of test tubes (one set for each level of enzyme).
2. Adjust the final volume to 1.0ml with 0.001 M HCl (reagent 3).
3. Add 1ml 0.1M borate buffer, P^H7.6 to all the test tubes and keep in a water bath at 37°C.
4. Add 6ml TCA reagent to one set of triplicate tubes- this serves as blank for other two sets.
5. Add 2ml casein solution (37°C) to all the tubes.
6. Allow the tubes to remain at 37°C for 10min.
7. Stop the reaction by adding 6 ml TCA reagent to the experimental tubes.
8. Vortex the tubes and incubate for 30min at room temperature.

9. Filter through no. 42 paper and read absorbance of filtrate at 275 nm against the blank.

Calculation

One CIA unit is arbitrarily defined as an increase of 0.01 absorbance units at 275 nm in 30 min per 10ml of reaction mixture.

Amylase inhibitor activity

Principle

The amylase inhibitor activity (AIA) is measured based on the inhibition of amylase activity.

Reagents and equipment

1. 0.02M Phosphate buffer: a) Dissolve 2.78g monobasic sodium phosphate in water and make upto one litre. b) Dissolve dibasic sodium phosphate(Na_2HPO_4 $2H_2O$) 7.17g slowly and make upto one litre. Mix both solutions and adjust p^H to 6.9 using a p^H meter.
2. 3:5 dinitrosalyscylic acid: Dissolve10g of 3:5 dinitrosalyscylic acd in 300 ml water and to this add 200 ml 2NNaOH, and 300 g sodium potassium tartarate and make upto one litre.
3. Dissolve 2.925 g sodium chloride in one litre 0.02M Phosphate buffer.
4. Starch: Prepare 1% starch solution in sodium chloride solution (reagent 3).
5. Amylase solution: a) Take human saliva and dilute 10 fold using reagent 3. Store it overnight at 5° C. Centrifuge at 3000g for 15min at 5°C. Use the supernatant for amylase assay. b) dissolve 10mg crude alpha amylase from porcine pancreas in 50ml of reagent 3.
6. Balance.
7. P^H meter.
8. Shaker.
9. Centrifuge.
10. Water bath.
11. Vortex mixer.
12. Spectrophotometer.

Procedure

1. Weigh 0.5 g defatted sample into a screw cap tube and shake with10ml phosphate buffer for 2h on a tube shaker.
2. Centrifuge at 3000 g for 15min.
3. Incubate the supernatant at 70°C for 10min and filter. Use the filtrate as inhibitor extract.
4. 1ml inhibitor extract is incubated at 30°C for 30 min.
5. Add 1ml starch solution. After 3min add 2ml 3:5 dinitrosalyscylic acid reagent to stop the reaction.
6. Place the tubes in boiling water bath for 5min
7. Cool the tube, add 10ml water and read absorbance at 550nm.
8. Repeat the study for the enzyme inhibitor blank. Boil enzyme inhibitor extract for 8min and cool. This can be used as enzyme inhibitor blank.
9. Repeat the study for enzyme blank. For this add enzyme after 3:5 dinitrosalyscylic acid reagent is added
10. Take different concentrations of maltose standard and add 2 ml 3:5 dinitrosalyscylic acid.
11. Place the tubes in a boiling water bath for 5min, cool and read absorbance at 550 nm.
12. Plot a standard graph of maltose for calculation.

Calculation

$$\text{Amylase units inhibited / g} = \frac{= (y - z - x) \times 100 \times 10}{500}$$

Where Y = mg maltose liberated by the enzyme

Z = mg maltose present in enzyme blank

X = mg maltose liberated in the presence of enzyme inhibitor.

Glucosinolates-Quick Test

(Institute of Agronomy and Plant Breeding, University Gottingen, Germany)

Principle

Glucosinolates are thio-esters of glucose and other sugars. They become antinutritional after hydrolysis by thioglucosidases, which are present in the plant.This is a quick test method for determination of glucosinolates.

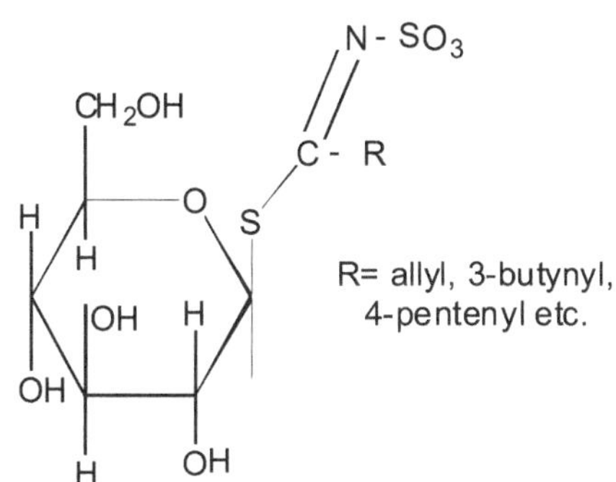

General structure of glucosinolates

Reagents and equipment

1. Polypropylene tubes.
2. Water bath.
3. Centrifuge.
4. Microtitre tray.
5. Vibrator.
6. Hot air oven.
7. Multi- channel photometer.
8. Methanol.
9. Diazobenzene sulphonic acid.
10. O- Phosphoric acid.
11. Linetta standard.

Procedure

1. Take 200 mg finely powdered seeds into poly propylene test tubes.
2. Add 300 µl 60% methanol. Keep for 5min at 80°C (water bath).
3. Add 2ml water, mix in a Vortex mixer for 15min at 80°C.
4. Mix and centrifuge.
5. Take 50 µl of the above extract in a well of a microtiter tray.
6. Add 200 µl reagent solution: 20mmol = 1.535 g 4-diazobenzene sulfonic acid. Dissolve in 250 ml water + 250 ml 85% O-phosphoric acid. Solution is slightly brown. Stable at 4°C for several weeks.

7. Mix in a vibrator.
8. Coupling reaction: Keep for 30 min at 70 °C.
9. Photometry against a blank (300µl reagent) with a multi-channel photometer at 510 nm.
10. Conversion to the concentration unit "µmol/g seed": Concentration (µmol/g seed = A_{510}/0.2g seeds –0.4 X 546).
11. The control value "A_{510}/0.2g seed" for the standard "Linetta" should be between 1.15 and 1.26 (O =1.20).

Gluco test method for glucosinolates (Thies, 1985)

Principle

The glucosinolates are estimated after their reaction with the enzyme myrosinase by reflectometry using the test papers.

Reagents and apparatus

1. Reflectometer.
2. GLUCOTEST paper.
3. Myrosinase(100 units enzyme activity).
4. Charcoal filter papers.
5. Filter supports.
6. Hamilton syringes.
7. 1, Chlorohexidine: 10% in methanol.

Procedure

1. Weigh 5g of finely powdered seed into a 250 ml glass beaker.
2. Add boiling water over the sample filling up to the 200ml mark.
3. Stirr the suspension for 1 min to speed up the extraction of the glucosinolates.
4. Cool 1ml of the extract to less than 60°C in a polystyrol tube of 3.5 ml voume.
5. Pour the cooled extract into a tube containing 100 units of myrosinase. Close with thumb and mix by 4 quick turns. After a reaction period of 5 min, add 5 ml of a protein precipitating agent (10% chlorohexidin diacetate in methanol) and shake the tube again thoroughly by hand.

6. Draw the contents of the tube into a 1ml syringe and press the same tightly onto the filter support armed with activated charcoal paper.
7. Pull out the piston from the syringe: the syringe must thereby be kept horizontal to facilitate the passage of the extract into the upper part of the filter unit
8. Place filter unit with syringe onto a 3.5 ml polystyrol tube and collect at least 5 drops of a clear filtrate. Discard opaque or coloured liquid.
9. Tranfer 17 µl of the filtrate onto a clean slide forming 1cm long strip.
10. Calibrate the reflectometer with a 17 mm long piece of GLUCOTEST paper and place the same immediately afterwards onto the liquid strip on the slide. Cover with cardboard box to exclude light effects. After 1 minute repeat.
11. Calibration of the reflectometer with a second dry test-paper strip (necessary because the commercial reflectometer is programmed with 1minute reaction time only).
12. Insert the treated test paper with a forceps into the reflectometer after a total reaction time of 2 min (in the dark) and note the result of the measurement in digital form.
13. Repeat steps 10 to 13 and calculate the mean X of both measurements.

Calculation

Use the formula: $C = (X - 90)\ 0.5$

Where C= the content of glucosinolates in the sample in ì mol per g air dry seeds. The value 90 is obtained as a mean when extracts are measured which are prepared as above but without myrosinase addition.

Reference

Thies, W. 1985. Glucosinolatgehalt in Handelspartien von 00-Raps – Schnelle Bestimmung durch Taschen-Reflektometer. Raps 3: 112-114.

Glucosinolates- Desulphation and HPLC (Buchner,1987)

Principle

Crude glucosinolate extracts are prepared without defatting of the meal and addition of internal standards before extraction. Glucosinolates in the crude extracts are enzymatically desulphated on-column. The resulting solution of desulphoglucosinolates is separated and measured using reversed-phase high-performance liquid chromatography (HPLC) with a solvent gradient. Calculations are based on HPLC peak areas, use of relative response factors and internal standard of known purity.

Reagents and equipment

1. 15 m mol/l glucotropaeolin, 15 mmol/l sinigrin and 3 m mol/1 sulfanilic acid as internal standards.
2. 20 mM/l tris buffer pH 7-8, 20 mM/1 sodium carbonate buffer pH 6.5-7, 20mM/1 sodium acetate buffer pH 4.
3. 20 mM/l potassium phosphate buffer pH 4.
4. Anion exchangers: DEAE-Sephadex A-25, Servacel Ecteola 23, Serdolit AW-2, AW-4, AW-14, Duolite A 30 B. Swollen in 2 M acetic acid, filled in the pasteur pipettes, rinsed with the 10 fold volume 6 M imidazole formiate, rinsed again with the 20 fold volume of water.
5. Sulfatase type H-1.
6. DEAE-Sepharose C1-6B formate form .
7. 0.2 mol/l sodium acetate.
8. Ultra pure water and acetonitrile for HPLC.
9. Pasteur pipettes, glass wool placed in the tips as ion exchange columns.
10. Water bath or heating plate.
11. Small laboratory centrifuge.
12. Millipore dip-in-filter type Immersible C X-10.
13. Low temperature oven.
14. U.V. spectorphotometer.
15. HPLC chromatograph: HPLC equipment with C-18 column, a UV-Visible Spectrophotometric detector(SPD) and a data recorder.

Procedure

Purification of the sulfatase

1. 200 mg sulfatase is dissolved in 20 ml water.
2. The solution is passed through 20 ml Sepharose column (diameter 26 mm).

3. Rinse the same with 40 ml water.
4. Elute with 60 ml of 0.2 mol/litre sodium acetate.
5. Concentrate the eluate to about 4 x 100 µl using four dip-in-filter.
6. Dilute to 4 x 2 ml with water.
7. Concentrate to 4 x 100 µl once more.
8. Rinse the filters, combine and dilute the filtrates to 20 ml sulfatase stem solution.
9. Store in a deep freeze.
10. Regeneration of column: rinse with 10 ml 6 M/l imidazole formiate and 50 ml water subsequently (one column filling can be used up to 10 times).
11. 50 µl of sulfatase stem soulution + 2 ml 0.15 m mol/l sinigrin (stem solution 1:10 diluted) should result at least in an absorption decrease at 227 nm of 0.0338 absorption units/minute.

Sample preparation and HPLC analysis

1. Weigh 0.2 g of finely powdered seed.
2. Add 1.5 l hot 70% methanol/water mixture and immediately afterwards 200 µl of internal standard glucotropaeolin (1 part standard to 2 parts final volume diluted rapeseed extract). Shake.
3. Extract 10 minutes at 80°C.
4. Centrifuge 3 minutes at 3000 rpm.
5. Decant supernatant and repeat step (2) to (5) for the precipitate adding more internal standard two times.
6. Combine supernatants and fill up to 5ml.
7. Put 500 to 1000 µl of the extract to 20 mg sephadex columns.
8. Rinse two times with 1ml 20 mmol/l sodium acetate buffer pH 4.0.
9. Add 75 µl sulfataste solution (1 part sulfatase stem solution + ½ part water).
10. Cap the columns.
11. Desulfation takes about 6 hrs at 39°C.
12. Elute with 1.5 ml water. Shake the samples.

13. If necessary filter the samples.
14. Inject 30-80 µl into HPLC. Programme: 0-16.5 min 0-16.5% acetonitrile, 16.5-18 min 16.5-0% acetonitrile (rest: water). Flow speed: 2.0 ml/min.
15. Detection at 229 nm.
16. Calculate the amounts of glucosinolates using the area of the internal standard and the relative response factors.

Reference

Buchner, 1987. Comparison of Procedures for Optimum Extraction, Purification and Analysis of Desulfo Indolyl Glucosinolates. Glucosinolates in Rapeseeds: Analytical Aspects. Martinus Nijhoff Publishers. Dordrecht. pp 77-79.

9

TOXINS

Certain constituents associated with plants are toxic. When such items are consumed by men and animals, they become hazardous. Mold growth on food items may be the cause of toxicity. Some of them are called mycotoxins. Aflatoxins are the best examples of such a group of compounds. Sometimes processing also produces some toxins. Fumigants, smoke and some metals (mercury, lead, cadmium, arsenic etc.) used for such purposes are some such toxic constituents. Pesticides used in protecting the crops and grains during storage remain on the items and the residues are hazardous.

Plants are capable of synthesizing a multitude of chemicals that cause toxic reactions when consumed. Pulses contain a number of toxic substances like enzyme inhibitors and lathyrogens. Goitrogens (which cause hypothyroidism and thyroid enlargement) are present in some plant species. Some allergens are also present in plant foods.

Estimation of aflatoxins

Principle

Four of the aflatoxins namely B_1, B_2, G_1 and G_2 are the most widely occurring in *A. flavus* infested foods. They are very strong hepatocarcinogens. The aflatoxins are extracted into chloroform (from the methanolic crude fraction) and the same is separated on a TLC (thin layer chromatography) plate using Benzene:ethanol:water(46:35:19 v/v). The aflatoxins are identified under U.V.

light based on their fluorescence. The developed spots can be identified based on Rf values of known aflatoxin standards. They are estimated using photo-fluorimeter. Their quantitation can also be made based on visual intensities of the spots.

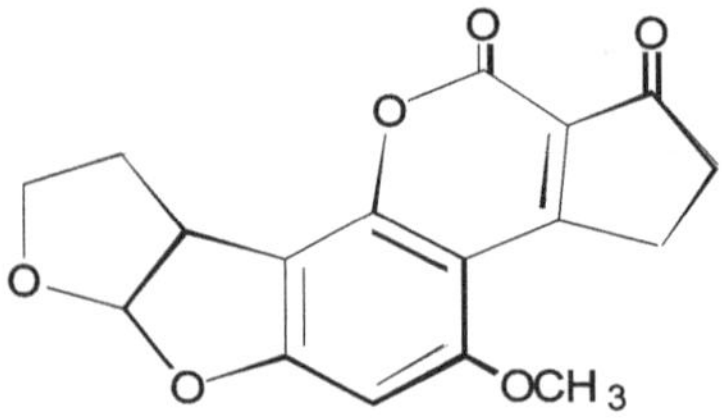

Aflatoxin B_2

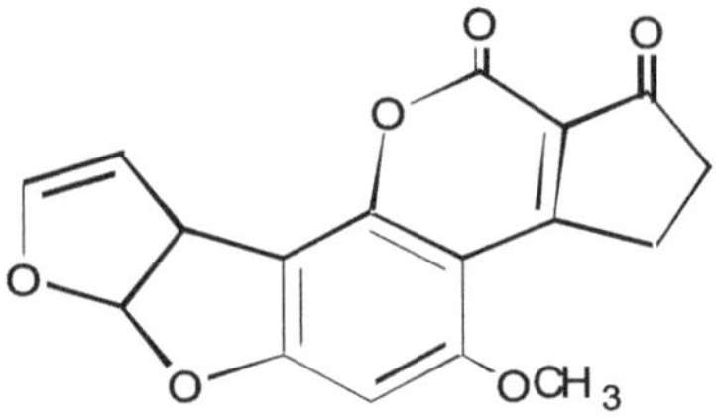

Aflatoxin B_1

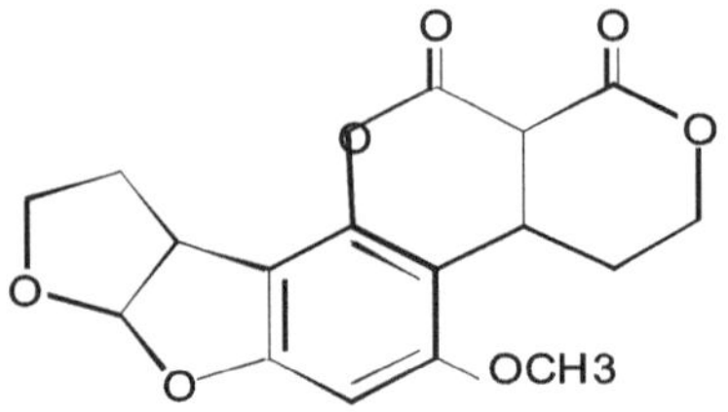

Aflatoxin G_2

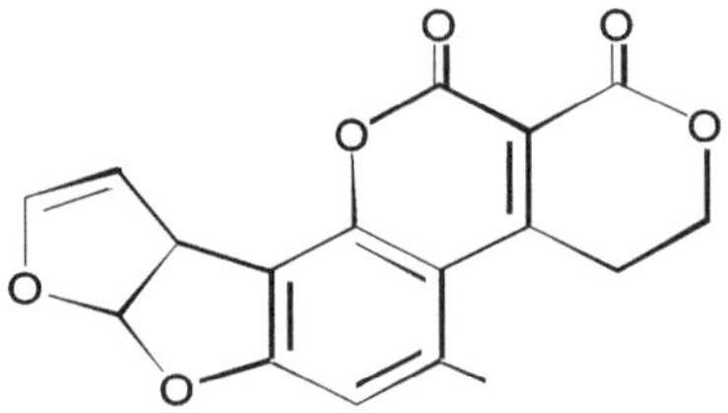

Aflatoxin G_1

Reagents and equipments

1. Methanol: water (55:45).
2. Hexane.
3. Chloroform.
4. Toluene: acetonitrile(98:2).
5. Silica Gel G.
6. TLC Plates, spreader, chamber etc.
7. U.V. cabinet.
8. Photofluorimeter.
9. Hot water bath.
10. Hot air oven.
11. Benzene:ethanol:water(46:35:19 v/v).

12. Benzene: methanol:acetic acid(90:5:5).
13. Aflatoxin standards,B_1, B_2, G_1 and G_2.

Procedure

a) Extraction

1. Powdered, defatted plant sample (for Ex. groundnut meal), 50 g is accurately weighed into 500 ml conical flask. Add to it 250 ml methanol: water (55:45 v/v), 100 ml hexane and 2.0 g NaCl.
2. Shake thoroughly for about one hr and filter/centrifuge.
3. Extract the methanol layer twice with hexane to remove all the oil.
4. Take out the methanolic layer and extract thrice with about 10ml volume of chloroform each time.
5. Pool all the chloroform layers in a 100 ml beaker and evaporate the solvent on a boiling water bath to minimum volume.
6. Then add to it 1.0 ml toluene: acetonitrile (98:2) to dissolve aflatoxins.
7. Transfer the liquid to a small sample tube and dry the solvent again and add 1.0 ml Toluene: Acetonitrile (98:2) solvent to it. Use this solution for TLC.

b) TLC plate preparation

1. Take 40 g silica gel g in 82 ml distilled water and pour the suspension through a spreader on to a TLC plate at 0.25 mm thickness. Incubate for 4hr at 180^{0}C for activation.

c) TLC separation

1. Spot 10 µl of the above extract on silica gel G TLC plates (0.25nm) along with standard mixture (B_1, B_2, G_1 and G_2, 10µl each) from a 20 ppb solution.
2. Develop the plate in a Benzene: ethanol : water (46:35:19 v/v) solvent system or Benzene: Methanol: Acetic acid (90:5:5).
3. After developing the plate view the spots under 366mm U.V. radiation and mark the corresponding spots of B_1, B_2, G_1 and G_2.
4. Scrape the the spots of silica gel and take them separately into centrifuge tubes.

5. Add 5ml methanol and dissolve the toxins.
6. Centrifuge the mixture.
7. Decant the supernatant and estimate the concentration using a photofluorimeter. Use 10 µl of 20 ppb solution of standard B_1 or any other aflatoxin in 5 ml methanol for standard reading.

Calculation

$$\text{Concentration of unknown (ng or ppb)} = \frac{\text{U-B}}{\text{S-B}} \text{X conc. of std.} = \text{(A) ng (µg/kg)}.$$

B = Blank reading

S = Standard reading

U = Unknown sample reading

µg/kg aflatoxin in unknown sample = 2 x A

Aflatoxins (TLC- Method)

Principle

The aflatoxins are extracted into chloroform and separated by TLC. Aflatoxins fluoresce under U.V. light. The spots after extraction are estimated spectrphotometrcally. The developed spots can be identified based on Rf values of known aflatoxin standards. Their quantitation can also be made based on visual intensities of the spots.

Reagents and equipment

1. Acetone.
2. Chloroform.
3. Methanol.
4. Lead acetate 10% solution.
5. Anhydrous sodium sulphate.
6. Silica gel G.
7. Aflatoxin standards.
8. Balance.
9. Thin layer applicator set.

10. Wrist action shaker.
11. Water bath.
12. Hot air oven.
13. Hot air blower.
14. TLC solvent tank.
15. Chromatography sprayer.
16. Forced draught air oven.
17. U. V. Cabinet with 365 nm U.V. light.
18. Spectrophotometer (U.V-Visible) with cuvettes.

Procedure

1. Take 50 g of the sample (for E.x. groundnut seeds) and grind to a fine powder.
2. Place the 50g sample in a stoppered conical flask, add 250 ml of 70% aqueous acetone, and seal the stopper with cellotape.
3. Shake the flask for 30 min at 300 rpm.
4. Filter the extract through No. 4 Whatman paper.
5. Transfer 150 ml of the filtrate to a 250 ml beaker.
6. Add 30 ml of 10% lead acetate solution and boil on a steam bath to reduce the volume to 150 ml and then cool to room temperature.
7. Filter the extract through N0.4 Whatman paper.
8. Transfer the filtrate to a 250 ml separating funnel.
9. Extract the filtrate twice with chloroform (each time with 50ml of chloroform).
10. Drain the chloroform layers through a layer of 15–20 g of anhydrous sodium sulphate in a funnel lined with a filter paper.
11. Wash the sodium sulphate with 20ml chloroform.
12. Combine the chloroform extracts in a 150 ml beaker.
13. Evaporate the chloroform extracts to near dryness on steam a bath.
14. Dissolve the residue in 500 ml chloroform.
15. Detect aflatoxins by TLC.

Preparation of TLC plates

1. Weight out silica gel G (30 g) into a stoppered flask, add 60 ml distilled water and shake for 1 min.
2. Coat the plates with a slurry of slica gel G (250 μ thick) and leave the silica gel plates until gelled (15 min).
3. Heat the plates for one hour in a forced draught oven at 110^0C.
4. Cool the plates in a dust –free atmosphere for 30 min and store them in a plate cabinet.

Development of TLC plates

1. Apply 20 μl extract of samples to the chromatoplates using a micropipette. Spot the extract samples in a line 2 cm from the bottom of the plate and at least 2 cm inside from either side. Spotting should be done as quickly as possible and in subdued light. Also spot a qualitative standard of aflatoxin.
2. Develop the chromatoplate in a solvent system of chloroform: acetone (1:9) in a developing tank. Develop the plates until the solvent front has run 10-12 cm from the base line.
3. Remove the plates from the developing tank and allow them to dry in air in subdued light.
4. Examine plates in the dark room at a distance of 30 cm from the ultraviolet lamp (peak emission 365nm) and observe the presence or absence of blue or green fluorescent spots corresponding to aflatoxin B and G.
5. Compare the fluorescent spots from the sample extracts with those from the standards which have similar R_f values.
6. Scrape the the spots of silica gel and take them separately into centrifuge tubes.
7. Add 5ml methanol and dissolve the toxins.
8. Centrifuge the mixture.
9. Decant the supernatant and estimate the concentration using a photofluorimeter. Use 10 μl of 20 ppb solution of standard B_1 or any other aflatoxin in 5 ml methanol for standard reading.

Reference

Pons, W.A., Cuculu, A.F., Lee L.S., Franz, A.O. and Goldblatt, L.A. 1966. Determination of Aflatoxins in Agricultural Products: Use of Aqueous Acetone For Extraction. Journal of Association of Official Analytical Chemists. 49: 554-562.

Aflatoxin content (Spectrophotometric method) (Nabney and Nesbitt, 1965)

Principle

The aflatoxins are extracted from the TLC plate after their fractionation into methanol. It is later estimated spectrophotmetrically by measuring the absorbance at 363 and 420nm.

Reagents and equipment

1. TLC plates, spreader, chamber, srayer etc.
2. U. V. Camber.
3. Spectrophotometer with cuvettes.
4. Methanol.
5. Microsyringe

Procedure

1. Extraction and fractionation of the toxins can be carried out as per procedures described earlier.
2. Determintion of aflatoxins, particularly aflatoxin B1, is based on the intensity of the ultraviolet light at 363 nm.
3. After separation by thin-layer chromatography, scrape the silica gel 'G' containing the aflatoxin B_1 spot from the plate and extract with cold methanol for 3 minutes.
4. Filter off the methanol into a 50 ml beaker and wash the silica gel 'G' three times with methanol, making the combined filtrates up to 5 ml.
5. Record the ultraviolet absorption specturm of the methanolic solution in a 2 cm cell and calculate the amount of aflatoxin B_1 present in the sample.
6. The optical density at 363 nm, minus that at 420 nm, is used for calculations.

Calculation

The aflatoxin B_1 concentration is given by $\dfrac{D \times M \times 10^6}{C \times 200 \times 2}$ μg /kg

Where D is the corrected optical density at 363 nm, M is the molecular weight (312) of aflatoxin B_1, and C is the molar extinction coefficient (22,000).

Reference

Nabney, J. and Nesbitt, B.R. 1965. A spectrophotometric method for determining the aflatoxins. Analyst 90:155-160.

Aaflatoxins (Visual method)

Principle

The spectrophotometric method described above is useful in the case of fairly high concentrations of aflatoxins. However, in most of the samples, and in particular when dealing with natural contamination, levels may be low. For such samples, the visual method of quantitation is preferred.

Reagents and equipment

1. TLC plates, spreader, chamber, sprayer etc.
2. Microsyringe.
3. Chloroform:methanol (1:9).
4. U.V. Cabinet.
5. Methanol.
6. Chloroform:methanol (1:9).

Procedure

1. Extraction and fractionation of the toxins can be carried out as per procedures described earlier.
2. Spot successively 5 μl, 10 μl, and 20 μl aliquots of the sample extract on the TLC plates. On the same plate, spot 5 μl, 10 μl, 15 μl, and 20 μl, of aflatoxin standard containing 0.02 μg, 0.05 μg, and 0.1 μg of aflatoxin B_1 respectively.
3. Develop the plate in a solvent system of chloroform:methanol (1:9) in a developing tank. Develop the plates until the solvent front has run 10-12 cm from the base line.

4. Remove the plates from the developing tank and allow them to dry in air in subdued light.
5. Examine the plates in the dark room under long wave UV light and observe the presence or absence fo aflatoxins.
6. Compare the fluorescent intensities of the spots of sample extracts with those of the standards. If the spot of smallest aliquot (5μl) of the sample is too intense to match the standard spot, the sample extract can be diluted.
7. The process of dilution and spotting is repeated over and again if the spot of smallest portion of the sample extract is again too intense to match the standard.

Calculation

$$\text{Aflatoxins in } \mu g/kg = \frac{S \times Y \times V}{X \times W}$$

S = μl aflatoxin standard which matches the unknown (spot from the sample extract)

Y = concentration of aflatoxin standard, μg/ml

V = μl of final dilution of sample extract

X = μl of sample extract spotted giving fluorescent intensity

equivalent to S (standard aflatoxin)

W = weight of sample (g)

Aflatoxins by ELISA (Trucksess, *et al.*, 1990)

Principle

Aflatoxins are estimated by ELISA method after their normal extraction and separation on TLC plates as per regular procedure (mentioned earlier).

Reagents and equipment

1. High-speed blender; micropipet and tips; filter paper; test tube (12 x 75 mm, borosilicate); timer; syringe (20 ml) etc.
2. Fluorodensitometer-366 nm excitation, 420 nm emission.
3. TLC plates-Silica Gel 60.
4. Fluorimeter- with 360 nm excitation filter, 450 nm emission filter.

5. Affinity column-Aflatest.
6. LC System- with injector, pump, UV detector, and C18 column.
7. ELISA test kit- Immuno Dot Screen Cup consisting in part of the following: antibody-coated cup; aflatoxin-enzyme conjugate; phosphate-buffered solution (pH 7.2); tetramethylbenzidine aqueous solution (0.4 g/l); hydrogen peroxide solution (0.02% in 0.13 aqueous citric acid).
8. Chloroform.
9. Methanol: water (1:4).
10. 34 mg quinine sulfate/ ml 0.1N H_2SO_4.
11. 0.001 % bromine solution.
12. Methanol-water (8:2).

Procedure

TLC methods

1. Aflatoxin contaminated samples (corn, peanuts, and peanut butter) are extracted with chloroform and water. All final extracts are analyzed and quantitated by a TLC-fluorodensitometric technique.

Monoclonal antibody column method

1. Extraction of peanuts and peanut butter (or other plant sample): Weigh 50g test portion into a blender. Add 5g NaCl and 125 ml methanol-water (6: 4). Blend for 1min at high speed.
2. Add 125 ml water, and blend at low speed for 2 seconds.
3. Filter and collect 5ml (equivalent to 1g test sample). Proceed with column chromatography.
4. Column chromatography: Couple monoclonal antibody column to 30ml syringe barrel with connector. Add test extract to syringe barrel, and use syringe plunger to push extract through column. Flow rate is about 6 ml/min. Remove syringe take plunger out of barrel, and reconnect syringe to column.
5. Add 20 ml water to barrel and apply pressure to push water through. Remove as much water from packing as possible. Place small test tube under column. Add 1ml methanol directly to column and use syringe to push it through column (1-2 drops/s).

6. Use fluorimeter cuvette as receiver to collect eluate and continue with quantitation by either bromination solution fluorometry or LC (eluate is equivalent to 1g test sample).

7. Bromination solution flouorometry: Set instrument excitation at 360 nm and emission at 450 nm. Insert blank and adjust zero knob to display 0.Insert standard solution (34 mg quinine sulfate/ ml 0.1N H_2SO_4. This solution gives same fluorescence intensity as 20ng B_1 /2 ml) and adjust span knob to display 20. Add 1ml 0.001 % bromine solution to eluate (total volume, 2 ml). Mix, and place test tube in instrument. Wait exactly 1 min and record reading (ng total aflatoxin/g test sample).

8. Quantitation by LC: Add 0.2 ml water to eluate, mix, and evaporate under stream of nitrogen to 100-200 µl. Apply entire test extract to C18 column. Mobile phase is methanol-water (45 : 55) and flow rate is 1ml/min. Set UV detector at 365 nm and 0.01 AUFS, Approximate retention times for aflatoxin B_1, B_2, and G_1 are 16, 12, and 10 min, respectively.

ELISA Screening method

1. Extraction of corn, raw peanuts, cottonseed, and poultry feed.- Weigh 50g test portion in blender jar. Add 100 ml methanol-water (8+2). Blend 3 min at high speed, filter, and collect 5ml.
2. Measure 200 µl filtrate. Add 400 µl dilution buffer, mix, and save for immunoassay. Add 150 µl aflatoxin extract to ELISA cup, wait for 1min, and apply additional 150 µl test solution to ELISA cup, wait for 1min and apply additional 150 µl test solution to cup.
3. Wait 1min and add 100 µl test aflatoxin conjugate solution. Wait for 1min and add 1.5 ml wash solution. Ad 1.0ml substrate solution. Wait 1min and observe color development.
4. Determination: Observe center of cup for blue color or no color development exactly 1min after adding substrate. Test samples containing $\geq$ 20 ng aflatoxins/g show no color development; test samples containing <20 ng aflatoxins/g develop blue or gray color in center of the cup.

Reference

Trucksess, M.W., Young, K., Donahue, K.F., Morries D.K. and Lewis, E. 1990. Comparisonof Two Immuno Chemical Methods with Thin-Layer Chromatographic Methods for Determination of Aflatoxins. Journal of the Associaltion of Official Analytical Chemistry 73(3): 425-428.

Aflatoxins by HPLC method (Park *et al.*,1990)

Principle

Aflatoxins are extracted, purified and derivatized with trifluoroacetic acid (aflatoxins B_1 and G_1 to B_2a and G_2a, respectively), separated by reverse-phase liquid chromatography, and detected using their fluorescence. Method can measure 0.1 ng of aflatoxins B_1, B_2, G_1, and G_2. Detection limit is 0.3 ng/g for each aflatoxin.

Reagents and equipment

1. Liquid chromatograph.- with fluorescence detector, excitation filters (360 nm) and glass emission filters (440 nm), and column.- 15 cm X 4.6 mm i.d., Supelcosil, 5 µm, or equivalent.
2. Adjustable pipets.- 10-1000 µl and 100-1000 µl with disposable tips (Eppendorf).
3. Filter tube.- Glass, 15 cm i.d., with coarse frit bed support (glass wool not useful).
4. Solvents.- Distilled-in-glass: methanol, hexane, methylene chloride, benzene, acetone, acetonitrile. Anhydrous ethyl ether stored in metallic container. Glass-bottled ether may form peroxides soon after opening which degrade aflatoxins.
5. LC elution solvent.- H_2O:CH_3CN:methanol (700 :150 : 150). Adjust ratio of water to obtain baseline resolution of aflatoxins B_2a and G_2a.
6. Silica gel for column chromatography: Activate silica gel by drying 4 hr at 100°C. Cool to room temperature. Weigh desired quantity (100g) in to glass-stopper container. Add 1 ml H_2O in small increments; agitate silica gel between additions. Shake or tumble mechanically 4-6 h. If mechanical tumbler or shaker is not available, shake manually 5 min/h over 8 h. Let stand 16 h. Perform suitability test (as mentioned below).
7. Trifluoroacetic acid (TFA): Transfer 1 or 2 ml TFA to a vial with Teflon-lined cap. Keep in freezer when not in use. Discard if discoloration appears.
8. Sodium sulphate.- Anhydrous, coarse. Heat 2-3 hr at 600 °C to remove organic impurities.
9. Aflatoxin standard solution.- 1). Aflatoxin stock solution.- 10 µg/ml. Prepare individual stock solution in benzene:CH_3CH (98:2) 2). Working standard solutions.- Use Eppendorf pipette to transfer appropriate quantity

of stock solutions to each of the vials to obtain the amounts of afltoxins in each vial.

Vial	$B_1$and B_2ng	B_2 and G_2ng
1.	250	125
2.	500	250
3.	1000	500
4.	2000	1000

10. Evaporate solutions to dryness under gentle stream of nitrogen (drying may be facilitated by warming to 40 °C). Using Eppendorf pipet, and 200 μl hexane and 50 μl TFA to each vial, cap, and vortex-mix for 30s. Let solutions stand 5 min; then add 10.0 ml H_2O:CH_3CN (9:1) and vortex-mix 30s. Let layers separate for 5-10 min, or centrifuge at 1000 rpm for 30s. Final concentration of aflatoxins is as follows:

Vial	$B_1$and B_2 μg/10.05ml	B_2 and G_2 μg/10.05ml
1.	0.25	0.125
2.	0.5	0.25
3.	1.0	0.50
4.	2.0	1.00

Silica gel suitability test

1. Prepare 2.0 g silica gel column (as in column chromatography below). Transfer appropriate quantity of aflatoxin stock solutions to 50 ml beaker to contain 500ng each of aflatoxisn B_1 and G_1 and 250 ng each of aflatoxins B_2 and G_2. Evaporate mixture to dryness under gentle stream of nitrogen. Add 2-3ml CH_2Cl_2 and swirl for 10s. Quantitatively transfer to silca gel column with about 1 ml portions of CH_2Cl_2:acetone (9:1). Derivatize (as in derivatization below) and perform LC analysis. Compare chromatogram with that of derivatized standard. Peak heights (areas) should be nearly identical to those for working standard, i.e., B_1, G_1= 0.25 ug/10.05 ml. Recovery should be $\geq$ 90%.

Extraction

1. Transfer 50 g corn or 50g peanut butter to 1litre blender jar, add 200 ml methanol followed by 50 ml 0.1N HCl, and blend 3 min at high speed. Filter through 24 cm Whatman No.1 paper. Filtrate may not be completely clear. Collect 50 ml filtrate.

Partition

1. Transfer 50 ml filtrate to 250 ml separatory funnel. Add 50ml 10% NaCl solution, swirl, add 50 ml hexane, and shake gently for 30s.
2. Let phases separate and drain lower aqueous layer into another 250 ml separatory funnel. Discard hexane layer.
3. Add 25 ml CH_2Cl_2 and shake moderately for 30s. If emulsion occurs, break up with clean Pasteur pipette. Let phases separate and drain lower CH_2Cl_2 layer through 4 cm coarse granular, anhydrous Na_2SO_4 in glass filter tube.
4. Collect eluate in 250 ml beaker. Repeat partition with 2 additional 25ml portions of CH_2Cl_2 and vigorously shake and drain as above.
5. Collect eluate in the 250 ml beaker. Evaporate eluate on steam bath under gentle stream of nitrogen to 2-3 ml.

Column chromatography

1. Slurry 2.0 g silica gel with 10 ml ether:hexane (3:1) in 30 ml beaker, pour slurry into clean up column, and wash beaker with additional 5ml ether-hexane solvent to effect transfer. Keep stopcock closed and let silica gel settle without tamping.
2. Wash sides of column with 2-3 ml ether:hexane (3:1), using wash bottle. After gel settles, open stopcock, and, while column drains, add 1 cm anhydrous granular Na_2SO_4.
3. Transfer extract from partition (above) to column. Wash tip of beaker with 0.5 ml CH_2Cl_2 using wash bottle, and collect wash in column. Wash beaker with 2 ml CH_2Cl_2 and add wash to column. Do not use more than 5-6 ml CH_2Cl_2 to transfer extract to column.
4. With stopcock fully open, add 25 ml benzene:acetic acid (9:1), and then add 30ml ether:hexane (3:1) to column, draining each wash to top of Na_2SO_4.
5. Discard washes. Elute aflatoxin with 100 ml CH_2Cl_2:acetone (90:10) and collect eluate in 250 ml beaker.
6. Evaporate solvent on steam bath under gentle stream of nitrogen to about 6ml. Quantitatively transfer to a vial, using 2-3 ml CH_2Cl_2 as wash.
7. Evaporate eluate almost to dryness on steam bath in an aluminum block under gentle stream of nitrogen.

8. Evaporate remaining 200 µl just to dryness under gentle stream of nitrogen by holding vial in palm of hand and slowly rotaing vial.

Derivatization

1. Add 200 µl hexane to column extract, Then, add 50µl TFA (using Eppendorf pipette), cap vial, and vortex-mix vigorously for 30 s exactly). This procedure must be followed closely to ensure consistent reaction yields. Let mixture stand 5 min.
2. Using Eppendorf pipette, add 1.950 ml $H_2O:CH_3CN$ (9:1). Vortex-mix vigorously for 30s (exactly) and let layers separate for 10 min or centrifuge at 1000 rpm for 30 s. Concentration is 10 µg/2 ml aqueous CH_3CN.

LC Determination

1. Using instrument parameters inject 25µl of derivatized standard solutions.
2. Prepare standard curve to check linearity of responses.
3. Inject 25 µl TFA-treated sample solution (lower aqueous phase).
4. If sample peaks are outside linear range dilute aliquot of TFA-treated sample solution to suitable volume with H_2O-CH_3CN (9:1), re-mix on vortex mixer, and inject another 25 µl portion.

Calculation

Calculate individual aflatoxin concentrations as follows. Use responses of standard containing 500 ng B_1 and G_1, and 250 ng B_2 and G_2 for calculations.

Aflatoxins, ng/g = (P/P') X C X (2/10) X 1000 X D

Where P and P' = peak areas (integrator counts) or heights for sample and standard, respectively, per 25 µl injection; C = concentration of individual aflatoxins in standard solution (0.5 or 0.25 µg/10.5 ml); and D = dilution factor if 2 ml solution for injection is diluted.

Reference

Park, D.L., Nesheim, S., Truckess, M.W., Stack, M.E. and Newell, R.F. 1990. Liquid Chromatographic Method for Determination of Allatoxins B_1, B_2, G_1, and G_2, in Corn and Peanut Products: Collaborative Study. Journal of the Association of Official Analytical Chemistry. 73(3): 425-428.

Trichothecene mycotoxin (T-2) (Gas chromatographic method)

Principle

T-2 is a trichothecene mycotoxin. It is a naturally occurring mold by-product of *Fusarium* spp fungus which is toxic to humans and animals. The clinical condition it causes is alimentary toxic aleukia and a host of symptoms related to organs as diverse as the skin, airway, and stomach. Ingestion may come from consumption of moldy whole grains. T-2 can penetrate into and permeate through the human skin. Although no significant systemic effects are expected after dermal contact in normal agricultural or residential environments, local skin effects can not be excluded. Hence, the skin contact with T-2 should be limited.

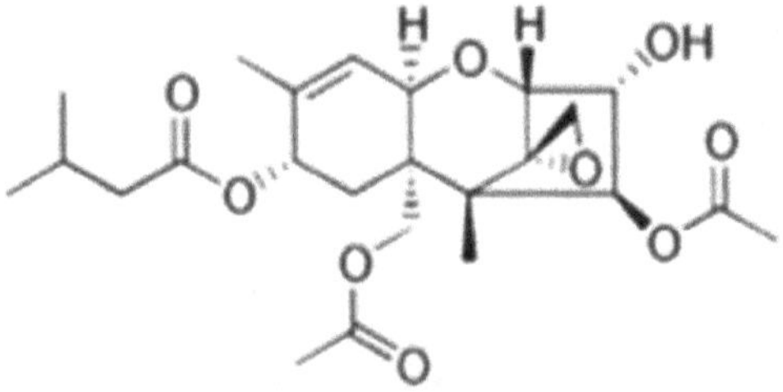

The T-2 toxin is extracted with methanol: water, purified and separated on a GLC column,Polydimethysiloxane (3 %). ECD is used for detection.

Reagents and equipment

1. *A. T-2 toxin standard solution*: Weigh accurately 10 mg of T-2 toxin [$C_{24}H_{34}O_9$] in a 100-ml amber volumetric flask. Dissolve in chloroform and make up to volume. (0.1mg/ml). Before use, dilute with chloroform to10 mg of T-2 toxin per ml.

   B. Internal standard solution: Weigh 0.25 g of methoxychlor in a 250-ml amber volumetric flask, dissolve by the addition of *n*-pentane, and make up to volume. Before use, dilute to get 0.1 μg of methoxychlor in 1 ml.

2. *Washing solution:* Dissolve 1.12 g of potassium hydroxide and 10 g of potassium chloride in water and make up to 1 litre.

3. *Silica gel:* Dry silica gel for column chromatography (particle size 74-149 μm 200-100 mesh) at 120°C for 2 hours.

4. *Diatomite:* Wash diatomite sequentially with warm water and methanol, and then air-dry.

5. Methanol/water (1:1).

6. Ammonium sulfate solution (30 w/v%).
7. Chloroform.
8. Separatory funnels.
9. Anhydrous sodium sulphate.
10. Glass column (10 mm inner diameter).
11. Water bath.
12. G. C. with Polydimethysiloxane (3 %)/ diatomite column and electron capture detector.

Procedure

a. *Extraction:* Weigh 10.0-50.0 g of sample, transfer it to a 500ml flask, add 250 ml of methanol/ water (1:1), extract by shaking for 30 min, and filter with filter paper. Transfer 60 ml of the extract to a 500 ml beaker, add 240 ml of ammonium sulfate solution (30 w/v%) and stir. Further add 25 g of diatomite and stir, and filter

b. *Purification*: Transfer 200 ml of the sample solution to the 300-ml separatory funnel A. Add 10 ml of chloroform, shake vigorously for 1 minute and leave at rest, and transfer the chloroform layer (lower layer) to the 300-ml separatory funnel B that contains 100 ml of the washing solution in advance. Add 10 ml of chloroform to the separatory funnel A, and treat similarly, and then transfer the chloroform layer to the separatory funnel B, shake mildly and then leave at rest. Pass the chloroform layer through a funnel that contains 25 g of sodium sulfate (anhydrous) in advance for dehydration, to obtain a sample solution to be subjected to column treatment.

c. *Column treatment*: Suspend 1g of sodium sulfate (anhydrous), 2 g of silica gel, and 1g of sodium sulfate (anhydrous) respectively in chloroform and sequentially pour into a column (10 mm inner diameter), wash the column by sequentially loading 50 ml of diethyl ether and 20 ml of chloroform, and elute so that the liquid level reaches to a height of 3 mm from the upper end of packing to prepare a column. Transfer 5 ml of the sample solution to a 50 ml recovery flask, add 20 ml of hexane, mix, and load on the column. Wash the recovery flask with a small amount of hexane three times, and add the washing to the column. Elute so that the liquid level reaches the height of 3 mm from the upper end of packing, and then sequentially load 20 ml of benzene and 30 ml of benzene/ acetone (19:1) and elute similarly. Place a 50 ml recovery flask under the column. Add 30 ml of diethyl ether to the column to elute T-2 toxin.

Concentrate the eluate under vacuum on a water bath at 50°C to an almost dry stage, and finally dry up under nitrogen gas. Add accurately 2 ml of benzene to dissolve the residue, and transfer to a 10 ml screw-capped amber test tube to be a sample solution to be subjected to derivatization. At the same time, transfer several amounts between 50-400 μl of the T-2 toxin standard solution to 10-ml screw-capped amber test tubes, respectively, dry up by nitrogen gas flow, and add accurately 2 ml of benzene to dissolve the residue to be the standard solutions to be subjected to derivatization.

d. *Derivatization*: Add 1 g of sodium sulfate (anhydrous) to the sample solution, shake, and add 100 μl of *N*-heptafluorobutyrylimidasole. Seal the test tube and shake for 1 min, further add 2 ml of sodium bicarbonate solution (5 w/v%), seal again and shake for 2 minutes, and then leave at rest. Take accurately a certain amount of the benzene layer (upper layer), dilute accurately 5-fold with the internal standard solution to be a sample solution to be subjected to gas chromatography. At the same time, operate using respective standard solutions in the same way as the sample solution, to be the derivatized T-2 toxin standard solutions to be subjected to gas chromatography.

e. *Gas chromatography*: Inject a certain amount each of the sample solution and respective derivatized T-2 toxin standard solutions to a gas chromatograph to obtain chromatograms. G.C. conditions may be as follows:

Detector: Electron capture detector

Column: Glass, 3 mm in inner diameter, 1.1 m in length

Column packing: Polydimethysiloxane (3%)/ diatomite for gas

chromatography (particle size 125-149 μm (120-100 mesh)

Carrier gas: N_2 (30 mL/min)

Column oven temperature: 220°C

Injector temperature: 325°C

Detector temperature: 325°C

f. *Calculation*: Obtain peak heights or areas of derivatized T-2 toxin and methoxychlor from the resulting chromatograms to prepare a calibration curve by the internal standard method, and calculate the amount of T-2 toxin in the sample.

Note

1. *Solvents to be used are reagents for residual agrochemicals or equivalents.*
2. *When a column is newly prepared, after column aging, set column oven temperature at 150-180°C, and inject 200 μl of a silanization reagent (Silyl-8 (Pierce Chemical) or equivalents) to the column, react for 1 hour, and then connect the column to the detector to stabilize the baseline. On and after the second use, inject 100 μl of the silanization reagent and operate similarly before use.*

Crude ricin (Funastu, *et al.,* 1971).

Principle

Ricin is present in castor seed. It is the most toxic constituent. Ricin is a two polypeptide chain linked with a single disulfide bond. It agglutinates red blood cells at 100-2000 mg/100 ml. Ricin being an albuminous protein is extracted into water and fractionated. Crude ricin is precipitated using ammonium sulphate and weighed gravimetrically.

Reagents and equipment

1. Petroleum ether (40-60°C).
2. Diethyl ether.
3. HCl.
4. NaCl.
5. Tris- HCl buffer.
6. Ammonium hydroxide 2% solution.
7. Ammonium sulphate.
8. Buchner funnel.
9. Vacuum pump.
10. Balance.
11. Dialysis membrane.

Procedure

Isolation, purification and quantification

1. Ricin from the castor seeds can be isolated as per the procedure of Funastu, *et al.,* (1971). The method, slightly modified, is given below.
2. The outer covering of the seeds (hulls) are removed, the kernel is powdered in a Knifetec sample mill. It is completely defatted using ice-cold petroleum ether (40-60°C) and diethyl either.
3. The defatted powder is suspended in double distilled water and adjusted to pH 4.0 using HCl. The suspension is agitated for 7-8 hrs.
4. The solution is filtered under vaccum. The yellow supernatant liquid is collected and the residue is decanted.
5. The proteins in the supernatant are precipitated by adding solid NaCl till its saturation.
6. The precipitate is dissolved and extensively dialysed for 72hrs in double distilled water and in Tris-HCl buffer, pH 7.5 at 4°C.
7. The insoluble particles and other impurities separated in dialysis are decanted. The clear protein solution is precipitated using 2% ammonium hydroxide. The resulting precipitate is decanted.
8. The clear protein solution is finally precipitated with saturated ammonium sulphate salt solution.
9. Crude ricin is thus separated and weighed for purposes of quantification.
10. The ricin content is expressed as per cent of the seed.
11. The ricin can be dissolved in either distilled water or in pH 4 aqueous HCl, for any further studies.

Reference

Fanatsu, M., Fanatsu, G., Ishiguro, M., Nanno, S. and Hara, K. 1971. Structure and Toxic Function of Ricin D. Proceedings of Japan Academy of Science. 47:715.

Ricin: RID assay (Pinkerton *et al.*,1999)

Principle

A RID (radial immunodiffusion) assay is used to quantify levels of specific antigens in a mixture of several antigens (Garvey et al., 1977). A specific antibody developed against ricin is incorporated into a thin layer of agar. Seed extracts containing the antigens for ricin and RCA are placed into wells cut into the agar layer. The antigens diffuse evenly outward and bind with specific antibodies to form an antigen-antibody precipitation ring around the well. The square of the diameter of the precipitation ring is plotted on a standard curve to estimate antigen concentration (Kemeny and Chantler, 1988). The RID technique described by Stansfied (1981) and Mancini *et al*., (1965) is modified slightly and presented below.

Reagents and equipment

1. Anti ricinus communis lectin.
2. Agarose.
3. PEG.
4. R.I.D. Plates.
5. Incubator.

Procedure

1. One hundred sixty seven microliters of rabbit anti-*Ricinus communis* lectin diluted with 34 mg/ ml of distilled water.
2. This solution is placed in 9.5 ml of a 1.2 % (v/v) agarose/3% (v/v) PEG 6000 solution which has been cooled to 50°C.
3. Antibody-agar solution is added to RID plates and allowed to solidify before the 2 mm wells are cut.
4. Ricin solutions of 3.5 and 2.9 mg/l are used to develop six standard concentrations. Ten microliters of these standards are added to the wells.
5. The plates are allowed to incubate at 26°C for 24 h, and then placed at 4°C for another 24 h.
6. The precipitate ring around each well is measured on two different axes and the square of the average diameter is used to estimate the ricin + RCA concentration.

References

Garvey, J.S., Cremer, N.E. and Sussdorf, D.H. 1977. Radial Immuno- Diffusion. Methods in Immunology. W.A. Benjamin, Inc., Boston.

Pinkerton S.D., R Rolfe, D.L., Auld, V., Ghetie and Kaytervacg, B.F. 1999. Selectionof Castor for Divergent Concentrations of Ricin and Ricinus Communis Agglutinin. Crop Science 39:353-357.

Kemeny, D.M. and S. Chantler, 1988. An introdction to ELISA. ELISA and Other Solid Phase Immunoassays. John Wiley and Sons, Ltd., London. Stansfield, W.D. 1981. Serology and Immunology. Macmillan Publushing Co., NewYork.

Mancini, G., Carbonara, A.O. and Heremans, J.F. 1965. Immunochemical Quantification of Antigens by Single Radial Immunodiffusion. Immunochemistry 2: 235-241.

Ricin, HPLC method

Principle

A solution containing proteins are chromatographed on a non-polar column. A more polar liquid that serves as the mobile phase is passed over the matrix and solute molecules (proteins) get eluted in proportion to their solubility (Lottspeich and Henschen 1981).

Reagents and equipment

1. HPLC equipment with C-18 column, a UV-Visible spectrophotometric detector(SPD) and a data recorder.
2. Mobile phase: Tris-HCl buffer, pH 7.5
3. Methanol 100% (HPLC grade).
4. Standard proteins like BSA and trypsin inhibitor
5. Nylon membrane 0.45μm

HPLC separation procedure

1. Ricin extraction can be carried out as described earlier(modified Funastu, *et al.,* (1971).
2. The crude and pure ricin fractions are evaluated to check their purity using reverse phase HPLC (Shimadzu LC-10AT, Japan) under isocratic conditions on C-18 column (4.6 x 25cm. Supelco products, USA).
3. Before applying the ricin samples, the column is calibrated using standard proteins like BSA, Trypsin inhibitor etc.
4. The samples are dissolved in 100% mobile phase (Tris-HCl buffer, pH 7.5) and then filtered using 0.45μm nylon membrane (Millipore, USA).
5. 20μl of the protein sample is injected into the column at a flow rate of 0.5ml per minute and monitored at 280nm using SPD.

6. The chromatogam shows a major ricin peak, wthin 5 min of elution. Both identification and quantification can be carried out using this HPLC separation. BSA can be used for quantification.

Reference

Lottspeich, F. and Henschen, A. 1981. HPLC in Proteins and Peptide Chemistry, Ed. Hupe. Walter Degrothy.

Hydrocyanic acid (Cyanogenic glycosides)

Principle

The cyanogenic glucosides present in lima bean and linseed are not good for health. However, they are lost during drying and processing. HCN, released from their glucosides, is measured and estimated.

Reagents and equipment

1. 0.02N silver nitrate.
2. Nitric acid.
3. 0.02N Potassium thiocyanate.
4. Ferric ammonium alum solution as indicator.
5. Kjeldahl distillation set up.
6. Gooch crucible.

Procedure

1. Thoroughly shake in an 800 ml Kjeldahl flask 10-20 g of finely powdered sample with 100 ml water for 20 min.
2. Add 100 ml water and carry out steam distillation. The condenser tube should be dipped into a mixture of 20 ml 0.02N silver nitrate and 1ml nitric acid in the receiver.
3. When 150 ml distillate is collected, it is centrifuged or filtered through a Gooch crucible.
4. The contents are titrated with 0.02N Potassium thiocyanate using ferric ammonium alum solution as indicator.
5. This gives a measure of excess silver nitrate. That which is consumed is computed from the volume (20 ml) taken.

Calculation

1ml of 0.02N silver nitrate solution = 0.54 mg of hydrocyanic acid.

Hydrocyanic acid (Alkaline titration method)

Principle

The HCN is collected into an alkali and the same is titrated with $AgNO_3$.

Reagents and equipment

1. NaOH solution (2.5 g in 100 ml water)
2. 6N NH_4OH solution
3. 0.02N silver nitrate
4. 5% potassium iodide solution.
5. Kjeldahl flask (distillation set up)

Procedure

1. Thoroughly shake in an 800 ml Kjeldahl flask 10-20 g of finely powdered sample with 100 ml water for 20 min.
2. Add 100 ml water and carry out steam distillation. The condenser tube should be dipped into a mixture of 20 ml of NaOH solution in the receiver.
3. Make up distillate to 250 ml in a volumetric flask.
4. Pipette 100 ml into a beaker.
6. Add 8 ml 6N ammonium hydroxide and 2ml 5% potassium iodide solution
7. Titrate with 0.02N silver nitrate solution from a microburette. The end point is a faint but permanent turbidity.

Calculation

1ml 0.02N silver nitrate = 1.08 mg of HCN

10

ALKALOIDS

Alkaloids are of plant origin and are alkaline in reaction. They contain N as part of a hetercyclic ring. They exhibit pharmacological activity. Presently more than 2000 alkaloids are known. They have a lot of uses in medicine and as flavoring agents, psychoactive agents, pain killers and as pesticides.

Estimation of Nicotine

Principle

Nicotine is alkaline in nature. Initially it is salted out using an alkali and then extracted into an organic solvent. Later the same is titrated with an acid to estimate the nicotine content.

Reagents

1. Glacial acetic acid.
2. Barium hydrate octahydrate.
3. Perechloric acid.
4. Potassium acid phthalate.
5. Benzene: chloroform (9:1) solution.
6. Saturated aqueous barium hydroxide solution.

7. Crystal violet indicator: 1% in glacial acetic acid.
8. Perchloric acid solution: dilute 2.1ml of 72% perchloric acid to one litre.
9. Wrist action shaker.

Procedure

1. Accurately weigh about 1.5g of finely powdered sample into a glass stoppered Erlenmeyer flask.
2. Add about 1g of granular barium hydroxide and 7.5 ml of barium hydroxide solution. Swirl the flask till the sample gets thoroughly wetted. If needed, add more barium hydroxide solution.
3. Pipette 50 ml benzene: chloroform solution into the flask. Stopper tightly and shake vigorously for 15 min in a shaking machine.
4. Allow the two layers to separate and filter the majority of the organic solution using Whatman no.2 filter paper.
5. Pipette 25 ml of the filtrate into a conical flask. Add one drop of indicator and titrate with 0.025N $HClO_4$.

Calculation

$$\%\text{ Total alkaloids (as nicotine)} = \frac{V \times N \times 16.23}{\text{Weight of the sample(MF basis)}}$$

Where V = ml of $HClO_4$ required to neutralize 25 ml sample extract taken,

N = Normality of $HClO_4$. MF = moisture free

Note

1. *Standardization of $HClO_4$: Accurately weigh 0.1g potassium acid phthalate into an Erlenmeyer flask, add 50ml glacial acetic acid and heat to dissolve. Cool and add 2 drops of crystal violet indicator. Titrate to a blue-green end point. Perform blank titration using 50ml glacial acetic acid and deduct from the above value.*

$$N = \text{Weight of KH phthalate X} \frac{4.897}{\text{ml of HClO4}}$$

Reference

Cundiff, R.E. and Markunas, P.C. Anal.Chem. 27: 1650-1653, 1955

Estimation of solanine

Principle

Solanine is a steroidal alkaloid and is present in potato. It is extracted into a weak acid and then precipitated with NH_4OH. The precipitate is dissolved in ethanol:H_2SO_4 mixture and treated with formaldehyde. The colour is measured at 565-570 nm.

Reagents

1. 5% acetic acid.
2. Conc. H_2SO_4 and 1% NH_4OH.
3. 96% ethanol.
4. 20% and 60% H_2SO_4.
5. 0.5% formaldehyde in 60% H_2SO_4.
6. 1M H_2SO_4.
7. Standard solanine solution: 0.2-0.3mM Solanine in 96% ethanol-20% H_2SO_4(1:1) mixture (molecular weight of solanine =868.04).
8. Centrifuge.
9. Spectrophotometer with cuvettes.

Procedure

1. Extract the sample (plant tissue) with 5% acetic acid (15-20 ml/ tissue) with grinding and filter through a cheese cloth.
2. Warm the filtrate to 70^0 C on a water bath and add conc. NH_4OH drop wise till the p^H is 10.
3. Centrifuge at low speed and discard the supernatant.
4. Wash the precipitate with 1% NH_4OH and centrifuge again.
5. Collect the precipitate, dry and weigh the crude solanine obtained.
6. Dissolve a known amount (50 mg) of crude solanine in a mixture of ethanol-20% H_2SO_4(1:1).

7. To one ml of the alkaloid solution add 5ml of 60% H_2SO_4. After 5min add 5ml of formaldehyde reagent. Allow to stand for 3hr at room temperature.
8. Measure the absorbance at 565-570nm against a blank.
9. Calculate the amount of solanine in the sample using a standard curve of pure solanine (prepared by following the steps as above).

Capsaicin content (Colorimetric Method)

Principle

Capsaicin is a protoalkaloid which is responsible for the pungency of chillies. The quality of chilli fruits, extracts or oleoresins is determined by the capsaicin content. The phenolic group in capsaicin reduces phosphomolybdic acid to lower acids of molybdenum. The resulting component is blue in color and is read at 650nm. The color intensity is directly proportional to the concentration of capsaicin.

Reagents and equipment

1. 0.4% Sodium Hydroxide.
2. 3% Phosphomolybdic Acid.
3. Dry Acetone (Add about 25g anhydrous sodium sulphate to 500ml acetone of analytical grade at least one day before use).
4. *Stock Standard Capsaicin Solution* :Dissolve exactly 50mg capsaicin in 50ml of 0.4% sodium hydroxide solution (1,000 µg / ml).
5. *Working Standard* : Dilute 10ml of the stock standard to 50 ml with 0.4% sodium hydroxide solution (200 µg / ml).

Procedure

1. Weigh 0.5g dry chilli powder into a glass-stoppered test tube or volumetric flask.
2. Pipette out 10ml dry acetone into the flask and shake it for 3hr in a mechanical shaker. Let the contents settle down or centrifuge (10,000rpm for 10 min).
3. Pipette out 1ml of the clear supernatant into a test tube and evaporate to dryness in a hot water bath.

4. Dissolved the residue in 5 ml of 0.4% sodium hydroxide solution.
5. Add 3 ml of 3% phosphomolybdic acid.
6. Shake the contents and let stand for 1hr.
7. Filter the solution quickly into centrifuge tubes to remove any floating debris.
8. Centrifuge at about 5,000 rpm for 10-15 min.
9. Transfer the clear blue colored solution directly into the cuvette and read the absorbance at 650 nm.
10. Run a reagent blank along with the test samples.
11. Prepare a standard graph using 0-200 μg capsaicin simultaneously i.e., pipette out 0.2, 0.4, 0.6, 0.8 and 1ml of working standard solution and proceed as above.

References

Quagliotti, L (1971) Hortic Res 11:93.

Theymoli Balasubramanian, Raj, D, Kasthuri, R and Rengaswami, P (1982) Indian J Hortic. 39:239.

Capsaicin content (Spectrophotometric method)

Principle

Capsaicin is a proto-alkaloid which is responsible for the pungency in chillies. The capsaicin from red chillies is extracted with ethyl acetate and made to react with ethyl acetate solution of vanadium oxychloride. Then it is read at 720nm. This method is sensitive and is useful to measure small quantities (less than 0.05%).

Reagents

1. Vanadium oxychloride (0.5%) in ethyl acetate.
2. Pure capsaicin (0.01%) in ethyl acetate (10mg in 100ml).
3. Ethyl acetate.
4. Spectrophotometer with cuvettes.

Procedure

1. Grind the sample well to pass through No. 40 sieve.
2. Place 2 g sample in a 100 ml volumetric flask. Add 50 ml ethylacetate
3. Let it stand for 24 h to extract (otherwise reflux the contents for 2.5h) and then make up to volume.
4. Dilute 1 ml of extract to 5 ml with ethyl acetate.
5. Add 0.5 ml vanadium oxychloride solution (just before reading) and shake.
6. Read at 720 nm in a spectrophotometer.
7. Subtract the absorbance value.
8. Prepare a standard curve using 0.5,1.0,1.5, 2.0 and 2.5 ml of standard capsaicin solution containing 50, 100, 150, 200 and 250 µg capsaicin respectively.

Calculation

$$\% \text{ capsaicin} = \frac{\mu\text{g capsaicin x 100 x100}}{1000 \text{ x } 1000 \text{ x } 1 \text{ x } 2} = \frac{\mu\text{g capsaicin}}{200}$$

Reference

Palacio, J J R (1977) JAOAC 60 970.

11

PLANT NUTRIENTS

Sixteen chemical elements are known to be important to a plant's growth and survival. The sixteen chemical elements are divided into two main groups: non-mineral and mineral. The Non-Mineral Nutrients are hydrogen (H), oxygen (O), and carbon (C). These nutrients are found in the air and water.

In a process called photosynthesis, plants use energy from the sun to change carbon dioxide (CO_2 - carbon and oxygen) and water (H_2O- hydrogen and oxygen) into starches and sugars. These starches and sugars are the plant's food. Since plants get carbon, hydrogen, and oxygen from the air and water, there is little the farmers and gardeners can do to control how much of these nutrients a plant can use.

The 13 mineral nutrients, which come from the soil, are dissolved in water and absorbed through a plant's roots. There are not always enough of these nutrients in the soil for a plant to grow healthy. This is why many farmers and gardeners use fertilizers to add the nutrients to the soil. The mineral nutrients are divided into two groups: macronutrients and micronutrients.

Macronutrients

Macronutrients can be broken into two more groups: primary and secondary nutrients. The primary nutrients are nitrogen (N), phosphorus (P), and potassium (K). These major nutrients usually are lacking from the soil because plants use large amounts for their growth and survival. The secondary nutrients are calcium (Ca), magnesium (Mg), and sulfur (S). There are usually enough of these secondary nutrients in the soil. Hence fertilization (external supply) is not always needed. Also, large amounts of calcium and magnesium are added when lime is applied to acidic soils. Sulfur is usually found in sufficient amounts from the slow decomposition of soil organic matter, an important reason for not throwing out grass clippings and leaves.

Micronutrients

Micronutrients are those elements essential for plant growth which are needed in only very small (micro) quantities. These elements are sometimes called minor elements or trace elements, but use of the term micronutrient is more widely accepted. The micronutrients are boron (B), copper (Cu), iron (Fe), chloride (Cl), manganese (Mn), molybdenum (Mo) and zinc (Zn). Recycling organic matter such as grass clippings and tree leaves is an excellent way of providing micronutrients (as well as macronutrients) to growing plants.

In general, most plants grow by absorbing nutrients from the soil. Their ability to do this depends on the nature of the soil. Depending on its location, a soil contains some combination of sand, silt, clay, and organic matter. The make up of a soil (soil texture) and its acidity (pH) determine the extent to which nutrients are available to plants.

Nitrogen is a part of all living cells and is a necessary part of all proteins, enzymes and metabolic processes involved in the synthesis and transfer of energy. Nitrogen is a part of chlorophyll, the green pigment of the plant that is responsible for photosynthesis. It helps plants in rapid growth, increasing seed and fruit production and improving the quality of leaf and forage crops. Nitrogen often comes from fertilizer application and from the air (legumes get their N from the atmosphere, water or rainfall contributes very little nitrogen).

Phosphorus (P) is an essential part of the process of photosynthesis. It is involved in the formation of all oils, sugars, starches, etc. It helps in the transformation of solar energy into chemical energy, proper plant maturation and withstanding stress. It also effects rapid growth, encourages blooming and root growth. Phosphorus often comes from fertilizer, bone meal, and superphosphate.

Potassium is absorbed by plants in larger amounts than any other mineral element except nitrogen and, in some cases, calcium. It helps in the building of protein, photosynthesis, fruit quality and reduction of diseases. Potassium is supplied to plants by soil minerals, organic materials, and fertilizer. It is a highly mobile element in the plants.

Calcium, an essential part of plant cell wall structure, provides for normal transport and retention of other elements as well as strength in the plant. It is also thought to counteract the effect of alkali salts and organic acids within a plant. Sources of calcium are dolomitic lime, gypsum, and superphosphate.

Magnesium is part of the chlorophyll in all green plants and essential for photosynthesis. It also helps activate many plant enzymes needed for growth. Soil minerals, organic material, fertilizers, and dolomitic limestone are sources of magnesium for plants. It is an essential plant food for production of protein. Also it promotes activity and development of enzymes and vitamins. It helps in chlorophyll formation and improves root growth and seed production. It has an important role in vigorous plant growth and resistance to cold.

Sulfur may be supplied to the soil from rainwater. It is also added in some fertilizers as an impurity, especially the lower grade fertilizers. The use of gypsum also increases soil sulfur levels.

Boron helps in the use of nutrients and regulates other nutrients. Also it aids in the production of sugar and carbohydrates. It is essential for seed and fruit development. Sources of boron are organic matter and borax.

Copper is important for reproductive growth. Aids in root metabolism and helps in the utilization of proteins.

Chloride aids plant metabolism. Chloride is found in the soil.

Iron is essential for formation of chlorophyll. Sources of iron are the soil, iron sulfate and iron chelate.

Manganese functions with enzyme systems involved in breakdown of carbohydrates and nitrogen metabolism. Soil is a source of manganese.

Molybdenum helps in the use of nitrogen. Soil is a source of molybdenum.

Zinc is essential for the transformation of carbohydrates. It regulates consumption of sugars. It is part of the enzyme systems which regulate plant growth. Sources of zinc are soil, zinc oxide, zinc sulfate and zinc chelate.

Mineral Analysis

Ash or mineral content

Ash is the inorganic residue remaining after the water and organic matter have been removed by heating in the presence of oxidizing agents. It provides a measure of the total amount of minerals within a sample material. The three main types of analytical procedures used to determine the ash content are 1.*dry* ashing, 2.*wet* ashing and 3. *low temperature plasma dry* ashing. Ash contents of organic material rarely exceed 5%. However, in some cases the content may be as high as 12% (*e.g.*, dried beef, sesame seed).

Sample Preparation: Typically, samples of 1-10g are used in the analysis of ash content. Solid samples are finely ground and then carefully mixed to facilitate selection of a representative sample. Samples high in moisture are often dried prior to ash analysis in a muffle furnace. High fat samples are usually defatted by solvent extraction, as this facilitates the release of the moisture. Other possible problems include contamination of samples by minerals in grinders, glassware or crucibles which come into contact with the sample during the analysis. For the same reason, it is recommended to use de-ionized water when preparing samples.

Dry Ashing: Dry ashing procedures use a high temperature muffle furnace capable of maintaining temperatures of 500°C to 600°C. Water and other volatile materials are vaporized and organic substances are burned in the presence of the oxygen in air to CO_2, H_2O and N_2. Most minerals are converted to oxides, sulfates, phosphates, chlorides or silicates. Although most minerals have fairly low volatility at these high temperatures, some are volatile and may be partially lost, *e.g.*, iron, lead and mercury. The sample is weighed before and after ashing to determine the concentration of ash present. The ash content can be expressed on either a *dry* or *wet* basis:

$$\% \text{ Ash (dry basis)} = \text{weight of ashx} \frac{100}{\text{weight of dry sample}}$$

Wet Ashing: Wet ashing breaks down and removes the organic matrix surrounding the minerals so that they are left in an aqueous solution. A dried ground sample is usually weighed into a flask containing strong acids and oxidizing agents (*e.g.*, nitric, perchloric and/or sulfuric acids) and then heated. Heating is continued until the organic matter is completely digested, leaving only the mineral oxides in solution. The digestion takes from 10 min to a few hours at temperatures around 350°C. The resulting solution can then be analyzed for specific minerals. Little loss of volatile minerals occurs because of the lower temperatures used in wet ashing. However, it is labour intensive and requires a special fume-cupboard (because of caustic acid fumes).

Low Temperature Plasma Ashing: A sample is placed into a glass chamber which is evacuated using a vacuum pump. A small amount of oxygen is pumped into the chamber and broken down to nascent oxygen (O_2 β 2O·) by application of an electromagnetic radio frequency field. The organic matter in the sample is rapidly oxidized by the nascent oxygen and the moisture is evaporated because of the elevated temperatures. The relatively cool temperatures (< 150°C) used in low-temperature plasma ashing cause less loss of volatile minerals than other methods. There is less chance of losing trace elements by volatilization. However, plasma ashing is more expensive (costly equipment).

Extractable Minerals

Principle

The sample is digested with nitric acid and perchloric acid mixture and the mineral content is estimated. The minerals extracted can be used fro further analysis by standard methods.

Reagents and equipment

1. HCl (0.03N) : To 99.74 ml distilled water, 0.26 ml concentrated HCl is added.
2. Di-acid mixture: Nitric acid and perchloric acid mixed in the ratio of 5:1.
3. Incubator.

Procedure

1. One g sample is taken in a conical flask. To this 50ml of 0.03N HCl is added and incubated at 37°C in a shaker –cum – water bath for 3 hr to simulate conditions that occur in human stomach.
2. The mixture is filtered through ashless filter paper (whatman 42) and the filtrate is oven dried.
3. The residue is digested in the di-acid mixture and proceed for the determination of individual minerals.
4. The extractable mineral content of Calcium, Iron, Phosphorus and Potassium are estimated in the samples using standard methods.

Calculations

$$\text{Minerals extractable} = \frac{\text{Minerals extractable in 0.03 N HCl} \times 100}{\text{Total minerals}}$$

Ash content

Principle

The sample is burnt (wet or dry) to remove completely the organic compounds. Only the mineral components remain which are estimated gravimetrically.

Reagents and equipment

1. Burner.
2. Muffle furnace.
3. Kjeldahl flask, 100 ml.
4. Glass beads.
5. Polythene bottles.
6. Concentrated HCl.
7. Concentrated H_2SO_4.
8. Concentrated HNO_3.
9. Concentrated $HClO_4$.
10. Silica crucible.
11. Desiccators.
12. Water bath.
13. Centrifuge.

Procedure

1. Two (2) g dried sample is taken in a weighed silica crucible.
2. The crucible is placed on a burner and heated till the material is completely charred.
3. Then this is placed in the muffle furnace and heated to 550^0-700^0C for 4 hrs.
4. The crucible is transferred to a dessicator to cool to room temperature and the weight of the crucible is later recorded.

Calculations

$$\text{Per cent ash} = \frac{\text{weight of crucible with ash- weight of empty crucible X 100}}{\text{weight the sample X 100-M}}$$

M = moisture percent in the sample

a. Ash solution

1. The ash is moistened with a small amount of glass distilled water (0.5 1.0ml) and 5 ml of distilled hydrochloric acid is added to it.
2. The mixture is evaporated to dryness on a boiling water bath.
3. Another 5ml of hydrochloric acid is added again and the solution evaporated to dryness as before.
4. Four ml of hydrochloric acid and a few ml of water are then added and the solution warmed over a boiling water bath and filtered into a 100 ml volumetric flask using whatman No 40 filter paper.
5. After cooling, the volume is made upto 100ml and suitable aliquots are used for the estimation of minerals like phosphorus, iron and calcium.

b. Wet digestion

Organic matter can be destroyed and oxidised by boiling with sulphuric, perchloric and nitric acids.

1. Weigh about 3 g of dry ground organic material in a 100 ml Kjeldahl flask.
2. Add 25 ml 3:2:1 (conc. nitric acid-conc. perchloric acid-conc. sulphuric acid) tri-acid mixture and shake well ensuring that no dry lumps (solids) remain. Add a clean glass bead (acid washed).
3. Leave aside for 3-4 hr or preferably overnight in a fume cupboard. (watch for foaming during the first hour especially with samples high in fat. If foaming occurs and threatens to overflow immerse the bulb of the flask in cool water till froth subsides.
4. Heat for about 30 min cautiously until the initial vigorous reaction has subsided (dense yellow fumes will be evolved).
5. Heat more strongly for 4 hr until most of the nitrous fumes are removed.
6. Continue heating until white fumes of perchloric acid are evolved. If charring occurs or flask contents tend to become dry, remove from heat, cool and add 5ml of nitric acid and continue heating.

7. Carry a blank with each set of samples.
8. When cool, transfer quantitatively with 3-4 washings of de-ionised water the contents of the digestion flask to 15 ml graduated test tube. Make upto the 10 ml mark with de-ionised water and mix thoroughly. Allow the heated solution to cool and make up to the 10 ml mark again if necessary.
9. The tubes are centrifuged for 30 min.
10. Transfer the ash solution to acid-washed polythene bottles and store in a cool place prior to elemental analysis by atomic absorption spectrophotometry or flame photometry.

Note

1. *Use only glass distilled or de-ionised water.*
2. *Nitric acid diluted 1:100 in de-ionised water is used as glassware washing liquid.*
3. *All operations should be carried out in a fume cupboard with proper ventilation.*

Atomic absorption spectrophotometry

Principle

Representative sample in a suitable liquid form is sprayed into the flame of an atomic absorption spectrophotometer and the absorption or emission of the mineral to be analyzed is measured at a specific wave length.

Reagents and equipment

1. Standard stock solution (1000 µg/ml): Dissolve 1g of the mineral salt in a minimum amount of redistilled HCl (10ml). Dilute to one litre with de-ionised water.
2. Working standard solutions: Dilute aliquots of the stock solutions with de-ionised water to make at least 4 standard solutions within the range of determinations.
3. Atomic absorption spectrophotometer

Procedure

1. Depending upon the mineral to be determined standards, sample and blank solutions may be aspirated into the flame directly or after suitable dilutions to attain working range of the instrument.
2. The operating conditions recommended by the manufacturer should be followed.
3. Read at least 3-4 ranges of standard solutions before and after sample readings. Flush burner with de-ionised water between samples and check for zero setting.
4. Prepare calibration curve from the readings of standards.
5. Determine the concentration of samples from the standard graph.

Calculation

$$\text{Mineral content in ppm} = \frac{\mu\text{g mineral in one ml x dilution factor}}{\text{ml aliquots x g sample}}$$

Mineral	Wave length nm	Flame	Optimum working range µg/ml
Ca (dilute with 0.1% Lanthanum chloride sol)	422.7	Rich air- C_2H_2(oxidizing)	0.5-2.5
Cr	357.9	Rich air- C_2H_2(reducing)	2.5-10.0
Cu	324.7	Air- C_2H_2(oxidizing)	2.0-8.0
Fe	248.3	Rich air- C_2H_2(oxidizing)	2.0-8.0
Mg	285.2	Rich air- C_2H_2(oxidizing)	0.1-0.6
Mn	279.5	Air- C_2H_2(oxidizing)	1.0-4.0
Zn	213.9	Air- C_2H_2(oxidizing)	0.4-1.6

Boron (Dodd's method -titrimetry)

Principle

Ashed sample is utilized to estimate boron content by titrimetry. The boric acid in the ashed sample is titrated with CaOH(0.04N).

Reagents

1. Sodium hydroxide -1N:Dissolve 40g NaOH in water and make up to one litre.

2. CaOH - 0.04N: Prepare from lime water or CaOH salt and standardize with KIO_4.
3. $CaCl_2$: Dissolve 20g $Cacl_2.6H_2O$ or 13g dry $CaCl_2$ in water and make up to 100 ml.
4. Mannitol.
5. Methyl red indicator: Dissolve 1 g in one litre ethanol.

Procedure

1. Ash the sample in the presence of sodium hydroxide.
2. To the boric acid solution (in 100 ml volumetric flask) obtained, add 5 ml $CaCl_2$ solution. Make just alkaline with 1N Na OH. Make up to 100 ml.
3. Filter through a dry Whatman no.44 filter paper. Discard initial few drops. Collect the rest of the filtrate into a dry flask.
4. Transfer 75ml of the filtrate to a 250 ml conical flask. Add 3 drops methyl red and make just acid by adding a few drops of 1N H_2SO_4. Then add a few more drops of the acid and boil for 2 min to expel CO_2. Cover with watch glass to prevent re-absorption of CO_2 and cool under running water.
5. Titrate with 0.04N CaOH to the neutral point of methyl red. Read the burette reading.
6. Add 3-5 g mannitol and continue titration to the phenolphthalein colour range.
7. The amount of alkali between the methyl red and phenolphthalein end point corresponds to the boric acid present in 75 ml of the solution.
8. Carry out a blank titration using only reagents. Add potassium hydrogen phosphate and NaOH solutions to the blank. Deduct the value from the above titration to get the correct titre value.

Calculation

One ml of 0.04N CaOH solution corresponds to 0.432mg boron. Calculate the actual boron in the sample based on the sample weighed for ashing and the dilutions effected.

Reference

Dodd, A.S. Analyst, 52: 459-466, 1927 from C.S. Piper, Soil and Plant analysis.

Boron (Tintometry)

Principle

The boron in the sample is converted into a red coloured compound with quinalizarin. The intensity of the colour is measured using a tintometer and compared with that of the standards.

Reagents and equipment

1. Conc. H_2SO_4
2. Dilute H_2SO_4: Dilute 5 ml Conc. H_2SO_4 to 500 ml with water.
3. Ca $(OH)_2$: Prepare a saturated solution.
4. Quinalizarin solution: Dissolve 0.01 g quinalizarin in a mixture of 90 ml of conc. H_2SO_4 and 10 ml water.
5. Standard boric acid: Dissolve 2.857 g boric acid in water and dilute to one litre. (1 ml = 0.5 mg boron). Dilute this solution (20 ml to one liter) to get 10 µg boron per litre.
6. Tintometer.
7. Muffle furnace.
8. Centrifuge.
9. Platinum crucible.

Procedure

1. Ash one gram sample in a platinum basin in a muffle furnace at 450°C for a short time.
2. Cool and add 2 ml reagent 3 above and evaporate to dryness.
3. Re-ignite the material at 600°C.
4. When cool, dissolve the residue in 6-15 ml dilute H_2SO_4. Stir and centrifuge.
5. Transfer clear supernatant one ml to a test tube and add 9 ml Conc. H_2SO_4 and 0.5 ml reagent 4.
6. Place in a desiccator after stirring and keep overnight.
7. Next day determine the colour – red units in a tintometer.
8. Compare the colour in units with that of the standards prepared similarly.

9. Calculate the boron content taking into account the dilutions.

Reference

Mounsfell, P.W. New Zealand. J. Sci. Techn. (1940) 22: 100-111B.

Calcium (Titrimetric method)

Principle

Calcium is precipitated as oxalate and titrated with standard potassium permanganate to have quantitative estimate.

Reagents and equipment

1. 4% ammonium oxalate solution.
2. Dilute ammonia solution (2 ml of liquor ammonia + 98 ml water).
3. 1 N H_2SO_4.
4. 0.01 N Potassium permanganate solution.
5. 0.01 N Oxalic acid: Sodium oxalate is dried in an oven at 100-105°C for 12 hr. Exactly 0.67 g is dissolved in redistilled water, 5ml concentrated H_2SO_4 is added and the solution made up to 1litre after it has cooled down.
6. Standardization of potassium permanganate solution: 25 ml of 0.01 N oxalic acid is transferred to an Erlenmeyer flask. One ml of concentrated H_2SO_4 is added, warmed to about 70°C and titrated against $KMnO_4$ soultion till the faint pink colour persists.
7. Centrifuge.

Procedure

1. Two ml of the ash solution is taken into a 15ml centrifuge tube. Add 2 ml of distilled water and 1 ml of 4% ammonium oxalate solution. Mix thoroughly and leave overnight.
2. Again the contents are mixed and centrifuged for 5 min at 1500 rpm. The supernatant liquid is poured off and the centrifuge tube drained by inverting the tube for 5 min on a rack (care should be taken not to disturb the precipitate).
3. The mouth of the centrifuge tube is wiped with a piece of filter paper. The precipitate is stirred and the sides of the tubes are washed with 3 ml of dilute ammonia.

4. It is centrifuged again and drained as before. The precipitate is washed once more with dilute ammonia to ensure the complete removal of ammonium oxalate.

5. The precipitate is dissolved in 2 ml of 1N H_2SO_4. The tube is heated by placing it in a boiling water bath for 1 min and titrated against 0.01 N $KMnO_4$ solution to a definite pink colour persisting for at least 1 min.

Calculation

1ml of 0.01 N $KMnO_4$ is equivalent to 0.2004 mg of calcium.

$$\text{mg of calcium/100 ml} = (X\text{-}b) \times \frac{100}{2}$$

where, X= number of ml of 0.01 N $KMnO_4$ required to titrate the sample,

b= number of ml of 0.01 N $KMnO_4$ required to titrate 2ml of 1N H_2SO_4 (blank).

If the normality of $KMnO_4$ is N, the value obtained in the above formula should be multiplied by the factor, N/ 0.01.

Note

Potassium permanganate solution needs to be frequently standardized.

Calcium content (Colorimetric method)

Principle

Calcium forms a coloured complex with the O-cresolphthalein dye, which is made more specific in the presence of 8-quinolinol.

Reagents and equipment

1. Standard $CaCl_2$: 10µg Ca^{++}/ ml prepared by dissolving required amount of anhydrous $CaCO_3$ in dilute HCl and making up to a definite volume with water.

2. Ammonia/ammonium chloride buffer, pH 10.5: Dissolve 0.24 g NH_4Cl in 100 ml of 5% NH_4OH solution (v/v).

3. O-Cresolphthalein solution, 0.1%: 100 mg of O- cresolphthalein dissolved in 28 ml of ammonia/ammonium chloride buffer and diluted to 100 ml with water.

4. 8-Quinolinol: 1% solution in absolute ethanol.

5. Spectrophotometer with quartz cuvettes.

Procedure

1. Different aliquots of the $CaCl_2$ solution with Ca++ content varying from 2-10 µg are taken in graduated stoppered tubes and volume in each case is made up to 5.5 ml with water.
2. Subsequently, 1ml of 8-quinolinol solution and 2.5 ml ammonia/ ammonium chloride buffer are added and the volume adjusted to 9.0 ml.
3. The contents are finally treated with 1ml of O-cresolphthalein solution.
4. After addition of each reagent mentioned above, the contents are thoroughly mixed.
5. Water blank in the place of $CaCl_2$ solution is simultaneously carried throughout the procedure.
6. Standards are read against the blank in a spectrophotometer at 565 nm. Readings should be read after 5min (stable up to 3hr).
7. Concentration of sample in µg is read from the standard graph using the absorbency reading.
8. Further calculations are made based on the weight of the sample and related dilutions etc.

Phosphorus content (Fiske and Subba Row method)

Principle

Organic matter is destroyed by digestion with H_2SO_4 (or fuming nitric acid). The phosphate containing solutions are treated with molybdic acid to produce phosphomolybdic acid. This is reduced by the addition of 1,2,4-aminonaphtholsulphonic acid reagent to obtain a blue colour, the intensity of which is proportional to the amount of phosphate present.

Reagents and equipment

1. 10 and 5N H_2SO_4.
2. 30% H_2O_2 or fuming nitric acid.
3. Aminonaphtholsulphonic acid reagent: 195 ml of 15% sodium bisulphite solution is taken in a stoppered cylinder and 0.5 g of 1,2,4-aminonapholsulphonic acid is added to it followed by 5ml of 20% sodium sulphite. It is stoppered and shaken until the powder is dissolved. If the solution is not complete, more of sodium sulphite, 1ml at a time is added with shaking. The solution is then transferred to a brown glass bottle and

stored in the cold. This solution is stable for four weeks. Prepare afresh for each analysis.

4. Standard phosphate: 351 mg of pure dry monopotassium phosphate (KH_2PO_4) is dissolved in water and transferred quantitatively to a 1 litre volumetric flask. Ten ml of 10 N H_2SO_4 is added and the solution diluted to the mark with water and mixed. This solution contains 0.4 mg of phosphorus in 5ml. It is stable indefinitely.
5. 2.5% Ammonium molybdate: 2.5 g of reagent grade ammonium molybdate is dissolved in water, transferred to 100ml volumetric flask, filled to the mark and mixed.
6. Spectrophotometer with quartz cuvettes.
7. Kjeldahl digestion and distillation assemblies.

Procedure

1. Two ml of ash extract containing 2-20 μg of phosphorus is transferred to a test tube or micro-Kjeldahl flask and 2.5 ml of 5N H_2SO_4 is added; also add a quartz chip to minimise bumping. The contents of the flask are digested on a micro burner.
2. After evaporation is complete and the mixture turns brown or black with no further change, the tube is removed, cooled slightly and 1 drop of 30% H_2O_2 (if it gives too high a blank then 1 or 2 drops of fuming nitric acid) added to fall directly into the digestion mixture.
3. The heating is then continued until the contents of the tube become colourless. When the mixture is clear, the tube is cooled and a few ml water is added and the solution is heated to boiling momentarily.
4. The contents are then transferred to a 25 ml volumetric flask quantitatively with washings. To the flask is added 2.5 ml of 2.5% ammonium molybdate solution, followed by 1ml of amino-naphtholsulphonic acid reagent. The solution is made up to 25 ml with water and mixed well.
5. It is allowed to stand for 5 min, then the intensity of the colour is read in a colorimeter or a spectrophotometer.
6. For photometric measurements, the most accurate procedure is to run a digested blank and a series of digested standard solutions along with each series of unknowns.
7. The photometer is set to zero OD with blank and colour of standard and unknown solutions are read at 660 nm. The standard curve is plotted and the unknown read off from the curve.

Calculation

mg of total acid soluble phosphorus per 100ml solution =

$$\frac{\text{mg P in unknown} \times 100}{\text{ml of solution taken}}$$

Phosphorus (Vanadomolybdate method)

Principle

Phosphrus is converted to phosphomolybdic acid and the colour is measured spectrophotometrically.

Reagents and equipment

1. Tri-acid mixture of HNO_3:H_2SO_4:$HClO_4$(9:4:1).
2. HNO_3: vanadomolybdate reagent: a) solution A: dissolve 25 g ammonium molybdate in 400 ml water. Solution B: dissolve 1.25 g ammonium meta vanadate in 300 ml boiling water. Cool and add 250 ml conc. HNO_3 and again cool to room temperature. Finally solution A is added and the contents diluted to one litre.
3. Standard P solution: Dry KH_2PO_4 (0.2195 g) is dissolved in 400 ml distilled water in a 1000 ml volumetric flask. Add 25 ml 7N H_2SO_4. Make up to volume (50 ppm P solution). Dilute 50 times to get 1ppm solution.
4. Spectrophotometer with quartz cuvettes.

Procedure

1. Take 1 g sample (plant) into a 100 ml volumetric flask. Add 10 ml tri-acid mixture of HNO_3:H_2SO_4:$HClO_4$(9:4:1). The flask is placed on low heat initially followed by high temperature digestion. After cooling add 20 ml water. Make up to volume. If necessary filter the contents through no.1 filter paper.
2. Take 2 ml of the ash solution in a 50 ml volumetric flask. Add 10 ml vanadomolybdate reagent, make up to volume and mix well.
3. After 10 min measure absorbency at 470 nm.
4. Run a blank alongside and measure the colour.
5. Deduct blank value.

6. Take 2, 4, 6, 8, 10, 12 and 14 ml of 1ppm standards into 50 ml volumetric flasks. Add 10 ml vanadomolybdate reagent. Make up to volume. Measure absorbency at 470 nm. Plot a graph of absorbency against concentration.
7. Read sample concentration from the graph.
8. Make calculations based on the sample weight and dilutions effected.

Potassium content (Flame photometric method)

Principle

Potassium in the sample is determined using a flame photometer. It is based on the principle that atoms of some specific elements take energy from flame and get excited to the higher orbit. Such atoms release energy of a wavelength which is specific for that element and is proportional to the concentration of atoms of that element.

Reagents and equipment

1. Tri-acid mixture of HNO_3:H_2SO_4:$HClO_4$(9:4:1).
2. Standard "K" solution: Dissolve 1.0609 g KCl (A.R. grade) in distilled water and make up to 1000ml (1000 ppm conc.). Dilute 10 times to get 100 ppm solution.
3. Flame photometer:

Procedure

1. Take 1g sample (plant) into a 100ml volumetric flask. Add 10 ml tri-acid mixture of HNO_3:H_2SO_4:$HClO_4$(9:4:1). The flask is placed on low heat initially followed by high temperature digestion. After cooling add 20ml water. Make up to volume. If necessary filter the contents through no.1 filter paper.
2. Dilute the digest suitably to the desired measurable level in instrument.
3. Standard curve: Set the instrument at highest concentration of 5 ppm using the standard filter (549 nm).
4. Measure the concentration of sample by loading to the flame photometer at 549 nm.

Calculation

Make calculations based on the sample weight, dilution and volumes taken for measurement

$$K\ (\%) = R \times \frac{5}{100} \times \frac{100}{\text{sample}} \times \frac{100}{1000000}$$

R= Reading of flame photometer, where 5 ppm K=100 R and if further dilution is made, then appropriate modification is required in the calculation.

Iron (Wong's method)

Principle

Iron is determined colorimetrically using the fact that ferric iron gives a blood red colour with potassium thiocyanate.

Reagents and equipment

1. 30% H_2SO_4
2. 7% Potassium persulphate soultion: 7 g potassium persulphate is dissolved in glass distilled water and the solution made up to 100 ml.
3. 40% Potassium thiocyanate solution: 40g KCNS is dissolved in 90 ml glass distilled water 4ml acetone added and the volume made upto 100 ml.
4. Standard iron solution: 702.2 mg ferrous ammonium sulphate is dissolved in 100 ml glass distilled water. Add 5ml of 1:1 HCl, make up to 1 litre and mix thoroughly (0.1 mg Fe/ml). The standard solution is prepared fresh once in 6 months.
5. Working standard: (10 mg Fe/ml) is prepared by diluting the above solution 10 times.
6. Spectrophotometer with quartz cuvettes

Procedure

1. To an aliquot (6.5 ml or less) of the mineral solution enough water is added (if necessary) and make up to a volume of 6.5 ml followed by 1.0ml of 30% H_2SO_4, 1.0 ml potassium persulphate solution and 1.5 ml 40% KCNS soultion.
2. The red colour that develops is measured within 20 min at 540 nm.

3. A standard (10-100 mg) is run similarly and a standard graph is prepared.
4. Calculations are made based on the concentration read from the graph.

Reference

Jackson, M.L. 1967. Soil Chemical Analysis.

Copper (Iodimetric method)

Principle

Copper present in the solution is estimated through iodimetric titration.

Reagents and equipment

1. Concentrated HCl, HNO_3, H_2SO_4.
2. 1:1 H_2SO_4: water.
3. 1:3 ammonium hydroxide.
4. KI salt.
5. Potassium thiocyanate.
6. Saturated bromine water.
7. 1:3 ammonium hydroxide.
8. Ammonium acid fluoride.
9. Glacial acetic acid.
10. Hypo: 25 g sodium thiosulfate penta hydrate in one litre water (0.1N): Add 2 g sodium carbonate. Standardize against potassium dichromate solution.

Procedure

1. Weigh about 2 g of the finely ground sample containing copper (plant, soil or ore) into a 250 ml beaker.
2. Add slowly 5 ml of conc. HCl and 10 ml of conc. HNO_3. Cover with a watch glass and heat over a low flame until only a white residue of silica remains.
3. Remove the watch glass and add 10 ml 1:1 H_2SO_4.
4. Evaporate the solution in a fume hood. Cool and transfer to a conical flask with 25 ml water.

5. Add 5 ml saturated bromine water and boil gently to remove excess bromine.
6. Add 1:3 ammonium hydroxide. Add 5 ml glacial acetic acid and 2 ml ammonium acid fluoride and stir.
7. Dissolve 3 g KI in 10 ml water and add to the copper solution.
8. Titrate the liberated iodine immediately with hypo solution.
9. Towards the end of titration (disappearance of brownish colour), add 5ml starch solution and 2 g potassium thiocyanate.
10. Swirl the flask and complete the titration with hypo (dis-appearance of blue colour).

Calculation

Based on the sample weighed and the titre value calculations are made. Take into account any dilutions effected during the digestion.

Copper (Colorimetric method)

Principe

Copper forms a colored complex with sodium diethyl dithiocarbamate which is extracted and measured colorimetrically.

Reagents

1. EDTA solution: 25 g EDTA sodium salt is dissolved in 500 ml of glass distilled water. It is well shaken with solid diethyl dithiocarbate (0.5 g) and 100 ml $CHCl_3$ till the colour disappears. 8 ml of 0.1% cresol red is added. The mixture is shaken well and the EDTA solution kept in a stoppered bottle.
2. Copper standard: Dissolve 0.982 g $CuSO_4$, $5H_2O$ in water and 5ml conc. H_2SO_4 is added and made up to 250 ml in a volumetricflask (1mg/ml copper).
3. 1% Sodium diethyl dithiocarbamate.
4. Kjeldahl digestion assembly.

Procedure

1. 2 g of dry sample is digested in a pyrex Kjeldahl flask with 5 ml glass distilled water and 4 ml conc. H_2SO_4 and 5ml distilled HNO_3. The digestion is carried out with constant addition of HNO_3 till a clear digest is obtained. One ml of $HClO_4$ is then added and the digestion continued for another 15-20 min.
2. The digest is then quantitatively transferred to a separating funnel using glass distilled water to make up to 50 ml. 5 ml EDTA solution is then added and the solution neutralized with distilled NH_3 (to get a pink color).
3. It is cooled and 1ml of 1% sodium diethyl dithiocarbamate is added followed by 10 ml distilled CCl_4. The mixture is shaken for exactly 2min.
4. After CCl_4 layer separates, it is run into a colorimetric tube through a plug of cotton and the intensity of colour is read at 440 nm against a reagent blank extracted similarly. (The blank extraction does not contain sample with all other steps being the same as above).
5. Copper content is read from a standard graph prepared using 0-40 μg levels of copper standard. Calculations are made based on the dilutions and the weight of sample taken.

Magnesium (Gravimetric method)

Principle

Magnesium is converted to magnesium pyrophosphate and estimated gravimetrically.

Reagents and equipment

1. 10% Ammonium phosphate solution.
2. 10% Ammonium citrate solution.
3. 0.1% Methyl red indicator in ethanol.
4. 1:4 and 1:10 NH_3: H_2O solution.
5. Muffle furnace.
6. Desiccator.

Procedure

1. To the Ca free filtrate (obtained from the filtrate after precipitation of Ca as oxalate) is added 30ml of conc. HNO_3.

2. It is evaporated on a water bath. 5 ml conc. HCl and 100 ml of water are then added and the solution stirred well with a glass rod.
3. Then add 10 ml of 10% ammonium phosphate and 5 ml of 10% sodium citrate. The contents are stirred well.
4. Add 2-3 drops of methyl red indicator and neutralize with the addition of 1:4 dilute ammonia.
5. Strong NH_3 (25 ml) is then added, stirred well and mixture allowed to stand overnight.
6. The contents are filtered through Whatman no. 44 or 40 filter paper and washed free from chlorides using 1:10 dilute NH_3.
7. The funnel with precipitate on the filter paper is dried in an oven.
8. The filter paper is transferred to a weighed crucible and ashed slowly on a burner.
9. It is then kept in a muffle furnace at 900^0 C for 2 hr.
10. The crucible and the contents are cooled in a desiccators and weighed to get Mg as its pyrophosphate.

$$\text{mg of } \frac{\text{Magnesium}}{\text{100 g sample}} =$$

$$\text{weight of ash x } \frac{48.64}{222.6} \text{ x } \frac{100}{\text{ash sol.taken}} \text{ x } \frac{100}{\text{wt of sample}} \text{ x } 1000$$

Barium content (Titrimetric method)

Principle

The barium in the sample is titrated with EDTA solution and the barium content is estimated

Reagents and equipment

1. 1M NaOH
2. Methyl thymol blue-potassium nitrate mixture (30-50 mg of a1% solid mixture of the indicator in potassium nitrate). Satisfactory colour is achieved with this mixed indicator.
3. 0.01M EDTA solution.

Procedure

1. Pipet 25 ml of wet digestion solution (containing 0.01 M barium only) into a 250 ml beaker and dilute to about 100 ml with de-ionised water.
2. Adjust the p^H to 12 by adding 3-6 ml of 1M NaOH. The p^H must be above 11,5.
3. Add 50 mg of methyl thymol blue- potassium nitrate mixture .
4. Titrate with 0.01 M EDTA solution till the colour changes from blue to grey.

Calculation

1mol EDTA = 1mol Ba^{++}

Chloride (as silver chloride)

Principle

The chloride is quantitatively precipitated as AgCl. The amount of silver nitrate consumed during titration with potassium thiocyanate is used to measure the chlorides in the sample.

Reagents and equipment

1. Water bath.
2. 0.1N NaOH.
3. N/50 $AgNO_3$.
4. Ferric indicator: 20 g ferric ammonium sulphate is dissolved in 50 ml water to which 1 or 2 drops of nitric acid are added.
5. N/100 potassium thiocyanate.

Procedure

1. 2-5 g sample is weighed into a boiling tube and 60 ml of 0.1N NaOH is added.
2. The tube is kept in a water bath and digested. When complete, the solution is made up to 100 ml.
3. A 30 ml aliquot of the digest is taken in another boiling tube and add 5ml of N/50 $AgNO_3$ and 5 ml HNO_3. Continue digestion in the boiling water bath for another 30 min. nitric acid fumes are allowed to escape.

4. The tube is then cooled and 40 ml of distilled water and 20 drops of ferric indicator are added and solution is kept overnight in the cold.

5. The next day, the contents are titrated against N/100 potassium thiocyanate.

6. A reagent blank is run simultaneously to obtain the blank titer value.

Calculation

1ml of 0.01 potassium thiocyanate = 0.335 mg of Cl

$$\frac{\text{mg Chloride}}{\text{100 g sample}} = \text{blank-sample titer value} \times \frac{0335}{0.01} \times \text{N of KSCNS} \times \frac{100 \times 100}{30 \times \text{Wt}}$$

Sulphur (Gravimetric method)

Principle

Sulphur is precipitated with barium chloride as sulphate and is estimated gravimetrically.

Reagents and equipment

1. 0.1% Methyl orange.
2. 40% Na OH.
3. 5% Barium chloride.
4. HCl.
5. Conc. HNO_3.
6. Water bath.
7. Sand bath.
8. Kjeldahl digestion assembly.
9. Hot air oven.
10. Muffle furnace.

Procedure

1. To 2-3g sample in a micro Kjeldahl flask is added 5ml conc. HNO_3.
2. The flask is then kept in a boiling water bath for 10hr. and then on a sand bath and digested till a liquid digest results. (Additional HNO_3 is added if needed.)
3. 2 ml $HClO_4$ is added digested further (10-16 hr).
4. The digest is transferred to a 500ml beaker with washings and solution is made up to 200 ml.

   1-2 drops of methyl orange indicator are added. The solution is neutralized with 40% NaOH
5. The solution is again acidified with HCl and boiled for 5 min.
6. 10 ml barium chloride is added and the solution boiled for 5 min.
7. It is kept overnight and filtered through no. 40 or 44 filter paper and washed till free from barium chloride.
8. The precipitate along with filter paper is kept in an oven for drying.
9. The filter is then taken into a weighed crucible and heated at 600°C (in a muffle furnace) for 3-4 hr and weighed.

Calculation

$$\frac{\text{g sulphur}}{\text{100 g sample}} = \text{wt of ash} \times 0.1374 \times \frac{100}{\text{wt of sample}}$$

Sulphur (Turbidimetric method)

Principle

The material is digested with nitric and perchloric acids and the sulphur content is measured turbidimetrically as barium sulphate (using barium chloride).

Reagents and equipment

1. Nitric acid 70%: mix 70 ml HNO_3 in water and make upto 100 ml.
2. Perchloric acid 70% (v/v).
3. HCl 6N: Mix 500 ml conc. HCl in in water and make upto one litre.
4. Barium chloride.

5. KCl standard solution: dissolve 5.43g K_2SO_4 in water and dilute to one litre (1 ml = 1 mg sulphate).
6. Balance.
7. Digester.
8. Spectrophotometer.

Procedure

1. Weigh 1 g defatted sample into a macro digestion tube.
2. Add 5 ml nitric acid (reagent1) to the tube.
3. Digest the sample at 100^{0}C for 30min.
4. Cool and add 6 ml perchloric acid.
5. Digest the sample again at 235^{0}C for one hour.
6. After cooling add 10 ml 6NHCl.
7. Make up the volume to 250 ml with water. Shake well and allow to stand overnight.
8. Transfer 15 ml to 50 ml flask and add 250 mg $BaCl_2$.
9. Shake for 10 min in a shaker.
10. Record the absorbance of the turbid suspension at 420 nm.
11. The sulphate content is read from graph prepared using standard K_2SO_4.

Calculation

$$\text{S content } \frac{\text{g}}{\text{100g sample}}$$

$$\text{total} \frac{\text{volume}}{\text{wt of sample}} \text{ x conc of } \frac{\text{unknown}}{\text{volume of sample}} \text{ x}$$

$$\frac{100}{1000 \text{ x } 1000} \text{ x absorption of } \frac{\text{unknown}}{\text{std}} \text{ x conc of std}$$

Iodine

Principle

Iodine estimation is based on catalytic reduction of ceric (Ce^{++++}) ion to cerous (Ce^{+++}) ion by iodine.

Reagents equipment

1. Sodium carbonate – potassium chlorate solution: 212 g of anhydrous Na_2CO_3 and 20 g of $KClO_3$ are dissolved in water and the volume made up to one litre.
2. 0.02 N Arsenious acid reagent : 0.986 g AsO_3 is dissolved in 10 ml of 0.5 N NaOH in a beaker while heating. This is added to 850 ml of water and to this 20 ml HCl and 20.6 ml conc. H_2SO_4 are added. The volume is then made up to one litre.
3. 0.03N ceric ammonium sulphate: 48.6 ml conc. H_2SO_4 is added to about 600 ml of water and then 20 g of ceric ammonium sulphate is dissolved in this solution. The volume is then made up to one litre with water.
4. Standard KI solution: Initially a stock solution containing 4 μg of iodine is prepared. From this suitable aliquots can be taken to have 0.04, 0.08, 0.12 and 0.16 μg iodine/ml fo preparation of standard curve based on colorimetric readings.
5. Muffle furnace.
6. Centrifuge.
7. Spectrophotometer with cuvettes.

Procedure

1. *Digestion*: Weigh enough quantity of sample (containing approximately 0.04–0.08 μg of iodine) and take into a pyrex test tube (15 × 125 mm). 0.5ml of Na_2CO_3- $KClO_3$ reagent and dry the contents at 110 °C, which may take about 2 hr.
2. *Incineration*: Incinerate the contents of the tube in a muffle furnace at 610°C for 30 min.
3. *Extraction*: The ashed material is extracted with 10 ml of arsenious acid at room temperature. After 15 min the contents are centrifuged at 2000 rpm.

4. *Colorimetry*: 5ml of supernatant is pipetted into a test tube and keep at 37°C. Later add 1ml of ceric ammonium sulphate. After 20 min (reduction would be complete) read the colour in a colorimeter or spectrophotometer at 420 nm.
5. *Standard curve*: Standard curve is prepared using different concentrations of KI mentioned above.
6. Concentration of sample is read from the standard curve based absorbence value obtained from the colorimeter. Calculations are made based on sample weights, dilutions etc.

Note

The standard curve needs to be prepared afresh for each estimation, when they are on different dates.

Cobalt (Nitroso-R-salt method)

Principle

Cobalt forms an intensely coloured soluble compound with sodium salt of 1-nitroso-2-naphthol 3:6-disulphonic acid. This cobalt compound is stable in nitric acid. This is the basis for the colorimetric determination of cobalt.

Reagents and equipment

1. Conc. H_2SO_4.
2. Conc. HNO_3.
3. Perchloric acid (sp.gr. 1.54).
4. 0.2M citric acid: Prepare citric acid solution and standardize it with NaOH solution.
5. 1N Na OH: Prepare using carbonate free conc. NaOH solution.
6. Buffered soda: Dissolve 6.184 g boric acid and 35.62 g $Na_2HPO_4.2H_2O$ in water. Add 500 ml 1N NaOH and dilute to one litre.
7. Dithizone in CCl_4 (0.05%): Dissolve 0.25 g dithizone in CCl_4. Dilute to one litre and store in a refrigerator.
8. Dithizone in $CHCl_3$ (0.2%): Dissolve dithizone in $CHCl_3$ containing 1% alcohol.
9. Bromophenol blue: Dissolve 0.04 g bromophenol blue in 50 ml alcohol. Add 1.49 ml of 0.1N NaOH. Make up to 100 ml with water.

10. Cresol red: 0.04 g cresol red in 50 ml alcohol. Add 2.62 ml 0.1NNaOH. Make up to 100 ml with water.
11. Methyl red: Dissolve 0.025 g in 100 ml 50% alcohol.
12. Phenolphalein: Dissolve 0.05 g in 100 ml 50% alcohol.
13. Nitroso – R - salt : Dissolve 35 g sodium salt of 2-naphthol 3:6 disulphonic acid (R –salt) in 400 ml water. Acdify with 10 ml conc. HCl, cool to 10^0C. Add drop by drop a solution of sodium nitrite (7 g in 25 ml water). Keep the reaction mixture dipped in ice. Filter yellow crystals. Wash with cold alcohol. Dissolve this salt (1 g) in 100 ml water and later make up to 500 ml.
14. Bromine solution (0.2N): Prepare using a saturated solution. Standardize with hypo using KI. Prepare afresh when needed.
15. Standard cobalt: Dissolve 0.4037 g $CoCl_2.6H_2O$ in water, add 10 ml HCl (1:2 aqueous). Dilute to one litre. One ml of this solution =100 µg cobalt. Prepare a working standard of 0.05 µg/ml from the above solution.
16. Kjeldahl digestion assembly.
17. Silica basin.

Procedure

1. Digest 10 g plant material in a 500 ml Kjeldahl flask. Add 80 ml HNO_3, 3ml H_2SO_4 and 5ml $HClO_4$. After digestion cool and dilute with 10 ml water. Boil and filter through a no.44 filter paper soaked in 10% HCl. Collect the filtrate and evaporate, till a small quantity of sulphuric acid remains.
2. Take up the residue in 7.5 ml of 0.2N citric acid and dilute to 30 ml
3. Add 5 drops of bromophenol blue and add 1N NaOH till a greenish-blue tint is reached.
4. Transfer to a separating funnel and extract with 20 ml lots of dithizone in chloroform thrice. This chloroform solution extracts copper.
5. Adjust the aqueous solution to p^H 8.3 by adding buffered soda.
6. Then extract the cobalt with dithizone in CCl_4. Collect the extracts into pyrex boiling tubes.
7. Distill off CCl_4 and digest with 1 ml HNO_3, 0.2 ml H_2SO_4 and 0.5 ml $HClO_4$ till colourless.
8. Rinse the contents into a silica basin and fume off at 350^0C.

9. Dissolve the residue in 1ml of 0.2 M citric acid. Transfer the solution to a boiling tube with minimum water.
10. Adjust to p^H 8.0 with buffered soda.
11. Develop the cobalt nitroso R- salt complex by adding 1 ml 0.2% nitroso R-salt. Boil and add 1 ml conc. HNO_3.
12. Then discharge the yellow colour of the reagent by adding 0.5 ml of 0.2N bromine.
13. After removing excess bromine, cool and dilute to 10 ml.
14. Measure the colour at 425 nm.
15. Compare the colour against a standard either visually or using a spectrophotometer.

Cobalt (Nitroso-R-salt method-modified)

Principle

Cobalt forms an intensely coloured soluble compound with sodium salt of 1-nitroso-2-naphthol 3:6-disulphonic acid. This cobalt compound is stable in nitric acid . This is the basis for the colorimetric determination of cobalt.

Reagents

1. Conc. HNO_3.
2. Conc. Ammonia.
3. Ammonium citrate (1M) : dissolve 210 g citric acid, add 400 ml water and 200 ml conc. ammonia. Add more ammonia till alkaline. Dilute to one litre. Make the solution metal free using dithizone in chloroform.
4. Nitroso –R- salt: Dissolve 35 g sodium salt of 2-naphthol 3:6 disulphonic acid (R –salt) in 400 ml water. Acidify with 10 ml conc. HCl, cool to 10°C. Add drop by drop a solution of sodium nitrite (7 g in 25 ml water). Keep the reaction mixture dipped in ice. Filter of yellow crystals. Wash with cold alcohol. Dissolve this salt (1 g) in 100 ml water and later make up to 500 ml.
5. Bromine water: Saturated solution in water.
6. Standard cobalt solution: Dissolve 0.4037 g $CoCl_2.6H_2O$ in water, Add 10 ml HCl (1:2 aqueous). Dilute to one litre. One ml of this solution =100 µg cobalt. Prepare a working standard of 0.05 µg/ml from the above solution.

Procedure

1. Make cobalt extract of the plant material acidic with excess H_2SO_4.
2. Add 5 ml conc. NH_3 and evaporate on a water bath to remove excess NH_3.
3. Take up residue in 5 ml 1M ammonium citrate, 5 ml water and 0.5 ml nitoso-R-salt.
4. Heat for 5-10 min, transfer to a boiling tube and dilute to 15 ml.
5. While still warm add 2.5 ml bromine water. After 15 min remove excess bromine.
6. When cold dilute to 15 ml.
7. Measure the colour at 425 nm.
8. Compare the developed colour with that of similarly prepared standards.
9. Calculate cobalt content based on sample taken and dilution of solutions etc.

Manganese (Periodate method)

Principle

The manganese is converted to permanganate and the colour is measured to quantify the mineral.

Reagents and equipment

1. Conc. Sulphuric acid
2. Conc. Nitric acid
3. Perchloric acid of specific gravity,1.54.
4. Ammonium persulphate: Dissolve 20 g ammonium persulphate in warm water and dilute to 100 ml. Prepare afresh prior to use.
5. Standard $MnSO_4$ solution: Take 0.5756 g dry $KMnO_4$ in a 2 litre volumetric flask and add 500 ml water. Add 40 ml conc. H_2SO_4 and reduce the permanganate by addition of sodium metabisulphite till the solution of manganese becomes colourless. Oxidize sulphurous acid by adding a little HNO_3. When cool dilute to 2000 ml. The solution contains 0.1 mg Mn/ml.
6. Hot plate.

Procedure

1. Digest 2-10 g of the plant sample with sulphuric (5 ml), nitic (4 ml) and perchloric (4 ml) acids. Later add a small quantity of nitric acid to completely oxidize the organic material. When cool add 5ml of ammonium persulphate solution, heat again to fuming for about 5 min.

2. When cool add 2 ml phosphoric acid and 35-50 ml warm water. Filter through 44 paper and wash with water. Collect filtrate in a 200 ml silica basin and evaporate to 25 ml.

3. Then add 0.3 g potassium periodate crystals. Boil cautiously over a low flame till the permanganate colour appears.

4. Dilute with 25-35 ml water and boil for 2 min till maximum colour develops.

5. Transfer the solution to a suitable volumetric flask and dilute. Keep in a boiling water bath for 15 min. Later make up to volume. Compare the colour to the standard solution prepared separately (545 nm).

6. Pipet different aliquots of standard Mn solution to volumetric flasks and add 10 ml conc. H_2SO_4 and 2 ml phosphoric acid and dilute each solution to about 60 ml. Add about 0.3 g potassium periodate and heat on water bath (5 min). Dilute/ make up to volume. Leave on the water bath for 15min. The colour developed can be used for obtaining the concentration of unknown sample.

Molybdenum content

Principle

Molybdenum (Mo) is determined as thiocyanate after reduction of Mo with $SnCl_2$. The molybdenum thiocyanate is extracted with ether. The red colour of ethereal solution is measured against a standard.

Reagents equipment

1. Conc. HCl.

2. Potassium thiocyanate: Dissolve 10 g potassium thiocyanate in water and dilute to 100 ml.

3. $SnCl_2$ solution: Dissolve 10 g $SnCl_2 . 2H_2O$ in dil HCl (1:9). Prepare afresh.

4. Ethyl ether.

5. Standard Mo solution: 0.552 g ammonium mlybdate - $(NH_4)_6Mo_7O_{24}$ $4H_2O$ in water and dilute to one litre. Dilute 10 ml 100 ml to have 30 μg/ml.
6. Muffle furnace.
7. Spectrophotometer with cuvettes.

Procedure

1. Take 2 g sample into a silica crucible and put in a muffle at 450-500°C.
2. Moisten the ash with water and add 10 ml HCl, warm and transfer to 100ml beaker and dilute to 40 ml
3. Boil and filter into a 100 ml volumetric flask and make up to mark.
4. Transfer 50 ml to a separating funnel. Add 7 ml conc. HCl. Mix and add 3ml potassium thiocyanate and 3 ml $SnCl_2$ while shaking.
5. Add 10 ml ether. Shake vigorously. Drain the lower aqueous layer.
6. Collect the ether layer and re extract the aqueous layer with 5 ml ether again twice. Pool the ether extracts. Measure the colour at 465 nm.
7. Prepare a standard : Take 5 ml Mo solution (30 μg/ml) in a separating funnel. Dilute to 50 ml and add 7 ml conc. HCl. Add 3 ml KCNS solution and 3 ml $SnCl_2$. Extract with ether as above and dilute ether extracts to 25 ml with ether.
8. Compare the unknown sample colour with the standard and determine the concentration of Mo.

Zinc content

Principle

The dithizone reacts with zinc to a coloured compound. It is measured at 535 nm.

Reagents and equipment

1. Conc. H_2SO_4.
2. Conc. HNO_3.
3. Perchloric acid (sp.gr.1.54).
4. Dithizone in $CHCl_3$: 5 g dithizone in 500 ml $CHCl_3$. Dissolve and store in a fridge. Prepare afresh.

5. Ammonium citrate buffer: Dissolve 5 g citric acid in 50 ml water and 200 ml 4N ammonia. Purify this solution by extraction with dithizone in $CHCl_3$. The buffer contains dithizone to extract 350 µg Zn from a plant digest.
6. 1N Ammonium hydroxide.
7. 0.02N HCl.
8. Carbon tetra chloride.
9. Dithizone in $CHCl_3$: Dissolve 0.25 g dithizone in 600 ml CCl_4, warming if needed. Transfer to a big separating funnel and shake with 350 ml water containing 3-4 ml conc. NH_3. Separate the aqueous layer and wash thrice with 75 ml portions of CCl_4. Reject the CCl_4 layers. Then add 600 ml CCl_4 and acidify the aqueous layer. The dithizone moves to CCl_4 layer. Separate the dithizone in CCl_4 and dilute to 2500 ml. store in a fridge.
10. 0.5M ammonium citrate : Dissolve 210 g citric acid in 1500 ml water. Add conc. Ammonia to a p^H of 8.5-8.7. Dilute to 2 litre. Purify by washing with dithizone reagent prior to use.
11. Sodium diethyl dithiocarbamate.: Dissolve 0.25 g sodium diethyl dithiocarbamate in water and dilute to 100 ml.
12. Mixed reagent A: Take 0.5 g ammonium citrate and 300 ml 1N ammonium hydroxide, dilute with 3200 ml water. Prior to use dilute 9 volumes of this solution with one volume of carbamate reagent (11 above).
13. Standard $ZnSO_4$: 0.1 g pure Zn foil is placed in a volumetric flask. Add 200 ml water and 10 ml conc. H_2SO_4. Dissolve Zn by heating and dilute to one litre. This solution contains 100 µg Zn/ml. This solution can be diluted 10 times to have working standard (1 µg/ml).
14. Spectrophotometer.

Procedure

1. Digest 1 g sample with nitric, suphuric and perchloric acids. After the digestion is over, cool and add 15 ml water. Boil, cool and 25 ml ammonium citrate buffer. Extract with chloroform thrice.
2. Rinse the aqueous phase with 2 ml CCl_4 twice to remove dithizone.
3. Add 50 ml mixed reagent A and 10 ml dithizone- CCl_4 reagent.
4. Pipet 5 ml CCl_4 extract to a 25 ml flask and make upto 25 ml.
5. Determine % transmittance at 535 nm.

6. Calculate concentration based on a standard curve prepared separately.
7. Take into account the sample weight and dilutions made during extractions.

Hardness of water (EDTA method)

Principle

Hardness is caused by dissolved polyvalent ions. The main constituents which cause hardness are Ca and Mg. Hardness is commonly expressed as $CaCO_3$ in mg/l. The degree of hardness of water is 0-60 mg/l is soft, 60-120 mg/l is medium, 120-180 mg/l is hard and > 180 mg/l is very hard.

Reagents and equipment

1. Buffer solution: Dissolve 16.9 g NH_4Cl in 143 ml NH_4OH. Add 1.25 g magnesium salt of EDTA to obtain a sharp change in colour of indicator. Dilute to 250 ml.
2. Inhibitor: Dissolve 4.5 g hydroxylamine hydrochloride in 100 ml 95% ethanol.
3. Eriochrome black T indicator: Mix 0.5 g dye with 100 g NaCl (dry powder).
4. Murexide indicator: Prepare a finely powdered mixture of 200 mg murexide with 100 g of solid NaCl.
5. NaOH (2N): Dissolve 80 g NaOH in 1000 ml water.
6. Standard EDTA (0.01M): Dissolve 3.723 g EDTA disodium salt to 1000 ml. standardize against standard Ca solution (1 ml= 1 mg $CaCO_3$).
7. Standard calcium solution: Weigh 1 g $CaCO_3$ and transfer to a 250 ml conical flask. Then add 1:1 HCl till $CaCO_3$ dissolves completely. Add 200 ml water and boil for 20-30 min. Cool and add methyl red indicator. add NH_4OH (3N) drop wise till intermediate orange colour develops. Dilute to 1000 ml to obtain 1 ml = 1 mg $CaCO_3$.
8. Titration assembly.

Procedure

Total hardness

1. Take 25 ml of water sample into a conical flask.
2. Then add 1-2 ml buffer followed by 1ml inhibitor.

3. Add a pinch of Eriochrome black T and titrate with standard EDTA (0.01M) till wine red colour changes to blue. Note the volume of EDTA required (A).
4. Run a reagent blank. Note the volume of EDTA required (B).
5. Calculate volume of EDTA required by sample, C = A-B.

Calcium hardness

1. Take 25 ml of water sample into a conical flask.
2. Add 1 ml NaOH to increase the p^H to 12 and add a pinch of murexide indicator.
3. Titrate immediately with EDTA till pink colour changes to purple. Note the volume of EDTA required (A_1).
4. Run a reagent blank. Note the volume of EDTA required (B_1).
5. Calculate volume of EDTA required by sample, $C_1 = A_1 - B_1$.

Calculations

$$\text{Total hardness as } CaCO_3, \frac{mg}{litre} = C \times D \times \frac{1000}{\text{volume of sample in ml}}$$

Where C= Volume of EDTA required by sample (with Eriochrome black T indicator)

D = mg $CaCO_3$ equivalent to 1ml EDTA titrant (1 ml EDTA = 1.00 mg of $CaCO_3$)

$$\text{or } D = 1 \times \frac{\text{Molarity of EDTA}}{0.01M}$$

$$\text{Calcium hardness as } CaCO_3, \frac{mg}{litre} = \frac{C1 \times D \times 1000}{\text{volume of sample in ml}}$$

Where C1 = Volume of EDTA required by sample (with murexide indicator)

D = mg $CaCO_3$ equivalent to 1ml EDTA titrant

Magnesium hardness $CaCO_3$, mg/litre

= total hardness- calcium hardness.

Organic carbon in soils

Principle

Organic matter in the soil gets oxidized by chromic acid. Excess dichromate is added and the un-reacted dichromate is measured to calculate the organic carbon.

Reagents and equipment

1. 1N potassium dichromate solution.
2. Conc. Sulphuric acid.
3. H_3PO_4 1N.
4. NaF.
5. Diphenyl amine indicator.
6. Ferrous ammonium sulphate 1N.
7. Titration assembly.

Procedure

1. Take 1 g finely powdered soil sample into a conical flask. Add 10 ml 1N potassium dichromate solution. Mix thoroughly.
2. Add 20 ml Conc. Sulphuric acid and again shake.
3. Add 10 ml H_3PO_4 and 0.5g NaF.
4. Add 1 ml dephenyl amine indicator.
5. Titrate the contents with ferrous ammonium sulphate solution till the colour changes from blue violet to green.

Calculation

$$\text{Organic carbon \%} = \frac{10(B\text{-}T)}{B} \text{ X } 0.003 \text{ X } \frac{100}{\text{wt of sample}}$$

Where B = volume of ferrous ammonium sulphate for the blank

T = titre value for the sample

Chemical oxygen demand

Principle

The amount of oxygen required to oxidize the organic material in impure water, (for example, sewage water), is known as chemical demand oxygen (COD). The water is treated with excess potassium dichromate and the dichromate consumed (= COD) is determined after titration with ferrous ammonium sulphate.

Reagents

1. Mercuric sulphate.
2. Potassium dichromate 0.00833M: Weigh 1.225 g potassium dichromate and dissolve in 500 ml deionized water.
3. Ferrous ammonium sulphate, $7H_2O$ solution (0.025M): Dissolve 4.9 g salt in 150 ml and add 2.5 ml conc. H_2SO_4. Dilute to 500 ml in a volumetric flask.
4. Ag_2SO_4- conc. H_2SO_4 solution: Dissolve 5 g silver sulphate in 500 ml conc. H_2SO_4
5. Diphenyl amine indicator.
6. Refluxion assembly.
7. Titration assembly.

Procedure

1. Take 50 ml water sample in a 250 ml conical flask with a ground glass neck.
2. Add 1 g mercuric sulphate and 80 ml silver sulphate-H_2SO_4 solution.
3. Add 10 ml of 0.00833M potassium dichromate solution.
4. Fit the flask with the refluxion set up and boil for 15min.
5. Cool and rinse the sides with 50 ml water.
6. Add diphenylamine indicator.
7. Titrate with 0.025M ferrous ammonium sulphate solution. The indicator changes from blue to green at the end point. Note the titre value (A ml).
8. Conduct a blank titration and note the titre value (B ml).
9. B-A ml is the amount of potassium dichromate used up in the oxidation.

Calculation

COD = B-A x 0.2 x 20 mg/ml

1ml corresponds to 0.2 mg O_2 required by 50 ml sample

Dissolved oxygen in water (Biological oxygen demand - B.O.D.)

Principle

The dissolved O_2 in water is made to oxidize potassium iodide to iodine. The liberated O_2 is titrated against hypo (sodium thiosulphate) to estimate iodine, which is a measure of O_2. Biological oxygen is important to support aqueous organisms. Higher oxygen has a negative effect relating to corrosion. Hence it is an important test to decide the quality of water.

Reagents

1. 0.02N $Na_2S_2O_3$: Dissolve 4.9642 g of $Na_2S_2O_3$. $5H_2O$ in a one litre boiled distilled water. Preserve by adding 0.2 g solid NaOH and dilute to one litre.
2. $KMnO_4$ solution 0.1N: Dissolve 3.1607 g of $KMnO_4$ in distilled water. Boil for one hour, cool, filter and make up to one litre.
3. Alkaline KI solution: Dissolve 10 g KOH and 5 g KI in 20 ml of boiled distilled water and filter.
4. $MnSO_4$ solution: Dissolve 10 g $MnSO_4$. $4H_2O$ in 20 ml boiled distilled water and add a few drops of formaldehyde.
5. 1% Starch solution: add 6-7 ml water to 1 g starch and make a paste. Pour the paste to 100 ml boiling water with constanr stirring. Cool and use for titration.
6. Conc. H_2SO_4 (sp. Gr. 1.84).
7. Titration assembly.

Procedure

1. Take 0.02N $Na_2S_2O_3$ into a burette for the titration.
2. Take 300 ml water sample into a glass stoppered bottle and add 0.9 ml Conc. H_2SO_4 and 0.2 ml $KMnO_4$ solution. Stopper and shake thoroughly. Add more $KMnO_4$ solution to maintain pink colour.
3. Add 0.5 ml $K_2C_2O_4$ (potassium oxalate) - 0.2%. Stopper and mix well (to discharge the $KMnO_4$ colour).

4. Add 2 ml of $MnSO_4$ solution followed by 3 ml of alkaline KI solution. Stopper and shake and allow the precipitate to settle.
5. Add 1 ml of conc. H_2SO_4 solution, mix till the precipitate gets dissolved. Measure 102.2 ml of this solution into a conical flask and titrate with hypo. Towards the end when the colour of the solution is light yellowish add 2ml starch solution and titrate till the blue colour disappears. Note the titre value.
6. Repeat a duplicate and triplicate titration to obtain to similar values.

Calculations

Total volume of sample = 300 ml

Volume of reagents used = 0.9 ml Conc. H_2SO_4 + 0.2 ml $KMnO_4$ solution + 0.5 ml $K_2C_2O_4$ + 2 ml of $MnSO_4$ solution +3 ml of alkaline KI solution + 6.6 ml

Volume of solution taken for titration = 102.2 ml

Volume of 0.02N $Na_2S_2O_3$ = V ml (from burette)

Normality of sample with respect to Dissolved Oxygen =

$$\frac{\text{Normality of hypo}\left(0.02\text{N}\right)\text{ x volume of hypo}}{\text{volume of sample equivqalent to 102.2 ml}\left(100\text{ ml}\right)}$$

$$=\frac{1 \text{ x V x } 8}{100 \text{ x } 50}\text{g/l}$$

$$=\frac{1 \text{ x Vx } 8 \text{ x } 1000}{100 \text{ x } 50}\text{mg/l} = 1.6 \text{ mg/l}$$

$$\text{D.O.}\left(\text{mg/l}\right) = \frac{\text{V x } 0.18 \text{ x } 1000}{\text{volume of sample}}$$

Carbon and Hydrogen Content

Principle

The carbon content of a sample is the amount of carbon contained in it. The hydrogen content is the total amount of hydrogen in a sample minus that due to the moisture content of the sample. A sample of known mass is burnt at 1350^0C

in a stream of oxygen such that complete combustion occurs. All the carbon is converted into CO_2 and all H_2 to H_2O. The CO_2 is absorbed by soda asbestos and H_2O by Mg $(ClO_4)_2$.

Reagents and equipment

1. $Al_2 O_3$.
2. O_2 cylinder.
3. Mg $(ClO_4)_2$.
4. Soda asbestos pellets.
5. Bomb calorimeter.
6. Oxygen supply system: The sample is burned in a rapid stream (300 ml/ min) of O_2. The O_2 is passed through a drying tube before introduction (glass tube filled with magnesium perchlorate). CO_2 in O_2, if any, needs to be removed.
7. Absorption train : The H_2O vapour and CO_2 released on the combustion of the sample are collected by assorbing them in three packed columns.
8. *Tube 1* : Magnesium perchlorate - absorbs all H_2O. From mass gain the H_2 content is calculated. The C-content is calculated from mass gain of other tubes.

   Tube 2 : Filled with soda asbestos pellets – CO_2 absorption

   Tube 3 : Also with soda asbestos + short section of magnesium perchlorate. This is to absorb H_2O produced during absorption of CO_2. Glasswool is used to prevent dust from the adsorbent materials being carried between tubes. The tubes are joined to combustion tube via a heat resistant stopper, and to each other by short lengths of rubber hose.
9. Air purge: After absorption the tubes are purged with air supplied through cylinder or drawn through a vaccum pump. Air must be passed through magnesium perchlorate and soda asbestos. Air has to be vented via the flow mater.

Procedure

1. To prepare the apparatus for use connect up the apparatus without combustion boat and raise the temperature of furnace to 1350^{0}C.
2. Pass O_2 through the apparatus at 300 ml/min for 10 min.
3. Disconnect the absorption train and purge with air for 2-3 min. disconnect the train and seal from atmosphere. Allow to cool in a dust-free atmosphere for 15-20 min.

4. When cool, and immediately before use, remove rubber tubing and wipe each absorption tube to remove dust and grease. Weigh each unit accurately.
5. Take an oven-dried combustion boat, allow to cool and weigh.
6. Add about 0.5 g sample to boat and weigh accurately.
7. Cover the sample with 0.5-1.0 g Al_2O_3. Reweigh and place in desiccator.
8. Reconnect the weighed absorption train.
9. Insert the combustion boat containing the sample to a point 240 mm from centre of furnace. Connect the O_2 supply and pass O_2 at 300 ml / min
10. During the test follow the schedule for boat position given below using silica rod to push the boat along and withdrawing the rod after each movement.

Time elapsed (Min)	Movement forward (mm)	Distance from centre off urnace (mm)
0	Insert boat	240
1	40	200
2	20	180
3	20	160
4	20	140
5	20	120
6	20	100
7	20	80
8	20	60
9	20	40
10	40	40
14	End of test	0

11. After the end of the 14th min shut off O_2.
12. Disconnect absorption train, purge with air for 2-3 min, seal and leave to cool as before but for one hr. When cool, wipe and weigh as before. Remove the boat and discard. It may be useful to keep a nickel- chrome wire hook for this purpose. If the moisture content of $Al_2 O_3$ has not been determined within a month before test, determine the same.

Calculation of results

M_1= mass of sample M_2 = increase of mass in tube 1.- less due to Al_2O_3

M_3= total increase in mass of tubes 2 & 3.

C_{ad}= % C content of sample wet basis H_{ad}= % hydrogen content of sample wet basis M_{ad}= % moisture content of sample wet basis.

Then $C_{ad} = \frac{27.29 \times M_3}{M_1}$ and $H_{ad} = \frac{11.19 M_2 - 0.1119 M_{ad}}{M_1}$.

Gross calorific value (GCV)

Definition

The gross calorific value (GCV) of a substance is the heat released on combustion of that substance in an oxygen-filled bomb calorimeter at a constant volume. The conditions of combustion are such that the products of combustion are ash, gaseous CO_2, liquid water, SO_2 and N_2. GCV is usually reported in units of kJ/kg.

Principle

A known quantity of sample is burned under strictly controlled conditions. The sample is burned in oxygen to ensure complete conversion to its cembustion products. The heat released by this combustion is determined on the basis of the equation.

Heat released = Mass of the apparatus x specific heat capacity of the apparatus x temperature rise.

Since the mass of th apparatus is constant from test to test (same calorimeter and bomb and same water), it need not be known quantitatively. Instead the effective heat capacity is used in this equation. The effective heat capacity of the apparatus is found by burning a sample of thermo-chemical standard benzoic acid in the apparatus and measuring the temperature rise. The effective heat capacity is then the heat released by the acid divided by the temperature rise and this value is used in every subsequent test until the apparatus is changed. So the GCV of a sample can be found by burning a known mass in an apparatus with a known effective heat capacity, and measuring the temperature rise.

Equipment

The sample is burned in a crucible, which is contained in a vessel called a bomb. During a GCV determination, the assembled bomb rests inside a metal container called a calorimeter. This calorimeter is filled with a quantity of water sufficient to cover the bomb, and this quantity of water is the same for each determination. The calorimeter vessel is completely enclosed by a water jacket, but separated from it by an air gap of approximately 10 mm. This water jacket can be of 3 types.

1. *Adiabatic* : As the name implies, this type of jacket does not allow any heat to flow from the calorimeter to the jacket. It does this by means of a control system which keeps th jacket at the same temperature as the calorimeter. Two thermistors detect any temperature differences between the calorimeter and the jacket, and a heater warms the jacket as neccesary.
2. *Isothermal* : This is a jacket which is thermostatically maintained at constant temperature. Heat flows from th calorimetor to the jacket and has to be compensated for in the calculation of the GCV.
3. *Static* : This type of jacket also remains at constant temperature and does so by being large enough to absorbb heat produced on firing the bomb without significant temperature rise. The effect of heat flow is compensated for by the same method.

Procedure

A general procedure for a GCV determination is outlined below:

1. Prepare a sample.
2. Weigh the crucible .
3. Introduce the sample into the crucible.
4. Weigh the filled crucible.
5. Assemble the bomb, and fill with O_2 to 30 bar.
6. Fill calorimeter with pre-determined quantiy of water.
7. Place the bomb in a calorimeter, and calorimeter inside jacket.
8. Assemble stirrers, thermometers etc.
9. Allow time for calorimeter and jacket to reach the same temperature.
10. Record firing temperaure (caution: type of jacket and temperature varies).
11. Fire the bomb.
12. After the required interval record the final temperature.
13. Dis-assemble the apparatus.
14. Carry out analysis of bomb contents, if required.

Calculations

$$GCV = \frac{\Delta T\ (ca) - e_1 - e_2 - e_3 - e_4}{M}$$

ΔT = Corrected temperature rise.

Ca = Eeffective heat capacity of apparatus.

e_1 = correction for heat of combustion of wire.

e_2 = correction for heat of combustion thread.

e_3 = correction for heat of formation of H_2SO_4.

e_4 = correction for heat of formation of HNO_3.

M = mass of sample.

Calorific Value (based on composition of sample)

Calorific value of an organic sample is obtained by multiplying the sum of carbohydrates and proteins by four and adding to it the figure obtained by multiplying the hexane extractives (fat/oil) by nine.

Calorific value = 4 (Carbohydrate + Protein) + 9 x Hexane extractives.

12

PLANT PIGMENTS

Plant pigments impart colour to the different components like leaves, flowers, fruits etc. They in turn are responsible for the colour of raw food, we eat. Pigments like chlorophyll, anthocyanins, flavonoids, tannins, xanthones etc. are the different pigments that are present in plants. The primary function of pigments in plants is photosynthesis, which uses the green pigment chlorophyll along with several red and yellow pigments that help to capture as much light energy as possible. Chlorophyll is green in colour. Carotenoids like α, β, and γ, carotenoids are yellow in colour. Lycopene is red in colour. Flavones and falvonols are also yellow in colour. Anthocyanins are purple and magenta in colour. Other functions of pigments in plants include attracting insects to flowers to encourage pollination. The assay of pigments is mostly based on their native colour.

Chlorophyll assay

Principle

Fresh plant tissue is extracted with an organic solvent like methanol or acetone. The clear filtrate is used for measuring the absorbance at 663 and 645nm. The chlorophyll content is then calculated based on the absorbances.

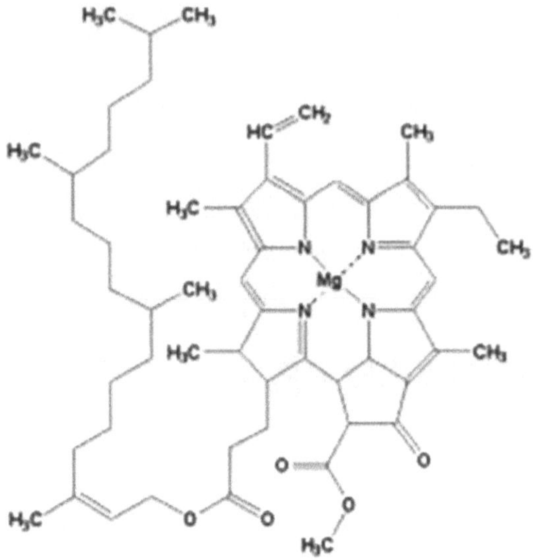

Chlorophyll A

Reagents and equipment

1. Acetone or methanol.
2. Mortar and pestle or a tissue grinder.
3. Buchner funnel.
4. Vacuum pump.
5. Spectrophotometer with cuvettes.

Procedure

1. Fresh tissue is ground in a mortar or macerator in the presence of excess acetone or methanol until all the colour is released from the tissue. $CaCO_3$ is added to prevent the formation of pheophytin.
2. The extract is filtered on a Buchner funnel and washed with acetone till the residue is colourless.
3. The extract along with washings are made up to a known volume. It can be stored in a refrigerator till analysis. Preferably use fresh extracts rather than stored ones. Work in dim light.
4. Chlorophyll a and b content can be obtained by measuring the absorbance at 663 and 645 nm in 1cm cells.

Calculation

Total chlorophyll mg/litre = $20.2A_{645} + 8.02\ A_{663}$

Chlorophyll a mg/litre = $12.7\ A_{663} + 2.69\ A_{645}$

Chlorophyll b mg/litre = $22.9\ A_{645} - 4.68\ A_{663}$

$$\text{Chlorophyll a mg/litre} = \frac{12.3A\ 663 - 0.86A645 \times V}{a \times 1000 \times W}$$

$$\text{Chlorophyll b mg/litre} = \frac{19.3A645 - 3.6A663 \times V}{a \times 1000 \times W}$$

Where A_{645} and A_{663} are absorbances at 645 and 663 nm.

V = the volume in ml of the total extract.

a= the length of light path in the cell, normally 1cm.

W = the fresh weight in grams.

Carotenes

Principle

The carotenoids are extracted and partitioned using organic solvents. Individual component fractionation is carried out on a column of activated magnesia. Carotene content is measured at 436 nm.

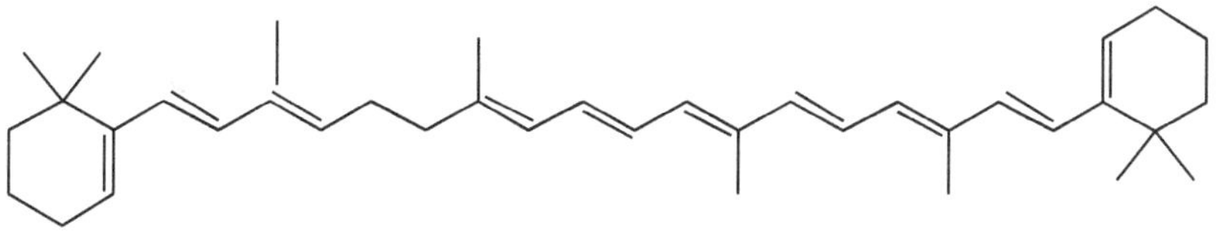

Beta carotene

Reagents and equipment

1. Dry acetone, alcohol free.
2. Pure hexane.
3. Activated magnesia.
4. Diatomaceous earth – Hyflo Supercel.
5. $MgCO_3$.

6. Glass column for column chromatography.
7. Spectrophotometer with cuvettes.

Procedure

1. Sample: Finely ground fresh plant sample should be blanched in boiling water. Sore in frozen condition.
2. Take 2-5 g sample into a blender. Add 40 ml acetone, 60 ml hexane and 0.1 g $MgCO_3$. Blend for 5 min.
3. Filter under suction. Wash the residue twice with 25 ml portions of acetone, then with 25 ml hexane and combine the extracts.
4. Wash acetone from the extract five times with 100 ml each of water. Transfer upper layer to 100 ml volumetric flask containing 9 ml acetone and dilute to volume with hexane.
5. Prepare a column of activated magnesia-diatomaceous earth (1:1). On top of it make 1cm layer of anhydrous sodium sulphate.
6. Pour the pigment extract into the column. Apply vacuum to elute the components.
7. Use 50 ml acetone-hexane (1:9) mixture to develop the chromatogram. Wash carotenes through the column.
8. Collect the entire eluate. Carotenes get eluted first.
9. Transfer the eluate and make up to 100 ml with acetone-hexane mixture.
10. Determine the absorbance at 436 nm.
11. Prepare a calibration curve with pure β-carotene. Read the sample concentration based on its absorbance.
12. Calculate the carotene content and express as mg/100 g sample.

Lycopene content

Principle

Lycopene imparts red colour to some fruits and vegetables. It is a carotenoid with the molecular formula $C_{40}H_{56}$. Lycopene has absorption maxima at 473 and 503 nm. One mole of lycopene in one litre of petroleum gives an absorption of 1.72×10^4 in 1cm cell at 503 nm. Thus a concentration of 3.1206 µg lycopene/ml gives a unit absorbance.

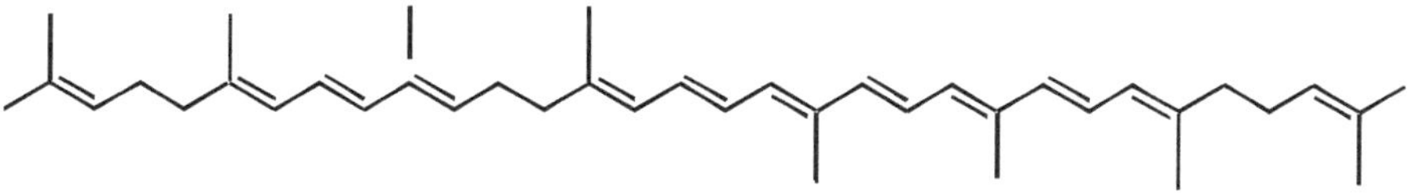

LYCOPENE

Reagents

1. Acetone.
2. Petroleum ether: P.E. (40-60° C).
3. Anhydrous sodium sulfate.
4. 5% sodium sulfate.
5. Spectrophotometer with cuvettes.

Procedure

1. Take 3-4 tomatoes and pulp them well to a smooth consistency in a blender
2. Weigh 5-10 g of the pulp.
3. Extract the pulp with acetone thoroughly till the residue is colourless.
4. Take the pooled acetone extract into a separating funnel containing 20 ml petroleum ether.
5. Add 20 ml of 5% sodium sulfate and shake. Add sufficient quantity of petroleum ether (20 ml).
6. Separate the P.E. layer. Re-extract the aqueous layer with P.E. till it is colourless
7. Pool the P.E. extracts and wash it with water.
8. Dry the P.E. extracts over anhydrous sodium sulfate. Keep it aside for 30 min or more.
9. Transfer the P.E. extract to 100 ml volumetric flask after complete washing of anhydrous sodium sulfate (till free of colour).
10. Make up to 100 ml with P.E.
11. Measure absorbance at 503 nm in a 1cm cuvette.

Calculation

Absorbance (1unit) = 3.1206 μg lycopene/ml

$$\text{mg lycopene/100 mg sample} = \frac{31.206 \text{ x absorbance}}{\text{wt of sample (g)}}$$

Reference

Ranganna, S. (1976) Manual of analysis of fruits and vegetable products. McGraw Hill, New Delhi

Curcumin

Principle

Curcumin is present in turmeric. It is an orange yellow coloured pigment. It is estimated using its absorbance at 425 nm in an alcoholic solution.

Reagents and equipment

1. Absolute alcohol.
2. Refluxion assembly.
3. Spectrophotometer with cuvettes.

Procedure

1. Dissolve 0.2-0.5 g of dry turmeric powder in 250ml alcohol.
2. Reflux the contents in a flask for 3-5hr.
3. Cool and decant the extract into volumetric flask and make up to volume.
4. Dilute a suitable aliquot, say 1-2 ml to 10ml with alcohol.
5. Measure the absorbance at 425 nm in a spectrophotometer.

Calculation

Curcumin content g/100g sample =

$$\frac{0.0025 \text{ x A425 x volume of final extract x dilution factor x 100}}{0.42 \text{ x wt of sample (x) 1000}}$$

0.42 absorbance at 425 nm = 0.0025 g curcumin

Anthocyanidins (flavon-4-ols)

Principle

Anthocyanins, the water-soluble flavonoid pigments occur in all tissues of higher plants, providing colour in leaves, plant stem, roots, flowers, and fruits, in small quantities. Anthocyanins are most visible in the petals of flowers, where they account for as much as 30% of the dry weight of the tissue. The anthocyanidins are determined by ionizing the middle ring of flavonoids by acid yielding a pink colour. The intensity of the pink colour is proportional to the concentration of flavon-4-ols. The absorbance produced is intense and is measured at 550 nm.

The colour of the pigments depend on the number of hydroxyl groups and or the methoxyl groups. The colour of the solution of an anthocyanidin also depends on the p^H. The pigments are generally red in acid solution, violet or purple in neutral and blue in alkaline solutions.

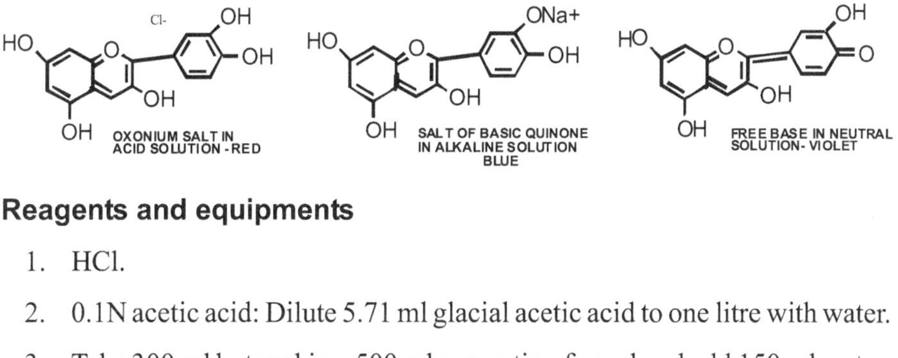

OXONIUM SALT IN ACID SOLUTION - RED

SALT OF BASIC QUINONE IN ALKALINE SOLUTION BLUE

FREE BASE IN NEUTRAL SOLUTION- VIOLET

Reagents and equipments

1. HCl.
2. 0.1N acetic acid: Dilute 5.71 ml glacial acetic acid to one litre with water.
3. Take 300 ml butanol in a 500 ml separating funnel and add 150 ml water. Shake vigorously and allow it to stand overnight. Remove the top layer (water -saturated butanol) and mix in a bottle with HCl in a ratio of 70:30.
4. Butanol.
5. Acetic acid.
6. Mix water saturated butanol, methanol and N/10 acetic acid in the ratio 70: 15;15. Use this for reagent blank.
7. Balance.
8. Constant temperature water bath.
9. Vortex mixer or tube rotator.
10. Spectrophotometer with cuvettes.

Procedure

1. Extraction: Grain or leaf samples are extracted with methanol. The phenolic compounds, if any, are adsorbed in poly vinyl pyrrolidine.
2. Take 0.5 ml sample extract.
3. Add reagent 3.
4. Prepare a blank with reagent 6 (reagent 3 and 6).
5. Put the tubes on a test tube rotator for 1hr.
6. Read the absorbance at 550 nm.

Calculation

Calculate the results (absorbance) at A_{550} g^{-1} dry weight.

Leghaemoglobin

Principle

Leghaemoglobin reacts with pyridine in strong alkali to produce haemochrome. It is measured at 556 nm.

Reagents and equipment

1. Diluent buffer: 0.1M sodium/potassium phosphate buffer p^H 7.4.
2. Alkaline pyridine reagent: Dissolve 0.8 g sodium hydroxide in 50 ml water and cool. Add 33.8 ml pyridine, dissolve and dilute to 100 ml with water (4.2M pyridine in 0.2M NaOH).
3. Sodium dithionate: Fine powder. Store in a desiccator.
4. Potassium hexacyanoferrate.
5. Centrifuge.
6. Spectrophotometer with cuvettes.

Procedure

1. Extract fresh or thawed nodules with 1-3 volumes of phosphate buffer and grind in a mixer. Filter and discard the nodule debris. Centrifuge the extract at 10,000 g for 10-30 min. dilute suitably.
2. Take 2-5 ml of the extract. Add equal volume of pyridine reagent and shake. The solution turns greenish yellow due to formation of ferric hemochrome.

3. Make it into two equal parts.
4. To one portion add a few crystals of sodium dithionate to reduce the hemochrome. Stir without aeration.
5. Read at 556 nm after 2-5 min against a reagent blank.
6. To the other portion add a few crystals of potassium hexacyanoferrate to oxidize the hemochrome and read at 539 nm.
7. Note the A_{556} and A_{539} absobencies.

Calculation

$$\text{Leghemoglobin concentration (m M)} = \frac{A556\text{-}A539 \times 2D}{23.4}$$

Where D is the initial dilution,

The calculation is based on the equation E= $23.4 \times 10^3\ mol^{-1}\ cm^{-1}$

Reference

Appleby, C.A. and Bergersen, F.J. 1980. In Methods for Evaluating Biological Nitrogen Fixation (Ed. by Bergersen, F.J.) John Wiley and Sons New York p 315.

13

PESTICIDE RESIDUE ANALYSIS

Chemicals in the form of insecticides , fungicides and herbicides are extensively used in large numbers in agricultural production. Commercial production of some crops would be impossible if chemicals are not used. Small quantities of chemical residues often remain in such crops and the harvested grains and other products. The residues of pesticides have been observed in small quantities (ppm level) in soil, water, agricultural produce and even in animal tissues and their products like meat and milk. They finally reach the consumer, namely, man (and even animals). Even the environment is now-a-days polluted with pesticides and other chemicals harmful to the humans. In view of the importance and seriousness of the problem the residues need to be estimated in the food and non-food products to assess their safety. Analytical details of a few pesticide residue procedures are presented below. Specialized texts may be consulted for more extensive information.

Organochlorine residues

Principle

The pesticide residues are extracted into hexane: acetone solvent. The residues are cleaned-up on alumina TLC plates. The same is further separated using argentation TLC for identification and quantification.

Reagents and equipment

1. Hexane: acetone (3:1).
2. Silica gel with $AgNO_3$ and alumina TLC plates.
3. Aldrin.BHC, DDT, endosulfan and heptachlor (1-10 ppm).
4. Cyclohexane.
5. Acetonitrile.
6. T.L.C. spreader chamber, sprayer etc. kit.
7. U.V. cabinet.
8. Vacuum pump.

Procedure

1. Add 8 ml water and 50 ml of 3:1 hexane, acetone mixture to 25 g grain or other plant sample in a stoppered flask. Shake for one hr on a flask shaker.
2. Decant and collect into a bottle. If necessary filter under suction. Take 10 ml and concentrate to 1ml.
3. Prepare an alumina TLC plate (0.5 mg thick). Activate in an oven for one hr in an oven.
4. Spot the sample (1 ml concentrated above) as a band at the bottom of the plate.
5. Develop with methyl cyanide (acetonitrile) to a height of roughly 10 cm.
6. Scrape out the alumina layer to a width of 2 cm just below the solvent front.
7. Extract the scraped alumina with 5 ml 3:1 hexane. Concentrate to 0.1 ml.
8. Spot this clean-up residue extract on silica gel G containing 100 mg $AgNO_3$ plates (0.25 mm) - argentation TLC.
9. Spot alongside some standard organo chlorine pesticides (aldrin. BHC, DDT, endosulfan and heptachlor)
10. Develop with cyclohexane.
11. Visualize the spots under U.V. light.
12. Identify based on Rf values. Visual identification also can be made. Quantitation is based on the area and density of the spots. Alternately densitometry can be used.

Carbaryl residues

Principle

Carbaryl residues are extracted into dichloromethane. After clean-up they are separated on a TLC plate and visualized under U.V. lamp.

Reagents and equipment

1. Dichloromethane.
2. Saturated sodium chloride.
3. Anhydrous sodium sulfate.
4. T.L.C. kit with spreader, plates, chamber, sprayer, etc.
5. Alumina TLC plates.
6. Chloroform.
7. U.V. cabinet.
8. Standard carbaryl (10 ppm solution).

Procedure

1. Weigh 25 g finely powdered sample into a 250 ml flask. Add 50 ml dichloromethane and shake for one hr.
2. Pour into a 500 ml separating funnel containing 250 ml water and 50 ml saturated sodium chloride. Shake well and allow to separate for 30 min.
3. Collect the lower dichloromethane through a plug of anhydrous sodium sulfate and store in a stoppered flask. Take 10 ml and concentrate to 1 ml.
4. Prepare an alumina TLC plate (0.5 mm thick). Activate in an oven for one hr in an oven.
5. Spot the sample (0.1 ml concentrated above) at the bottom of the plate.
6. Develop with chloroform.
7. Visualize the spot under U.V. light.
8. Spot a standard for identification and quantification (mostly based on visual area).

Carbaryl residues (Colorimetric method)

Principle

The carbaryl residues are extracted with methylene chloride. The color developed with p-nitrobenzenediazolium fluoborate is measured 477 nm.

Reagents and equipment

1. Acetone redistilled.
2. Aqueous acetone solution 10% v/v.
3. Methylene chloride redistilled.
4. Coagulating solution: Dissolve 0.5 g NH_4Cl in water containing 1 ml H_3PO_4 (85%).
5. Colour (chromogenic reagent) reagent: A mixture of 25 ml ethanol and 2ml acetic acid is saturated with p-nitrobenzenediazolium fluoborate (by mixing for several minutes). This solution is filterd and stored in the cold.
6. Diethylene glycol solution: Dilute 10 ml diethylene glycol to 100 ml with methylene chloride.
7. Alcoholic KOH (0.1 N in ethanol).
8. Carbaryl (1-naphthyl N- methylcarbamate): Prepare a solution in acetone 8μg/ml.
9. Blender.
10. Colorimeter or spectrophotometer.

Procedure

1. Take 25 g sample and add 150 ml methylene chloride and 100 g powdered anhydrous Na_2SO_4. Blend the contents (in a blender).
2. Decant and filter under suction. Rinse blender with 50 ml methylene chloride. Blend, decant and and filter as above. Repeat extraction and pool all the filtrates.
3. Concentrate the filtrate to 5 ml. Wash with acetone and transfer to a flask. Add 15 ml coagulating solution. Allow to stand for about 10 min. Filter (centrifuge) and make up to 25 ml in a volumetric flask.
4. Piet 5.0 ml sample solution into a beaker, add 2.0 ml alc. KOH and mix.

After 3 min add 10 ml acetic acid and 1.0 ml chromogenic reagent. Shake and allow to stand for 2 min.

5. Measure absorbance at 477 nm against water as control (blank).
6. Prepare a standard curve using carbaryl standard 8- 40 μg as in 4 above.
7. Read the concentration from the absorbance of sample from standard graph.
8. Calculate and express the residue levels as ppm taking into account sample weight and dilutions made in extraction.

Organophosphorus residues

Principle

The residues are extracted into dichloromethane and after clean-up are spotted on TLC plates for their fractionation. Identification and quantification are carried out using suitable spray reagents.

Reagents and equipment

1. Dichloromethane.
2. Saturated sodium chloride.
3. Anhydrous sodium sulfate.
4. Alumina TLC plates.
5. Standard malathion, methyl parathion sumithion and thimet (1-10 ppm).
6. Hexane: acetone (98:2).
7. Bromine liquid.
8. Bromocresol green indicator.
9. Wrist action shaker.
10. Densitometer.

Procedure

1. Weigh 25 g finely powdered sample into a 250 ml flask. Add 50 ml. dichloromethane and shake for one hr.
2. Pour into a 500 ml separating funnel containing 250 ml water and 50 ml saturated sodium chloride. Shake well and allow to separate for 30 min.

3. Collect he lower dichloromethane through a plug of anhydrous sodium sulfate. and store in a stoppered flask. Take 10 ml and concentrate to 1ml.
4. Prepare an alumina TLC plate (0.5 mm thick). Activate in an oven for one hr.
5. Spot the sample (0.1 ml concentrated above) at the bottom of the plate. Spot alongside malathion, methyl parathion sumithion and thimet.
6. Develop with 98:2 hexane: acetone solvent.
7. Dry the plates and expose to bromine vapours for 30 seconds.
8. Drive off the excess bromine and spray with bromocresol green.
9. Visualize the yellow spots over greenish blue background.
10. Identify based on R_f values and quantify based on area and intensity or using a densitometer.

Imidacloprid residues (HPLC method)

Principle

Imidacloprid belongs to group of compounds called nitroguanidines. It has excellent systematic activity. It shows low toxicity to warm blooded animals. Also it has low residual activity. A known mass of a solution containing imidacloprid residues and an internal standard is injected into a HPLC instrument equipped with U.V. detector to identify and quantify the residues.

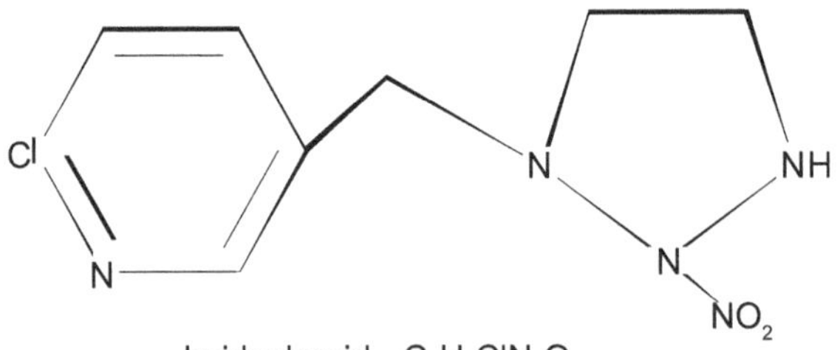

Imidacloprid - $C_9H_1ClN_5O_2$

Reagents

1. Internal standard: acenaphthene A.R. grade.
2. Acetonitrile.
3. Water : HPLC grade.

4. Sovent mixture: acetonitrile: water (60:20).
5. Internal standard: weigh 0.03 g acenaphthene into 100 ml volumetric flask. Add 40 ml solvent mixture and make up to volume.
6. Standard imidocloprid solution: Weigh 0.1 g imidacloprid into a 100 ml volumetric flask. Dissolve in 40 ml solvent mixture and make up to volume. Dilute this standard 10 times to obtain a dilute standard solution.
7. Internal standard mixture: 20 ml dilute standard + 20 ml internal standard solution in a 100 ml volumetric flask. Make up to 100 ml the solvent mixture.
8. Methanol : water (3:1) mixture.
9. 2N H_2SO_4 solution.
10. XAD-8 column: 25 g XAD resin suspended in 50 ml methanol. If turbid the methanol portion is rejected. The resin is re-suspended in methanol and packed into a column (20 mm diameter). A cotton plug is applied and the column is washed with methanol and water.
11. Acetonitrile: water (65:35) solvent mixture.
12. HPLC equipment with LC18 ODS column and U.V. detector.

Procedure

1. Extraction of residues: 10 g of finely powdered plant sample is weighed into a 250 ml conical flask. Add 75 ml methanol : water (3:1) and 1.25 ml 2N H_2SO_4. Keep it overnight.
2. Filter with suction and rinse with 80 ml methanol: water (3:1) mixture.
3. The filtrate is concentrated to about 20 ml. add 50 ml 5% NaCl. It is extracted thrice with hexane (10, 50 and 50 ml each time) to remove lipid constituents. Discard the hexane extract.
4. Transfer the aqueous phase to a 250 ml flask. Concentrate the same to about 1ml.
5. The aqueous extract is applied to the XAD-8 column. It is flushed with 50 ml water. The aqueous eluates are discarded. The active ingredient namely the pesticide residue is eluted with 150 ml methanol.
6. The methanol eluate is concentrated to dryness. Then add 10 ml acetonitrile: water mixture (1:1). 20 μl sample can be injected to the HPLC instrument.

7. The HPLC equipment is used for separation, identification and quantification of the imidacloprid residues. Mobile phase: actonitrile: water (35:65), flow rate; 1 ml/min, U.V. Detector at 270 nm, column: LC18 ODS, Injection volume : 20 μl.
8. The imidacloprid peak emerges at around 5.3 min. The identification and subsequent quantifications are based on the R_f values of the standard and that of the internal standard.

Calculation

$$\% \text{ Imidacloprid content} = \frac{H3 \times H2 \times m1}{H4 \times H1 \times m2} \times P$$

H1 = the height of imidacloprid peak in standard solution

H2 = the height of imidacloprid peak in sample solution

H3 = the height of acenaphthene peak in standard solution

H4 = the height of acenaphthene peak in sample solution

m1 = the mass in g of the standard

m2 = the mass in g of the sample

P = the % purity of the standard imidacloprid

Multi-residue analysis

Principle

The acetone extract of the sample is subjected to clean-up and separated using gel permeation and silica gel columns followed by HPTLC fractionation.

Reagents

1. Acetone.
2. Dichloromethane.
3. Polystyrene gel.
4. Cyclohexane.
5. Ethylacetate.
6. Silica gel mini-column.

7. Hexane.
8. Toluene.
9. HPTLC (silica gel) plates.
10. 2,6 dichlorophenol indophenols.
11. $AgNO_3$.
12. Densitometer.

Procedure

1.
   a) Clean-up: The sample is acetone extracted and partitioned into dichloromethane.
   b) The extract is fractionated using polystyrene gel permeation column chromatography. The solvent used for elution is cyclohexane: ethyl acetate (1:1)
   c) The eluent is further fractionated on a silica gel mini- column. The components are eluted with a) hexane: toluene 65:35, b) toluene, c) toluene: acetone 95:5, d) toluene: acetone 8:2, e) acetone.
2. HPTLC plates spotted with 8 mm start bands in a pre-saturated tanks:
   a) Organophosphorus pesticides: mobile phase : hexane acetone 8:2, dichloromethane, and ethyl acetate. Colour reagent 2,6 dichlorophenol indophenols. Detection and quantification is carried out at 570 nm on a densitometer.
   b) $AgNO_3$/U.V. for chlorinated insecticides/fungicides. Mobile phase : methanol: water 7:3, Detection 450 nm.
   c) For organophosphates, carbamates, triazines: silica gel HPTLC layers, mobile solvent = dichloromethane. Detection: 590 nm

Reference

Gardyan, C. and Their. 1991. Fresenius J. Anal. Chem. 339: 338-339.

14

APPENDIX

Important chemical terms

Percentage

It is expressed as either weight by volume or as volume by volume. It is the amount of solute in grams or ml in 100 ml of solvent. For ex. 5g solute in 95g solvent is 5% w/w solution. A 30 ml solute in 70ml solvent is 30% v/v solution. A 20g solute dissolved and made up to 100ml solvent is w/v solution.

Molarity or molar solution (M)

It is the number of moles (gram molecular weight) of solute dissolved in a litre of solution. For ex. if one gram molecular weight of a salt is dissolved in one litre of water, it is called 1M solution.

$$M = \frac{\text{number of moles of solute}}{\text{volume of solution in litre}} = \frac{\text{strength in g/l}}{\text{molecular weight}}$$

Normality

It is the number of gram equivalents (molecular weight/valency) of a solute dissolved in one litre of solution or number of milli equivalents of solute in one ml of the solution.

$$\text{Normality} = \frac{\text{strength in g/l}}{\text{equivalent weight}}$$

$$= \frac{\text{no of gram equivalents of solute}}{\text{no of litres of solution or volume of solution in litres}}$$

Mole fraction

It is the ratio of number of moles of solute to the total number of moles of both the solute and the solvent.

$$\text{Mole fraction} = \frac{\text{no of moles of solute}}{\text{no of moles of solute+no of moles of solvent}}$$

Indicators

Indicators are of different types. They are mainly 1. Oxidation reduction, 2.Self indicator, 3. Starch indicator and 4. External indicator.

Oxidation reduction: They exist in two different colours in the oxidized and reduced forms. They undergo redox reaction after the titration. A sharp change in colour occurs during oxidation reduction titrations. Some indicators are diphenyl amine, ferroin, N- phenyl anthranillic acid etc.

Self indicator: When $KMnO_4$ is used in titrations, the colour changes from pink to colourless and *vice- versa.*

Starch indicator: Iodine complexes with starch to form a blue coloured complex. The solution remains colourless till consumed. Excess iodine results in blue colour.

External indicator: This indicator is useful when the reactants are dark coloured and when the indicator forms insoluble precipitate. An example is potassium ferricyanide $K_3[Fe(CN)_6]$ in the titration of potassium dichromate ($K_2Cr_2O_7$). The final colour is deep blue.

Buffer solutions

A buffer solution (more precisely, pH buffer or hydrogen ion buffer) is an aqueous solution consisting of a mixture of a weak acid and its conjugate base, or vice versa. Its pH changes very little when a small amount of strong acid or base is added to it and thus it is used to prevent changes in the pH of a solution. Buffer solutions are used as a means of keeping pH at a nearly constant value in a wide variety of chemical applications. Many life forms thrive only in a relatively small pH range so they utilize a buffer solution to maintain a constant pH. One example of a buffer solution found in nature is blood.

Applications

Buffer solutions are necessary to keep the correct pH for enzymes in many organisms to work. Many enzymes work only under very precise conditions; if the pH moves outside of a narrow range, the enzymes slow or stop working and can denature. In many cases denaturation can permanently disable their catalytic activity. A buffer of carbonic acid (H_2CO_3) and bicarbonate (HCO_3) is present in blood plasma, to maintain a pH between 7.35 and 7.45.

Simple buffering agents

Buffering agent	pK_a	useful pH range
Citric acid	3.13, 4.76, 6.40	2.1 - 7.4
Acetic acid	4.8	3.8 - 5.8
KH_2PO_4,	7.2	6.2 - 8.2
CHES	9.3	8.3–10.3
Borate	9.24	8.25 - 10.25

Industrially, buffer solutions are used in fermentation processes and in setting the correct conditions for dyes used in colouring fabrics. They are also used in chemical analysis and calibration of pH meters. The majority of biological samples that are used in research are made in buffers, especially phosphate buffered saline (PBS) at pH 7.4.

For buffers in acid regions, the pH may be adjusted to a desired value by adding a strong acid such as hydrochloric acid to the buffering agent. For alkaline buffers, a strong base such as sodium hydroxide may be added. Alternatively, a buffer mixture can be made from a mixture of an acid and its conjugate base. For example, an acetate buffer can be made from a mixture of acetic acid and sodium acetate. Similarly an alkaline buffer can be made from a mixture of the base and its conjugate acid.

Universal buffer mixtures

By combining substances with pK_a values differing by only two or less and adjusting the pH, a wide range of buffers can be obtained. Citric acid is a useful component of a buffer mixture because it has three pK_a values, separated by less than two. The buffer range can be extended by adding other buffering agents. The following two-component mixtures have a buffer range of pH 3 to 8.

Buffer solution with a p^H range of 3 to 8

0.2M Na_2HPO_4 /ml	0.1M Citric Acid /ml	pH
20.55	79.45	3.0
38.55	61.45	4.0
51.50	48.50	5.0
63.15	36.85	6.0
82.35	17.65	7.0
97.25	2.75	8.0

A mixture containing citric acid, monopotassium phosphate, boric acid, and diethyl barbituric acid can be made to cover the pH range 2.6 to 12. Biological buffers cover 1.9 to 11 pH range.

Common buffer compounds useful to Biochemists and Biologists

Common name	Buffer range	Molecular weight	Chemical name
TAPS	7.7–9.1	243.3	3-{[tris (hydroxymethyl) methyl] amino} propanesulfonic acid
Bicine	7.6–9.0	163.2	N, N-bis (2-hydroxyethyl) glycine
Tris	7.5–9.0	121.14	tris (hydroxymethyl) methylamine
Tricine	7.4–8.8	179.2	N-tris (hydroxymethyl) methylglycine
TAPSO	7.0-8.2	259.3	3-[N-Tris (hydroxymethyl) methylamino]-2-hydroxypropanesulfonic Acid
HEPES	6.8–8.2	238.3	4-2-hydroxyethyl-1-piperazinee-than esulfonic acid
TES	6.8–8.2	229.20	2-{[tris (hydroxymethyl) methyl] amino} ethanesulfonic acid
MOPS	6.5–7.9	209.3	3- (N-morpholino) propanesulfonic acid
PIPES	6.1–7.5	302.4	piperazine-N, N2 -bis (2-ethanesulfonic acid)
Cacodylate	5.0–7.4	138.0	dimethylarsinic acid
SSC	6.5-7.5	189.1	saline sodium citrate
MES	5.5–6.7	195.2	2- (N-morpholino) ethanesulfonic acid
Succinic acid	7.4-7.5	118.1	2(R)-2- (methylamino) succinic acid

How Buffers Work

Only a small amount of a strong acid is necessary to drastically alter the pH. For certain experiments, it is necessary to keep a fairly constant pH while acids or bases are added to the solution either by reaction or by the experimenter. Buffers are designed to fill that role. Chemists use buffers routinely to moderate the pH of a reaction. Biology finds manifold uses for buffers which range from controlling blood pH to ensuring that urine does not reach painfully acidic levels.

A buffer is simply a mixture of a weak acid and its conjugate base or a weak base and its conjugate acid. Buffers work by reacting with any added acid or base to control the pH. For example, consider the action of a buffer composed of the weak base ammonia, NH_3, and its conjugate acid, NH_4^+. When HCl is added to that buffer, the NH_3 "soaks up" the acid's proton to become NH_4^+. Because that proton is locked up in the ammonium ion, the proton does not serve to significantly increase the pH of the solution. When NaOH is added to the same buffer, the ammonium ion donates a proton to the base to become ammonia and water. Here the buffer also serves to neutralize the base.

As the above example shows, a buffer works by replacing a strong acid or base with a weak one. The strong acid's proton is replaced by ammonium ion, a weak acid. The strong base OH^- is replaced by the weak base ammonia. These replacements of strong acids and bases for weaker ones give buffers their extraordinary ability to moderate pH.

Calculating the pH of Buffered Solutions

Buffers must be chosen for the appropriate pH range that they are called on to control. The pH range of a buffered solution is given by the Henderson-Hasselbach equation. For the purpose of derivation, imagine a buffer composed of an acid, HA, and its conjugate base, A. It is known that the acid dissociation constant pK_a of the acid is given by the expression:

$$K_a = \frac{(H+)(A-)}{(HA)}$$

The equation can be rearranged as follows:

$$(H^+) = K_a \frac{(HA)}{(A-)}$$

Taking the -log of this expression and rearranging the terms to make each one positive gives the Henderson-Hasselbach equation:

$$pH = pKa + \log\left(\frac{[base]}{[acid]}\right)$$

HA and A^- in the equation are generalized to the terms acid and base, respectively. To use the equation, place the concentration of the acidic buffer species where the equation says "acid" and place the concentration of the basic buffer species where the equation calls for "base". It is essential one uses the pK_a of the acidic species and not the pK_b of the basic species when working with basic buffers.

A buffer problem can be fairly simple to solve, provided one doen't get confused by the knowledge of all the other chemistry. For example, let's calculate the pH of a solution that is 0.5 M acetic acid and 0.5 sodium acetate both before and after enough SO_3 gas is dissolved to make the solution 0.1 M in sulfuric acid. Before the acid is added, we can use the Henderson-Hasselbach equation to calculate the pH.

$$pH = pK_a + \log\left(\frac{[base]}{[acid]}\right)$$

$$pH = 4.75 + \log\left(\frac{[0.5\ M]}{[0.5\ M]}\right)$$

$$pH = 4.75$$

The pK_a of acetic acid is 4.75

To calculate the pH after the acid is added, it is assumed that the acid reacts with the base in solution and that the reaction has a 100% yield. Therefore, 0.1 moles per liter of acetate ion reacts with 0.1 moles per liter of sulfuric acid to give 0.1 moles per liter of acetic acid and hydrogen sulfate. Here, the second dissociation of sulfuric acid is ignored because it is minor in comparison to the first. So the final concentration of acetic acid is 0.6 M and acetate is 0.4M. Plugging those values into the Henderson-Hasselbach equation gives a pH of 4.57. Note that a 0.1 M solution of strong acid would give to a pH of 1 but the buffer gives a pH of 4.57 instead.

To probe the useful range of the buffer, one needs to calculate the pH of the solution resulting from the same situation above but with different concentrations of the buffer. If the buffer is 1.0 M in both acetate and acetic acid, then the pH of the resulting solution after the introduction of acid is 4.66. However, if the solution is only 0.11 M in acetic acid and acetate, then a pH of

3.45 is obtained. Therefore, to make a more effective buffer, the concentration of the buffering agents is large in comparison to the added acid or base.

Simple indicators

Litmus: Litmus is a weak acid (in general all Indicators are weak acids). It has a seriously complicated molecule which we will simplify to HLit. The "H" is the proton which can be given away to something else. The "Lit" is the rest of the weak acid molecule. There will be an equilibrium established when this acid dissolves in water. Taking the simplified version of this equilibrium:

$$HLit_{(aq)} \rightleftharpoons H^{+}(aq) + Lit^{-}(aq)$$

The un-ionised litmus is red, whereas the ion is blue. Now use Le Chatelier's Principle to work out what would happen if hydroxide ions or some more hydrogen ions are added to this equilibrium.

Adding hydroxide ions

Hydroxide ions react with and remove these hydrogen ions

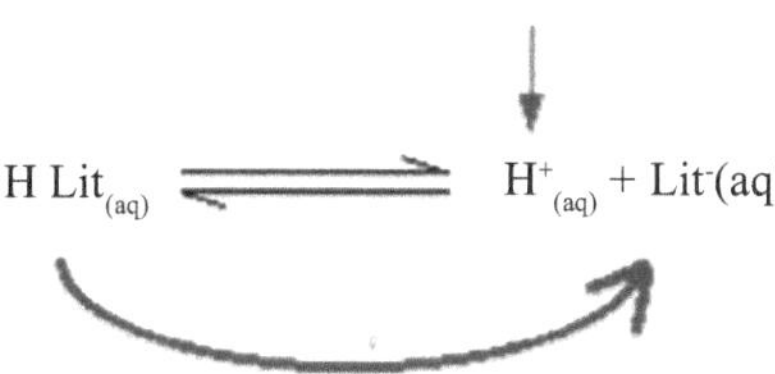

The equilibrium position moves to replace the lost hydrogen ions

Litmus turns blue

Adding hydrogen ions

Add extra hydrogen ions

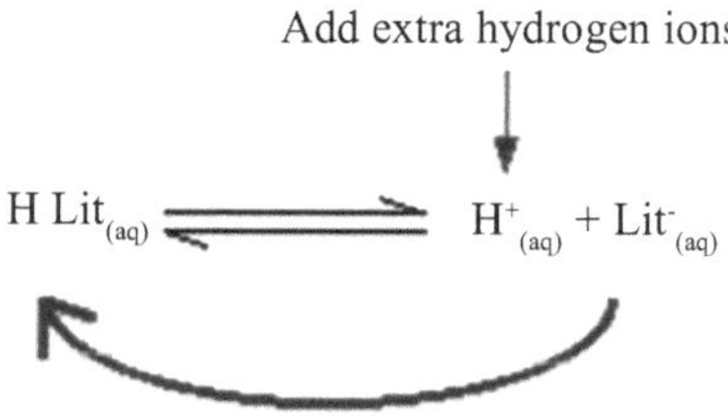

The equilibrium position moves to remove the extra hydrogen ions

Litmus turns red

If the concentrations of HLit and Lit⁻ are equal

At some point during the movement of the position of equilibrium, the concentrations of the two colours will become equal. The colour observed will be a mixture of the two. For litmus, it so happens that the 50 / 50 colour does occur at close to pH 7 - that's why litmus is commonly used to test for acids and alkalis.

Methyl orange: Methyl orange is one of the indicators commonly used in titrations. In an alkaline solution, methyl orange is yellow and the structure is:

the yellow form of methyl orange

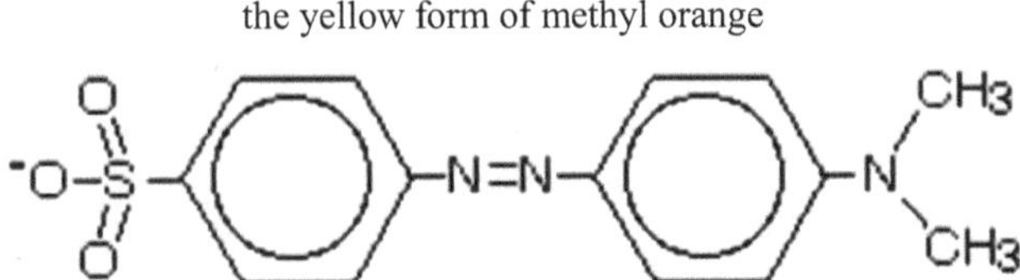

The hydrogen ion attaches to one of the nitrogen atoms in the nitrogen-nitrogen double bond to give a structure which is as follows:

the red form of methyl orange

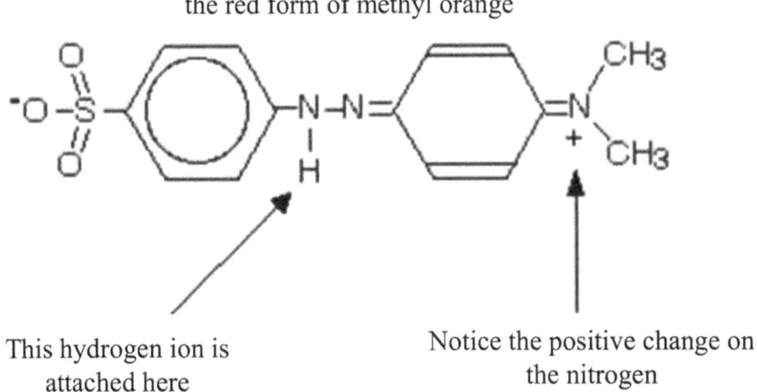

In the methyl orange case, the half-way stage where the mixture of red and yellow produces an orange colour happens at pH 3.7 - nowhere near the neutral.

$$\text{H-phph}_{(aq)} \rightleftharpoons H^{+}_{(aq)} + \text{phph}^{-}_{(aq)}$$

Phenolphthalein: Phenolphthalein is another commonly used indicator for titrations, and is another weak acid.

In this case, the weak acid is colourless and its ion is bright pink. Adding extra hydrogen ions shifts the position of equilibrium to the left, and turns the indicator colourless. Adding hydroxide ions removes the hydrogen ions from the equilibrium which tips to the right to replace them - turning the indicator pink. The half-way stage happens at pH 9.3. Since a mixture of pink and colourless is simply a paler pink, this is difficult to detect with any accuracy.

The pH range of indicators

Indicators don't change colour sharply at one particular pH (given by their pK_{ind}). Instead, they change over a narrow range of pH. Assume the equilibrium is firmly to one side, but now we add something to start to shift it. As the equilibrium shifts, we start to get more and more of the second colour formed, and at some point the eye will start to detect it.

For example, if we have methyl orange in an alkaline solution the dominant colour will be yellow. Now add acid so that the equilibrium begins to shift. At some point there will be enough of the red form of the methyl orange present that the solution will begin to take on an orange tint. As you go on adding more acid, the red will eventually become so dominant that one can no longer see any yellow. There is a gradual smooth change from one colour to the other, taking place over a range of pH. As a rough "rule of thumb", the visible change takes place about 1 pH unit either side of the pK_{ind} value.

The exact values for the three indicators are:

Indicator	pK_{ind}	pH range
litmus	6.5	5 - 8
methyl orange	3.7	3.1 - 4.4
phenolphthalein	9.3	8.3 - 10.0

The litmus colour change happens over an unusually wide range, but it is useful for detecting acids and alkalis in the laboratory because it changes colour around pH 7. Methyl orange or phenolphthalein would be less useful.

Some indicator solutions

1. Methylene-blue: Dissolve 1 g powdered methylene blue in water and make up to 500 ml.
2. Methyl- orange:.Dissolve 1 g methyl orange in 1.5 litres of boiling water.
3. Methyl-red: Dissolve 1 g in one litre ethanol.
4. Thymol- blue: dissolve 0.226 g in 50 ml ethanol and add 100 ml water.
5. Phenolphalein: Dissolve 1 g reagent in 100 ml methylated spirit.

6. Phenol- red: Grind 0.1 g phenol- red in 29 ml 0.01N NaOH and make up to 250 ml with water.

Names of phosphate salts used in buffer solutions

Formula	Name of salt	Other names
KH_2PO_4	potassium dihydrogenphosphate	potassium dihydrogen orthophosphatemonobasic potassium phosphatemonopotassium phosphateacid potassium phosphate potassium biphosphate
K_2HPO_4	potassium hydrogen phosphate	dipotassium hydrogen orthophosphate dipotassium' hydrogen phosphate dibasic potassium phosphatedi-potassium phosphate
K_3PO_4	potassium phosphate	tribasic potassium phosphate tripotassium phosphate

Tables of buffer solutions and their ranges

Buffering system	Useful buffering pH range @ 25°C
Hydrochloric acid/ Potassium chloride	1.0 - 2.2
Glycine/ Hydrochloric acid	2.2 -3.6
Potassium hydrogen phthalate/ Hydrochloric acid	2.2 - 4.0
Citric acid/ Sodium citrate	3.0 - 6.2
Sodium acetate/ Acetic acid	3.7 - 5.6
Potassium hydrogen phtaalate/ Sodium hydroxide	4.1 - 5.9
Disodium hydrogen phthalate / Sodium dihydrogen orthophospate	5.8 - 8.0
Dipotassium hydrogen phthalate / Potassium dihydrogen orthophospate	5.8 - 8.0
Potassium dihydrogen orthophosphate / sodium hydroxide	5.8 - 8.00
Barbitone sodium / Hydrochloric acid	6.8 - 9.6
Tris (hydroxylmethyl) aminomethane / Hydrochloric acid	7.0 - 9.00
Sodium tetraborate/ Hydrochloric acid	8.1 - 9.2
Glycine/ Sodium hydroxide	8.6 - 10.6
Sodium carbonate/ Sodium hydrogen carbonate	9.2 - 10.8
Sodium tetraborate/ Sodium hydroxide	9.3 - 10.7
Sodium bicarbonate / Sodium hydroxide	9.60 - 11.0
Sodium hydrogen orthophosphate / Sodium hydroxide	11.0 - 11.9
Potassium chloride/ Sodium hydroxide	12.0 - 13.0

Acetate buffer solutions pH 3 - 6

Prepare the following solutions (1) 0.1M acetic acid, (2) 0.1M sodium acetate (tri-hydrate) (13.6g / l). Mix in the following proportions to get the required P^H.

P^H	Vol. of 0.1Macetic acid	Vol. of 0.1Msodium acetate
3	982.3 ml	17.7 ml
4	847.0 ml	153.0 ml
5	357.0 ml	643.0 ml
6	52.2 ml	947.8 ml

Standardization buffers

For pH=7.00 : Add 29.1 ml of 0.1 molar NaOH to 50 ml 0.1 molar potassium dihydrogen phosphate.

Alternatively : Dissolve 1.20 g of sodium dihydrogen phosphate and 0.885 g of disidium hydrogen phosphate in 1 liter volume distilled water.

For pH= 4.00 : Add 0.1 ml of 0.1 molar NaOH to 50 ml of 0.1 molar potassium hydrogen phthalate.

Alternatively : Dissolve 8.954 g of disodium hydrogen phosphste.12 H_2O and 3.4023 g of potassium dihydrogen phosphate in 1 liter volume distilled water.

Methods to Prepare Buffer Solutions

1. 100 mM phosphoric acid (sodium) buffer solution (pH=2.1) Sodium dihydrogen phosphate dihydrate (M.W.=156.01)..50 m mol (7.8 g) Phosphoric acid (85 %, 14.7 mol/l).........................50 m mol (3.4 ml) Add water to make up to 1 litre.

2. 10 mM phosphoric acid (sodium) buffer solution (pH=2.6) Sodium dihydrogen phosphate dihydrate (M.W.=156.01)..5 m mol (0.78 g) Phosphoric acid (85 %, 14.7 mol/l).........................5 m mol (0.34 ml) Add water to make up to 1 litre (Alternatively, dilute 100 mM phosphoric acid (sodium) buffer solution (pH=2.1) ten times.)

3. 50 mM phosphoric acid (sodium) buffer solution (pH=2.8) Sodium dihydrogen phosphate dihydrate (M.W.=156.01)..40 m mol (6.24 g) Phosphoric acid (85 %, 14.7 mol/l).........................10 m mol (0.68 ml) Add water to make up to 1 litre.

4. 100 mM phosphoric acid (sodium) buffer solution (pH=6.8) Sodium dihydrogen phosphate dihydrate (M.W.=156.01)..50 m mol (7.8 g) Sodium dihydrogen phosphate 12-hydrate (M.W.=358.14)..50 m mol (17.9 g) Add water to make up to 1 litre.

5. 10 mM phosphoric acid (sodium) buffer solution (pH=6.9) Sodium dihydrogen phosphate dihydrate (M.W.=156.01)..5 m mol (0.78 g) Sodium

dihydrogen phosphate 12-hydrate (M.W.=358.14)..5 m mol (1.79 g)Add water to make up to 1 litre. (Alternatively, dilute 100 mM phosphoric acid (sodium) buffer solution (pH=6.8) ten times.)

6. 20 mM citric acid (sodium) buffer solution (pH=3.1) Citrate dihydrate (M.W.=210.14)...............16.7 m mol (3.51 g) Sodium citrate dihydrate (M.W.=294.10)..3.3 mmol (0.97 g) Add water to make up to 1 litre.
7. 20 mM citric acid (sodium) buffer solution (pH=4.6) Citrate dihydrate (M.W.=210.14)...............10 m mol (2.1 g) Sodium citrate dihydrate (M.W.=294.10)..10 mmol (2.94 g) Add water to make up to 1 litre.
8. 10 mM tartaric acid (sodium) buffer solution (pH=2.9) Tartaric acid (M.W.=150.09)..........................7.5 m mol (1.13 g) Sodium tartrate dihydrate (M.W.=230.08)........2.5 m mol (0.58 g) Add water to make up to 1litre.
9. 10 mM tartaric acid (sodium) buffer solution (pH=4.2) Tartaric acid (M.W.=150.09)..........................2.5 m mol (0.375 g) Sodium tart rate dihydrate (M.W.=230.08)........7.5 m mol (1.726 g) Add water to make up to 1 litre.
10. 20mM (acetic acid) ethanolamine buffer solution pH=9.6 Monoethanolamine (M.W.=61.87, d=1.017)...20 m mol (1.22 ml) Acetic acid (glacial acetic acid, 17.4 mol/l)..................10 m mol (0.575 ml) Add water to make up to 1 litre.
11. 100 mM acetic acid (sodium) buffer solution (pH=4.7) Acetic acid (glacial acetic acid) (99.5 %, 17.4 mol/l)..................50 m mol (2.87ml) Sodium acetate trihydrate (M.W.=136.08)........50 m mol (6.80 g) Add water to make up to 1 litre.
12. 100 mM boric acid (potassium) buffer solution (pH=9.1) Boric acid (M.W.=61.83)...........................100 m mol (6.18 g) Potassium hydroxide (M.W.=56.11)...........50 m mol (2.81 g) Add water to make up to 1 litre.
13. 100 mM boric acid (sodium) buffer solution (pH=9.1) Boric acid (M.W.=61.83)........... 100 m mol (6.18 g) Sodium hydroxide (M.W.=40.00)............... 50 m mol (2.00 g) Add water to make up to 1 litre.

Concentrations of acids and bases

Acid/base	Molecular weight	Specific gravity	% by weight	Molarity (M)
CH_3COOH	60.1	1.05	99.5	17.4
NH_4OH	35.0	0.89	28.0	14.3
HCOOH	46.0	1.20	90.0	23.4
HCl	36.5	1.18	36.0	11.3
HNO_3	63.0	1.42	71.0	15.9
$HClO_4$	100.5	1.67	70.0	11.6
H_3PO_4	98.0	1.70	5.0	13.7
H_2SO_4	98.1	1.84	96.0	18.0
HF	10.0	1.15	46.0	26.5

Some constants

Avogadro number (N) = 6.02214 x 10^{23}mol^{-1}

The gas constant (R) = 8.3145Jmol^{-1}K^{-1}

Charge on electron (e) = 1.602177 x 10^{-19}C

Faraday (F) = 96,485.3 C mol^{-1}

Speed of light (c) = 2.9979 x 10^{8} m s^{-1}

Electron volt (eV) = 1.6022 x 10^{8} m s^{-1}

Faraday's constant (F) = 9.6485 x 10^{4} C mol^{-1}

Electronic charge (e) = 1.6022 x 10^{-19}C

Preparation of some useful reagents

Dilute acids and bases

1. Dilute acetic acid: dilute 230 ml glacial acetic acid with water and make up to one litre (approximately 4M).
2. Dilute HCl: 1volume of conc.(concentrated) HCl is dluted to 3 volumes with water.
3. Dilute HNO_3: one volume of conc. HNO_3 is diluted to 10 volumes with water.
4. Dilute sulphuric acid: dilute 1volume of conc. sulphuric acid to 8 volumes of water.
5. Ammonium hydroxide dilute: dilute 1volume of conc. ammonia (density=0.880) to 3 volumes of water.

6. Potassium hydroxide, alcoholic: boil under reflux a mixture of 10 g of KOH and 100 ml rectified spirit for 30 min. Cool and filter through glass wool.
7. Sodium hydroxide, dilute(10% aqueous): 100 g NaOH dissolved in water, cool and dilute to one litre.

Salt solutions and some reagents

1. Calcium chloride: dissolve 100 g $CaCl_2,6H_2O$ (50 g anhydrous $CaCl_2$) in water and make up to one litre.
2. Ferric chloride (4.5%): dissolve 75 g $FeCl_3$, $6H_2O$ in water, add 10 nl conc. HCl and make up to one litre.
3. Ferrous sulphate: dissolve 10 g $FeSO_4$, $7H_2O$ in water, add 10 ml dilute H_2SO_4 and make up to 100 ml with water (prepare afresh prior to use).
4. Potassium iodide: 10% solution.
5. Potassium permanganate: 1% solution.
6. Silver nitrate: 2% solution.
7. Sodium bisulphite (Na HSO_3): dissolve 600 g of Na HSO_3 in water, make up to one litre.
8. Sodium carbonate: 20% solution in water.
9. Bromine water: shake 5 ml bromine with 100 ml water and decant off the clear aqueous solution.
10. Chlorine water: water saturated with chlorine gas (about 0.7%).
11. Fehling's solution: Solution A - dissolve 69.28 g $CuSO_4$ $5H_2O$ in water and makeup to one litre. Solution B – dissolve 346 g sodium potassium tartrate (Rochelle salt) and 120 g NaOH in water and make up to one litre. mix equal volumes of solutions A & B (prepare solutions afresh).
12. Schiff's reagent: dissolve 1 g rosaniline in 50 ml water. Saturate with SO_2 add about 1 g animal charcoal, shake and filter, make up to one litre with water.

Atomic weights of some elements

Element	Symbol	Atomic weight	Element	Symbol	Atomic weight
Arsenic	As	74.91	Manganese	Mn	54.93
Barium	Ba	137.36	Mercury	Hg	200.61
Bromine	Br	79.92	Nitrogen	N	14.01
Calcium	Ca	40.08	Oxygen	O	16.00
Carbon	C	12.00	Phosphorus	P	31.02
Chlorine	Cl	35.46	Platinum	Pt	195.23
Chromium	Cr	52.01	Potassium	K	39.10
Copper	Cu	63.57	Silver	Ag	107.88
Hydrogen	H	1.008	Sodium	Na	23.00
Iodine	I	126.92	Sulphur	S	32.00
Iron	I	126.92	Tin	Sn	118.70
Lead	Pb	207.22	Zinc	Zn	65.38
Magnesium	Mg	24.32			

SI Units and convenient Metric Equivalents

Dimension	SI base or Derived unit	SI or Metric unit	SI equivalent of metric unit
Time	Second(s)	Minute (min)	
	Hour (h), day (d)	Year (a)	
Length(depth)	Meter (m)	Hectare(ha)	10^4m^2
Area	m^2	Tone (t)	10^3kg
Mass	kilogram (kg)	Litre (L)	$10^{-3}m^3$
Volume	m^3		
Force	Newton (N)		
Pressure	Pascal (pa)		
Temperature	Kelvin (K)	Celsius (°C)	
Energy	Joule (J)		
Power	Watt (W)		
Photon fluxdensity	Mol $m^{-2}s^{-1}$	Mol m^{-2} d^{-1}	

Prefixes for multiples and sub-multiples of units

Prefix	Abreviation	Relative size of unit	Example
Nano	n	10^{-9}	ns nanosecond
Micro	µ	10^{-6}	µg microgram
Milli	m	10^{-3}	mL millilitre
Centi	c	10^{-2}	cm^3 cubic centimetere
Deci	d	10^{-1}	dm decimetre
Deca	da	10	dam decametre
Hecta	h	10^2	hm^2 square hectometre
Kilo	k	10^3	kw kilowatt
Mega	M	10^6	Mpa megapascal
Giga	G	10^9	GJ gigajoule
Tera	T	10^{12}	Tg tetragram

US, British and metric unit equivalents

Liters x 0.2642 = US gallons

US gallons x 3.785 = Liters

Imperial gallons x 1.201 = US gallons

Imperiol gallons x 4.546 = Liters

US gallons x 0.8327 = Imperial gallons

Cubic meters x 35.31 = Cubic feet

Cubic feet x 0.0283 = Cubic meters

Cubic meters x 264.2 = US gallons

US gallons x 0.0039 = Cubic meters

Metres x 3.281 = Feet

Feet x 0.3048 = Meters

Metres x 39.37 = Inches

Inches x 0.0254 = Meters

Centimeters x 0.3937 = Inches

Inches x 2.540 = Centimeters

Millimeters x 0.0394 = Inches

Inches x 25.4 = Millimeters

Kilograms x 2.2046 = Lbs.

Lbs. x 0.4536 = Kilograms

Tons (long) x 1016.05 = Kilograms

Tons (long) x 2240 = Lbs.

Tones (metr.) x 1000 = Kilograms

Tones (metr.) x 2204.6 = Lbs.

Tons (short) x 907.185 = Kilograms

Tons (short) x 2000 = Lbs.

Grams x 15.43265 = Grains

Grams x 0.0647989 = Grams

Grams x 0.0352740 = Ounces (US)

Ounces (US) x 28.349527 = Grams

Concentrations-conversion table

Proportion	potency	%	g/kg mg/g g/mg	ppm mg/kg pg/kg ng/mg	ppb pg/g ng/g pg/mg
1:100	1x 10^{-2}	1	10	10000	
1:1000	1x 10^{-3}	0.1	1	1000	
1:10000	1x 10^{-4}	0.0	10.1	1000	
1:100000	1x 10^{-5}	0.001	0.01	100	
1:1000000	1x 10^{-6}	0.0001	0.001	1	1000
1:10000000	1x 10^{-7}	0.00001	0.0001	0.1	100
1:100000000	1x 10^{-8}	0.000001	0.00001	0.01	10
1:1000000000	1x 10^{-9}	0.0000001	0.000001	0.001	1
1:10000000000	1x 10^{-10}				
1:100000000000	1x 10^{-11}				
1:1000000000000	1x 10^{-12}				

Energy dimensions-conversion factors

Given	Required dimension with conversion factor (1)Dimension					
Unit	J	kWh	MeV	mkp	$kcal_{15}^{o}$	erg
1 J	1	2.77778.10^{-7}	6.242.10^{12}	0.1019716	2.38920.10^{-4}	10^{7}
1 kWh	3600000	1		2.247.10^{19}	367097.8	860.11
3.6.10^{13}						
1 MeV	1.602.10^{-13}	4.45.10^{-20}	1	1.634.10^{-14}		3.827.10^{-17}
1.602.10^{-4}						
1 mkp	9.80665	2.72407.10^{-6}	6124-10^{13}	1		2.34301.10^{-3}
9.806665.10^{7}						
1 $kcal_{15}^{o}$	4185.5	1.16264.10^{-3}	2.613.1016	426.80		1
4.1855.10^{10}						
1 erg 10^{-7}		2.77778.10^{-14}	6.242.10^{5}	0.1019716.0^{-7}	2.38920.10^{-11}	1

(1) Examples: 1 J = 2.38920.10^{-4} kcal 1 MeV = 1.602.10^{-13} J

Estimated energy and fat requirements/day

Occupation	Work level	Energy requirement (Kcal)	Projected fat intake (g)	
			40%	20%
Male clerk (65kg)	Light Activity	2580	115	58
Subsistence farmer (58 kg)	Moderate activity	2780	124	62
Male heavy worker (65kg)	Heavy activity	3490	155	78
Retired male (60kg)	Light activity	1960	87	43
House wife (55 kg)	Light activity	1990	88	44
Female Dev. Country	Light activity	2235	99	50

FAO/WHO Recommended daily intake of energy

	Adult man (65 kg)	
	Per day	per kg/day
Energy (Kcal)	3000	46
(MJ)	12.55	0.19
Protein (Egg) (g)	37	0.57
(Animal source/ mixed)	46	0.71
Cereal diet (g)	62	0.95
Iron (mg)	5	
Calcium (mg)	400-500	
Vitamin A (μg)	750	
Vitamin D (μg)	2.5	
Thiamine (mg)	1.2	
Reboflavin (mg)	1.8	
Niacin (mg)	19.8	
Folic acid (μg)	200	
Vitamin B12 (μg)	2.2	
Ascorbic acid (mg)	30.0	

Vitamin Requirement

Vitamin	Qnt	Children		Males 0.5/4-6 yrs. 11-14;23-50		Females (Pregnant)
A	mg	420	500	1000	1000	1000
D	mg	10	10	10	5	10
E	mg	3	6	8	10	10
K	mg	12	20-40	50-100	70-140	70-140
Thiamine	mg	0.2	0.9	1.4	1.4	1.5
Riboflavine	mg	0.4	1.0	1.6	1.6	1.6
Niacin	mg	6	11	18	18	18
Vit. B6	mg	0.3	1.3	1.8	2.2	2.6
Pantothenic	mg	2	3-4	4-7	4-7	4-7
Biotin	mg	35	85	100-200	100-200	100-200
Folic acid	mg	30	200	400	400	800
B 12	mg	0.5	2.5	3.0	3.0	4.0
C	mg	35	45	60	60	80

Mineral element requirement

Element	Infants	Children		Males	
	Yrs - 0.5-1.0	4-6	15-18	25-30	51+
Sodium mg	500	900	1800	2200	2200
Calcium mg	540	800	1200	800	800
Potassium mg	1100	1550	3250	3900	3900
Phosphorous mg	360	800	1200	800	800
Magnesium mg	70	200	400	350	350
Copper mg	0.85	1.7	2.5	2.5	2.5
Iron mg	15	10	18	10	10
Zinc mg	5	10	15	15	15
Manganese	0.85	1.75	3.75	3.75	3.75
Iodine µg	50	90	150	150	150
Selenium mg	0.02	0.03-0.12	0.05-0.2	0.05-0.2	0.05-0.2
Molybdenum mg	0.04-0.08	0.06-0.15	0.15-0.5	0.15-0.5	0.15-0.5

FURTHER READING

1. Quantitative Analysis. 2000. R. A. De Jr. and A. L. Underwood, Prentice – Hall of India Pvt. Ltd. New Delhi.
2. Instrumental Methods of Chemical Analysis. 1985. Galen W. Ewing, McGraw Hill Book Company. New York.
3. Vogel's Text Book of Quantitative Chemical Analysis. 2000. 6th edition. J. Mendham, R.C. Denney, J.D. Barnes and M.J.K. Thomas. Pearson Education Ltd. U. K.
4. Practical Organic Chemistry. 2013. 4th edition. F. G. Mann and B.C. Saunders, Pearson Education Ltd. U. K.
5. A First Course in Food Analysis. 1999. A.Y. Sathe, New Age International (P) Ltd. Publishers, New Delhi.
6. Agricultural Plant Biochemistry. 2015. G. Nagaraj, New India Publishing Agency, New Delhi.
7. Oilseeds – Properties, Processing, Products and Procedures, 2009. G. Nagaraj, New India Publishing Agency, New Delhi.
8. Laboratory Manual in Biochemistry. 1981. Jayaraman, J. Wiley Eastern Limited, New Delhi.
9. Biochemical Methods. 1996. S. Sadasivam, and A. Manickam, New Age International (P) limited, New Delhi.
10. Standard Methods of Biochemical Analysis.2006. S.R. Thimmaiah, Kalyani Publications, India.

INDEX

C

D

K

L

M

N

S

T

U

V

W

Z

Zeitfracht Medien GmbH
Ferdinand-Jühlke-Straße 7
99095 Erfurt, Deutschland
produktsicherheit@kolibri360.de